Every one of your students has the potential to make a difference. And realizing that potential starts right here, in your course.

When students succeed in your course—when they stay on-task and make the breakthrough that turns confusion into confidence—they are empowered to build the skill and confidence they need to succeed. We know your goal is to create a positively charged learning environment where students reach their full potential to become active engaged learners. *WileyPLUS* can help you reach that goal.

WILEY **+**
PLUS

Wiley**PLUS** is a suite of resources—including the complete, online text—that will help your students:

- come to class better prepared for your lectures
- get immediate feedback and context-sensitive help on assignments and quizzes
- track their progress throughout the course

www.wileyplus.com

88% of students surveyed said it improved their understanding of the material. *

TO THE INSTRUCTOR

WileyPLUS is built around the activities you perform

Prepare & Present

Create outstanding class presentations using a wealth of resources, such as PowerPoint™ slides and image galleries. Plus you can easily upload any materials you have created into your course, and combine them with the resources *WileyPLUS* provides.

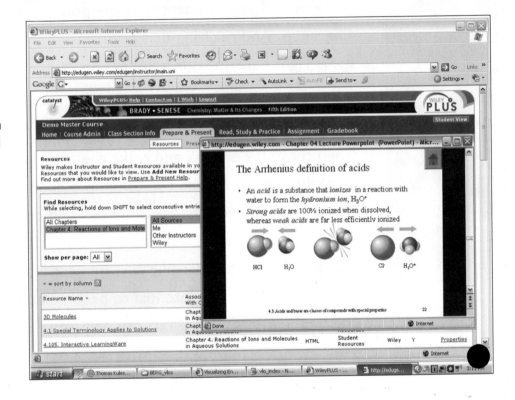

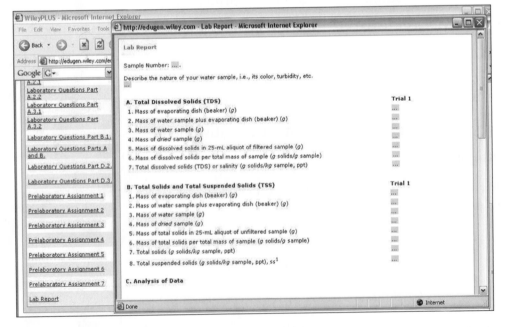

Create Assignments

Automate the assigning and grading of lab reports, homework, or quizzes by using the provided question banks. Student results will be automatically graded and recorded in your gradebook. *WileyPLUS* also links homework problems to relevant sections of the online text, hints, or solutions—context-sensitive help where students need it most!

Track Student Progress

Keep track of your students' progress via an instructor's gradebook, which allows you to analyze individual and overall class results. This gives you an accurate and realistic assessment of your students' progress and level of understanding.

Now Available with WebCT and eCollege!

Now you can seamlessly integrate all of the rich content and resources available with *WileyPLUS* with the power and convenience of your WebCT or eCollege course. You and your students get the best of both worlds with single sign-on, an integrated gradebook, list of assignments and roster, and more. If your campus is using another course management system, contact your local Wiley Representative.

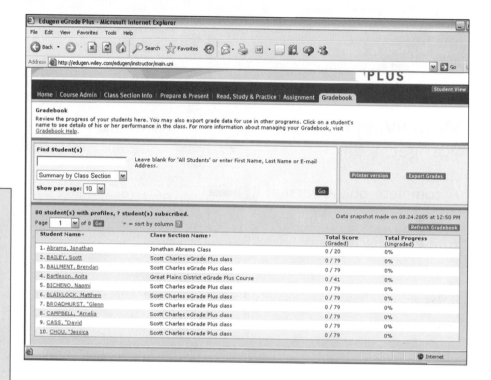

"I studied more for this class than I would have without *WileyPLUS*."

Melissa Lawler, *Western Washington Univ.*

For more information on what *WileyPLUS* can do to help your students reach their potential, please visit

www.wileyplus.com/experience

84% of students would recommend *WileyPLUS* to their next instructors *

TO THE STUDENT

You have the potential to make a difference!

WileyPLUS is a powerful online system packed with features to help you make the most of your learning potential, and get the best grade you can!

With Wiley**PLUS** you get:

A complete online version of your text and other study resources

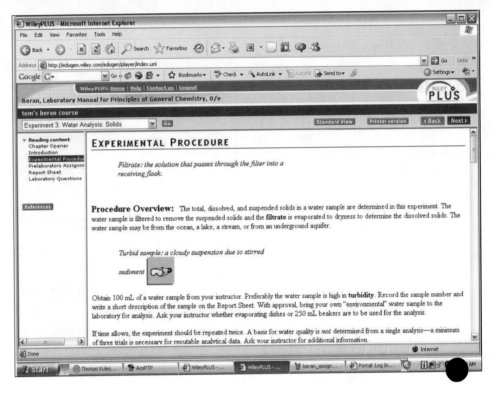

Problem-solving help, instant grading, and feedback on your homework and quizzes
You can keep all of your assigned work in one location, making it easy for you to stay on task. Plus, many homework problems contain direct links to the relevant portion of your text to help you deal with problem-solving obstacles at the moment they come up.

The ability to track your progress and grades throughout the term.
A personal gradebook allows you to monitor your results from past assignments at any time. You'll always know exactly where you stand.

If your instructor uses *WileyPLUS*, you will receive a URL for your class. If not, your instructor can get more information about *WileyPLUS* by visiting www.wileyplus.com

---

"It has been a great help, and I believe it has helped me to achieve a better grade."
Michael Morris, *Columbia Basin College*

74% of students surveyed said it helped them get a better grade.*

Laboratory Manual
for Principles of
General Chemistry

Laboratory Manual for Principles of General Chemistry

Eighth Edition

J. A. Beran

Regents Professor, Texas A&M University System
Texas A & M University—Kingsville

BICENTENNIAL
1807
WILEY
2007
BICENTENNIAL

John Wiley & Sons, Inc.

The author of this manual has outlined extensive safety precautions in each experiment. Ultimately, it is your responsibility to practice safe laboratory guidelines. The author and publisher disclaim any liability for any loss or damage claimed to have resulted from, or been related to, the experiments.

VICE PRESIDENT AND EXECUTIVE PUBLISHER   Kaye Pace
PROJECT EDITOR   Jennifer Yee
PRODUCTION EDITOR   Janet Foxman
EXECUTIVE MARKETING MANAGER   Amanda Wygal
DESIGNER   Hope Miller
PRODUCTION MANAGEMENT SERVICES   Jeanine Furino/GGS Book Services
PHOTO EDITOR   Tara Sanford
EDITORIAL ASSISTANT   Catherine Donovan
SENIOR MEDIA EDITOR   Thomas Kulesa
BICENTENNIAL LOGO DESIGN   Richard J. Pacifico
FRONT COVER PHOTO   ©Corbis Digital Stock

This book was set in Times Roman by GGS Book Services, and printed and bound by Courier/Westford. The cover was printed by Courier/Westford.

This book is printed on acid-free paper. ∞

To order books or for customer service, please call 1-800-CALL-WILEY (225-5945).

Library of Congress Cataloging-in-Publication Data:
Beran, Jo A.
Laboratory manual for principles of general chemistry / J. A. Beran—8th ed.
      p. cm.

ISBN 978-0-470-12922-7 (pbk.)
1.   Chemistry—Laboratory manuals.   I.   Title.
QD45.B475   2009
542—dc22   2007031692

Printed in the United States of America

10 9 8 7 6 5 4

Preface

Chemistry laboratories have changed with advances in technology and safety issues.

Welcome to the 8th edition! The immediate questions that come to mind are, "Why the 8th?" and "What's wrong with the 7th?" To respond to the latter question first: the 7th edition was the most well-received laboratory manual that Wiley has ever produced for general chemistry. The experiments were interesting, challenging, and had good pedagogy regarding laboratory techniques, safety, and experimental procedures. The reporting and analyzing of data and the questions (pre- and post-lab) sought to provide the intuitiveness of the experiment. The answer to "What's wrong with the 7th"—nothing really, except that chemistry like all sciences is constantly incorporating new instrumentation, new technology, and adapting to new developments and scientific trends.

In response to, "Why the 8th?": all of the "good" from the 7th has been retained, but because of these progressive trends in general chemistry, new experiments have added depth, relevance, and appreciation of the laboratory experience. Trends toward safer, more modern laboratory equipment, computer usage, and online information have been included. The open-endedness of each experiment is encouraged in "**The Next Step**" where, upon completion of the experiment, the student has the tools and experience to employ for studying additional chemical systems or topics of his/her interest. It is hoped that laboratory instructors and students will add their own "Next Step" for pursuing personal areas of interest/investigation. Reviewers have applauded the changes and challenges of the 8th edition.

While all comments of users and reviewers from the previous seven editions have been heavily weighed with each new edition, the task of presenting the "perfect" manual, like chemistry and science in general, is impossible. However, at this point in time, we feel it is the "best."

BREADTH (AND LEVEL) OF THE 8TH EDITION

This manual covers two semesters (or three quarters) of a general chemistry laboratory program. A student may expect to spend three hours per experiment in the laboratory; limited, advanced preparation and/or extensive analysis of the data may lengthen this time. The experiments were chosen and written so that they may accompany any general chemistry text.

FEATURES OF THE 8TH EDITION

Safety and Disposal. "Safety first" is again emphasized throughout the manual with recent advisories and guidelines being added. The introductory section outlines personal and laboratory safety rules and issues. Icons in the **Experimental Procedures** cite "**Cautions**" for handling various chemicals, the proper "**Disposal**" of chemicals, and the proper "**Cleanup**" of laboratory equipment. Pre-laboratory questions often ask students to review the safety issues for the experiment.

In addition, attention was focused on eliminating chemicals that have prompted disposal concerns. Experiments have been modified or combined to eliminate such chemicals as barium, bismuth, and thioacetamide, and to reduce the amount of silver ion from 7th edition experiments.

Laboratory Techniques. Numbered icons cited at the beginning of each experiment and within the Experimental Procedure are referenced to basic laboratory techniques that enable the student to complete the experiment more safely and efficiently. The **Laboratory Techniques** section provides a full explanation of seventeen basic general chemistry laboratory techniques (along with the corresponding icons) that are used repeatedly in the manual.

Organization. Experiments have been regrouped according to subject matter where, e.g., all redox experiments are grouped (Section 5) such that the sequential numbering for the experiments in the group indicates a greater degree of complexity. For example, Experiment 27, Oxidation-Reduction Reactions is the simplest of the reactions involving oxidation-reduction, not the 27th most difficult experiment in the manual, and Experiment 33, Electrolytic Cells: Avogadro's Number, is perhaps the most difficult of the oxidation-reduction experiments.

Dry Lab 2, Nomenclature, as well as a combined experiment (7th edition, Experiments 12 and 13) on Acids, Bases, and Salts (Experiment 6) have been moved forward in the 8th edition. Such organization appears to be the trend in general chemistry programs.

Report Sheets. They have been simplified! Data entries are now distinguished from calculation entries—the calculated entries are now shaded. Additionally, at the discretion of the instructor, the website, http://www.wiley.com/college/beran, provides downloadable Excel Report Sheet Templates for each experiment in which a numerical analysis is required.

Online References. A significant number of websites have been cited in various experiments and dry labs. An extensive list of online references is also provided in the **Laboratory Data** section of the manual.

NEW TO THE 8TH EDITION

New and Revised Experiments. Experiments of student interest and safety continue to be of importance in the 8th edition. Other than minor changes that appear in most experiments, the major changes and additions are:

- Experiment 3. Water Analysis: Solids. An expansion of the same experiment from the 7th edition, this experiment now introduces more details of the "chemistry" of the system investigated. While the experiment itself is no more difficult than before, an understanding of the salts and ions present is a beginning for understanding chemical systems.
- Experiment 6. Acids, Bases, and Salts. The combination of 7th edition, Experiments 12 and 13, has reduced some of the "drudgery (according to students)" of repetition in those two experiments. Being placed earlier in the manual encourages an earlier introduction to the properties of aqueous solutions.
- Experiment 8. Limiting Reactant. The handling and disposal of barium ion (7th, Experiment 8) had become an increasing problem. The barium phosphate limiting reactant system has been replaced by the calcium oxalate system.
- Experiment 20. New! Alkalinity. Both "T" and "P" alkalinity values are determined along with the interpretation and understanding of the significance of this water quality parameter.
- Experiment 30, Vitamin C Analysis. Simplified! A direct analysis of the vitamin C with potassium iodate is the revised procedure.
- Experiment 31. New! Dissolved Oxygen in Natural Waters. Water samples are obtained and the oxygen is "fixed" on site. The principles for the technique are explained with the subsequent analysis performed in the laboratory. Its significance is of interest to anyone in the biological or biochemical fields.
- Experiment 35. New! Spectrophotometric Metal Ion Analysis. Spectrophotometric analysis is so common in laboratories inside and outside of the general chemistry laboratory that additional exposure to its technique is important and useful to students in science. While a metal ion analysis is the focus of the experiment, the technique can be used and adapted for any student's interest.
- Experiment 38, Qual I: Na^+, K^+, NH_4^+, Mg^{2+}, Ca^{2+}, Cu^{2+}, and Experiment 39, Qual II: Ni^{2+}, Fe^{3+}, Al^{3+}, Zn^{2+}. The three cation qualitative analysis experiments from the 7th edition have been combined into two experiments, eliminating the use of thioacetamide, silver ion, bismuth ion, and manganese ion from the traditional qualitative analysis scheme.

The Next Step. "The Next Step" appears at the conclusion of the Experimental Procedure in each experiment. Based upon the tools and techniques gained upon completion of the experiment, The Next Step takes students from its completion to ideas for an independent, self-designed (and often open-ended) experience or experiment. Occasionally, tools and techniques learned from other experiments may be necessary for use and adaptation in the "project" that the student proposes. The initial title of this feature, "Where do we go from here?" was developed to answer the student's question of, "what more can I *now* do with what I just learned in the laboratory?" The thinking about chemistry just begins with "The Next Step" when the student leaves the laboratory, rather than ending with, "well, that experiment is over!"

Laboratory Equipment. Simple laboratory glassware and equipment, shown in the first sections of the manual, are necessary for completing most experiments. Where appropriate, the apparatus or technique is shown in the experiment with a line drawing or photograph. Analytical balances, spectrophotometers (Experiments 34 and 35), pH meters (Experiment 18), and multimeters (Experiments 32 and 33) are suggested; however, if this instrumentation is unavailable, these experiments can be modified without penalizing students. In general, hot plates have largely replaced Bunsen burners in the manual; however if not available, the Bunsen flame can still be safely used for heating.

CONTENTS OF THE MANUAL

The manual has five major sections:

- *Laboratory Safety and Guidelines.* Information on self-protection, what to do in case of an accident, general laboratory rules, and work ethics in the laboratory are presented.
- *Laboratory Data.* Guidelines for recording and reporting data are described. Sources of supplementary data (handbooks and World Wide Web sites) are listed. Suggestions for setting up a laboratory notebook are presented.
- *Laboratory Techniques.* Seventeen basic laboratory techniques present the proper procedures for handling chemicals and apparatus. Techniques unique to qualitative analysis (Experiments 37–39) are presented in Dry Lab 4.
- *Experiments and Dry Labs.* Thirty-nine experiments and four "dry labs" are subdivided into 12 basic chemical principles.
- *Appendices.* Seven appendices include conversion factors, the treatment of data, the graphing of data, names of common chemicals, vapor pressure of water, concentrations of acids and bases, and water solubility of inorganic salts.

CONTENTS OF EACH EXPERIMENT

Each experiment has six sections:

- *Objectives.* One or more statements establish the purposes and goals of the experiment. The "flavor" of the experiment is introduced with an opening photograph.
- *Techniques.* Icons identify various laboratory techniques that are used in the Experimental Procedure. The icons refer students to the Laboratory Techniques section where the techniques are described and illustrated.
- *Introduction.* The chemical principles, including the appropriate equations and calculations that are applicable to the experiment, and general interest information are presented in the opening paragraphs. New and revised illustrations have been added to this section to further enhance the understanding of the chemical principles that are used in the experiment.
- *Experimental Procedure.* The Procedure Overview, a short introductory paragraph, provides a perspective of the Experimental Procedure. Detailed, stepwise directions are presented in the Experimental Procedure. Occasionally, calculations for amounts of chemicals to be used in the experiment must precede any experimentation.
- *Prelaboratory Assignment.* Questions and problems about the experiment prepare students for the laboratory experience. The questions and problems can be answered easily after studying the Introduction and Experimental Procedure. Approximately 75% of the Prelaboratory questions and problems are new to the 8th edition.
- *Report Sheet.* The Report Sheet organizes the observations and the collection and analysis of data. Data entries on the Report Sheet are distinguished from calculated (shaded) entries. Laboratory Questions, for which students must have a thorough understanding of the experiment, appear at the end of the Report Sheet. Approximately 75% of the Laboratory Questions are new to the 8th edition.

INSTRUCTOR'S RESOURCE MANUAL

The *Instructor's Resource Manual* (available to instructors from Wiley) continues to be most explicit in presenting the details of each experiment. Sections for each experiment include

- an Overview of the experiment
- an instructor's Lecture Outline
- Teaching Hints

- representative or expected data and results
- Chemicals Required
- Special Equipment
- Suggested Unknowns
- answers to the Prelaboratory Assignment and Laboratory Questions
- a Laboratory Quiz.

Offered as a supplement to the *Instructor's Resource Manual* is a **Report Sheet Template** for those experiments requiring the numerical analysis of data. The format of the templates is based on Microsoft Excel software and is available from Wiley upon adoption.

The Appendices of the *Instructor's Resource Manual* detail the preparation of all of the solutions, including indicators, a list of the pure substances, and a list of the special equipment used in the manual *and* the corresponding experiment number for each listing. Users of the laboratory manual have made mention of the value of the *Instructor's Resource Manual* to the laboratory package.

REVIEWERS

The valuable suggestions provided by the following reviewers for this 8th edition are greatly appreciated:

Dennis J. Berzansky
Westmoreland County Community College

Maria Bohorquez
Drake University

David Neal Boehnke
Jacksonville University

Philip Delassus
University of Texas—Pan-American

Diana Glick
Georgetown University

Arlin Gyberg
Augsburg College

Alan Hazari
University of Tennessee—Knoxville

Newton Hilliard
Eastern New Mexico University

Steven C. Holman
Mississippi State University

Wendy S. Innis-Whitehouse
University of Texas—Pan-American

Susan Knock
Texas A&M University at Galveston

Walter Patton
Lebanon Valley College

Barbara Rackley
Tuskegee University

Michael Schuder
Carroll College

Kerri Scott
University of Mississippi

ACKNOWLEDGMENTS

The author thanks Dr. John R. Amend, Montana State University, for permission to use his basic idea in using emission spectra (without the aid of a spectroscope) to study atomic structure (Dry Lab 3); Dr. Gordon Eggleton, Southeastern Oklahoma State University, for encouraging the inclusion of the paper chromatography experiment (Experiment 4); the general chemistry faculty at Penn State University, York Campus for the idea behind the thermodynamics experiment (Experiment 26); and to Dr. Michael Schuder, Carroll College (MN), for his insightful chemical and editorial suggestions and opinions throughout the writing of the 8th edition.

What a staff at Wiley! Thanks to Jennifer Yee, Acquisitions Editor, for her keen insight, helpful suggestions, and unending commitment to see the manual through its birth; Janet Foxman and Jeanine Furino, Production Editors, for coordinating the production of the manual; Tara Sanford, Photo Editor at Wiley, for assistance in obtaining the new photographs for this edition; Jeanine Furino at GGS Book Services, for advancing the quality and detail of the line drawings.

Thanks also to the Chemistry 1111 and 1112 students, and laboratory assistants and staff at Texas A&M—Kingsville for their keen insight and valuable suggestions; also to my colleagues and assistants for their valuable comments.

A special note of appreciation is for Judi, who has unselfishly permitted me to follow my professional dreams and ambitions since long before the 1st edition of this manual in 1978. She has been the "rock" in my life. And also to Kyle and

Greg, who by now have each launched their own families and careers—a Dad could not be more proud of them and their personal and professional accomplishments. My father and mother gave their children the drive, initiative, work ethic, and their blessings to challenge the world beyond that of our small Kansas farm. I shall be forever grateful to them for giving us those tools for success.

James E. Brady, St. Johns University, Jamaica, NY, who was a co-author of the manual in the early editions, remains the motivator to review and update the manual and to stay at the forefront of general chemistry education. Gary Carlson, my *first* chemistry editor at Wiley, gave me the opportunity to kick off my career in a way I never thought possible or even anticipated. Thanks Jim and Gary.

The author invites corrections and suggestions from colleagues and students.

J. A. Beran
Regents Professor, Texas A&M University System
MSC 161, Department of Chemistry
Texas A&M University—Kingsville
Kingsville, TX 78363

Photo Credits

Preface: Granger Collection; Page 1 (**top**): Courtesy Fisher Scientific; Page 1 (**bottom**): Courtesy Fisher Scientific; Page 2 (**top**): Courtesy Fisher Scientific; Page 2 (**bottom**): Courtesy Fisher Scientific; Page 3: Courtesy Flow Sciences, Inc.; Page 5: Yoav Levy/Phototake; Page 7 (**top**): Courtesy Fisher Scientific; Page 7 (**bottom**): Courtesy Fisher Scientific; Page 8 (**top left**): Courtesy VWR Scientific; Page 8 (**top center**): Courtesy Fisher Scientific; Page 8 (**top right**): Courtesy Fisher Scientific; Page 8 (**bottom left**): Courtesy VWR Scientific; Page 8 (**bottom center**): Kristen Brochmann/ Fundamental Photographs; Page 8 (**bottom right**): Courtesy VWR Scientific; Page 9: Yoav Levy/Phototake; Page 10: Yoav Levy/Phototake; Page 11: Courtesy Mettler Instrument Corp.; Page 13 (**top**): Ken Karp; Page 13: (**center**): Courtesy Fisher Scientific; Page 13 (**bottom left**): Courtesy Fisher Scientific; Page 14 (**top**): Michael Watson; Page 14 (**bottom left**): Courtesy Fisher Scientific; Page 14 (**bottom right**): Courtesy VWR Scientific; Page 15: Peter Lerman; Page 16 (**top**): Courtesy VWR Scientific; Page 16 (**bottom**): Courtesy Fisher Scientific; Page 17 (**top left**): Courtesy Scientech, Inc.; Page 17 (**top center**): Courtesy VWR Scientific; Page 17 (**top right**): Courtesy Sartorius Co.; Page 17 (**bottom right**): Courtesy Corning Glass Works; Page 17 (**bottom far right**): Courtesy Fisher Scientific; Page 18 (**top left**): Ken Karp; Page 18 (**top right**): Ken Karp; Page 18 (**bottom left**): Ken Karp; Page 19 (**left**): Courtesy Professor Jo A. Beran; Page 19 (**right**): Courtesy Professor Jo A. Beran; Page 20 (**top**): Courtesy Professor Jo A. Beran; Page 20 (**bottom left**): Courtesy Professor Jo A. Beran; Page 20 (**bottom right**): Courtesy Professor Jo A. Beran; Page 21 (**left**): Ken Karp; Page 21 (**right**): Ken Karp; Page 22 (**top right**): Courtesy Professor Jo A. Beran; Page 22 (**bottom left**): Courtesy Professor Jo A. Beran; Page 22 (**bottom right**): Courtesy Professor Jo A. Beran; Page 23: Courtesy VWR Scientific; Page 24 (**left**): Courtesy VWR Scientific; Page 24 (**right**): Courtesy Professor Jo A. Beran; Page 25 (**left**): Courtesy Professor Jo A. Beran; Page 25 (**center**): Courtesy Professor Jo A. Beran; Page 26 (**left**): Ken Karp; Page 26 (**center**): Courtesy Professor Jo A. Beran; Page 26 (**right**): Courtesy Professor Jo A. Beran; Page 25 (**right**): Courtesy Fisher Scientific; Page 27 (**left**): Courtesy Professor Jo A. Beran; Page 27 (**right**): Courtesy Professor Jo A. Beran; Page 28 (**left**): Courtesy Fisher Scientific; Page 28 (**center**): Courtesy Fisher Scientific; Page 28 (**right**): Courtesy Fisher Scientific; Page 29 (**top left**): Ken Karp; Page 29 (**top center**): Ken Karp; Page 29 (**top right**): Ken Karp; Page 29 (**bottom left**): Courtesy Professor Jo A. Beran; Page 29 (**bottom right**): Courtesy Professor Jo A. Beran; Page 30: Courtesy Fisher Scientific; Page 31 (**left**): Courtesy VWR Scientific; Page 31 (**center**): Courtesy Professor Jo A. Beran; Page 31 (**right**): Courtesy Professor Jo A. Beran; Page 32 (**top left**): Ken Karp; Page 32 (**top center**): Courtesy Fisher Scientific; Page 32 (**top right**): Courtesy Professor Jo A. Beran; Page 32 (**bottom left**): Courtesy Professor Jo A. Beran; Page 32 (**bottom center**): Courtesy Professor Jo A. Beran; Page 32 (**bottom right**): Courtesy Professor Jo A. Beran; Page 33 (**top right**): Ken Karp; Page 33 (**bottom**): Courtesy Micro Essential Labs; Page 37: Courtesy Fisher Scientific; Page 38: Courtesy Fisher Scientific; Page 39: Courtesy Professor Jo A. Beran; Page 40: Terry Gleason/Visuals Unlimited; Page 42: Herb Snitzer/Stock Boston; Page 43 (**top**): NASA/GSFC; Page 43 (**center**): Courtesy VWR Scientific; Page 43 (**bottom**): Martin Bough/Fundamental Photographs; Page 44 (**top**): Angie Norwood Browne/Stone/Getty Images; Page 44 (**center**): Yoav Levy/Phototake; Page 44 (**bottom**): Courtesy Professor Jo A. Beran; Page 45: Richard Megna/Fundamental Photographs; Page 46 (**left**): Courtesy Fisher Scientific; Page 46 (**right**): Courtesy VWR Scientific; Page 47: Courtesy Professor Jo A. Beran; Page 48 (**left**): Courtesy Professor Jo A. Beran; Page 48 (**right**): Courtesy Professor Jo A. Beran; Page 49: Dean Abramson/Stock Boston; Page 50: Richard Megna/Fundamental Photographs; Page 53: Michael Watson; Page 54: OPC, Inc.; Page 56: Ken Karp; Page 61: Digital Vision/Getty Images; Page 63: Courtesy Professor Jo A. Beran; Page 69: Richard Megna/Fundamental Photographs; Page 71: Courtesy Norton Seal View; Page 72: Courtesy Fisher Scientific; Page 79 (**top**): Michael Watson; Page 79 (**bottom**): Courtesy Professor Jo A. Beran; Page 80: Courtesy Fisher Scientific; Page 81: Courtesy National Gypsum Company; Page 85 (**top**): Keith Stone; Page 85 (**top right**): Scott Camazine/Photo Researchers, Inc.; Page 88: Richard Megna/Fundamental Photographs; Page 89: Andy Washnik; Page 91: Peter Lerman; Page 92: Peter Lerman; Page 96: Kathy Bendo; Page 97: Peter Lerman; Page 98: Andy Washnik; Page 99 (**top right**): Kathy Bendo and Jim Brady; Page 99 (**bottom left**): Kathy Bendo and Jim Brady; Page 99 (**bottom right**): Peter Lerman; Page 100: Courtesy VWR Scientific; Page 109: Ken Karp; Page 111: Courtesy Professor Jo A. Beran; Page 113: Richard Megna/Fundamental Photographs; Page 117: Herring Laboratory; Page 120: Courtesy Professor Jo A. Beran; Page 121: Ken Karp; Page 124: L.S. Stepanowicz/Visuals Unlimited; Page 127: Michael Watson; Page 130 (**left**): Courtesy VWR Scientific; Page 130 (**right**): Courtesy VWR Scientific; Page 131: Courtesy Fisher Scientific; Page 137: Richard Megna/Fundamental Photographs; Page 140: Courtesy Borden Corporation; Page 143 (**top**): Richard Megna/Fundamental Photographs; Page 143 (**bottom**): Courtesy New York Public Library; Page 144 (**top**): Roger Rossmeyer/Corbis; Page 144 (**bottom**): Michael Watson; Page 148: Courtesy Professor Jo A. Beran; Page 155: Richard Megna/Fundamental Photographs; Page 157: Courtesy Bausch and Lomb; Page 158: Courtesy Library of Congress; Page 167: Courtesy VWR Scientific; Page 169 (**left**): Ken Karp; Page 169 (**right**): Courtesy Professor Jo A. Beran; Page 175 (**top**): Andy Washnik; Page 175 (**bottom**): Bruce Roberts/Photo Researchers, Inc.; Page 178: Hugh Lieck; Page 183 (**top**): Courtesy Fisher Scientific; Page 183 (**bottom**): Hugh Lieck; Page 186 (**top**): Courtesy Fisher Scientific; Page 186 (**bottom**): Hugh Lieck; Page 187: Courtesy Professor Jo A. Beran; Page 190: Michael Dalton/Fundamental Photographs; Page 193 (**top**): Michael Watson; Page 193 (**bottom**): Courtesy Professor Jo A. Beran; Page 195: Courtesy Professor Jo A. Beran; Page 196 (**left**): Courtesy Fisher Scientific; Page 196 (**right**): Ken Karp; Page 201: Peter Lerman; Page 202: Michael Watson; Page 203: Ken Karp; Page 204: Courtesy Center for Disease Control; Page 205: Ken Karp; Page 213 (**top**): Ken Karp; Page 213 (**bottom**): Kathy Bendo; Page 215 (**left**): Courtesy Professor Jo A. Beran; Page 215 (**right**): Courtesy Professor Jo A. Beran; Page 221: Courtesy Fisher Scientific; Page 222: Courtesy Fisher Scientific; Page 224: Richard Megna/Fundamental Photographs; Page 225: Courtesy Fisher Scientific; Page 231: Ken Karp; Page 233: Courtesy Professor Jo A. Beran; Page 234: Courtesy Professor Jo A. Beran; Page 239: PhotoDisc/Getty Images; Page 242 (**left**): Courtesy Professor Jo A. Beran; Page 242 (**right**): Courtesy Fisher Scientific; Page 248: Coco McCoy/Rainbow; Page 249: Courtesy The Permutit Co., a division of Sybron Corporation; Page 250: Bortner/National Audobon Society/Photo Researchers, Inc.; Page 257: Michael Watson; Page 258 (**left**): Michael Watson; Page 258 (**right**): Hugh Lieck; Page 261: Richard Megna/Fundamental Photographs; Page 265 (**top**): OPC, Inc.; Page 265 (**bottom**): Courtesy Fisher Scientific; Page 267 (**top left**): Courtesy OPC, Inc.; Page 267 (**top center**): Courtesy OPC, Inc.; Page 267 (**top right**): Courtesy OPC, Inc.; Page 267 (**bottom**): Courtesy VWR Scientific; Page 271: John Dudak/Phototake; Page 275: Ken Karp; Page 287: Andy Washnik; Page 290 (**left**): Courtesy Professor Jo A. Beran; Page 290 (**right**): Courtesy Fisher Scientific; Page 299: Andy Washnik; Page 302: Courtesy Fisher Scientific; Page 303: Courtesy Fisher Scientific; Page 309: Yoav Levy/Phototake; Page 311: Fundamental Photographs; Page 312: Andy Washnik; Page 314: Alaska Stock Images; Page 317 (**top**): Michael Watson; Page 317 (**bottom**): OPC, Inc.; Page 325: OPC, Inc.; Page 327 (**top left**): Andy Washnik; Page 327 (**top right**): Andy Washnik; Page 327 (**bottom**): Courtesy Professor Jo A. Beran; Page 328 (**left**): Courtesy VWR Scientific; Page 328 (**right**): Courtesy VWR Scientific; Page 329 (**top left**): Courtesy Professor Jo A. Beran; Page 329 (**bottom left**): Hugh Lieck; Page 329 (**right**): Courtesy VWR Scientific; Page 332: Ken Karp; Page 335 (**top**): Ken Karp; Page 335 (**bottom**): Michael Siluk/The Image Works; Page 336: Courtesy Professor Jo A. Beran; Page 337: Courtesy VWR Scientific; Page 343 (**top**): Courtesy Professor Jo A. Beran; Page 343 (**bottom**): Courtesy Fisher Scientific; Page 351: Michael Watson; Page 355 (**left**): Courtesy Fisher Scientific; Page 355 (**right**): Michael Watson; Page 363: Charles D. Winters/Photo Researchers; Page 364: Michael Watson; Page 365: Ken Karp; Page 371: Ken Karp; Page 372: Courtesy VWR Scientific; Page 375: Courtesy Fisher Scientific; Page 383: Andrew Lambert Photography/Photo Researchers, Inc.; Page 391: OPC, Inc.; Page 392 (**top**): National Audobon Society; Page 392 (**bottom**): Courtesy Fisher Scientific; Page 396: Ken Karp; Page 397: Courtesy Professor Jo A. Beran; Page 403: Courtesy VWR Scientific; Page 407: Peter Lerman; Page 409: Peter Lerman; Page 417: Yoav Levy/Phototake; Page 419: Andy Washnik; Page 421 (**left**): Ken Karp; Page 421 (**right**): Courtesy Professor Jo A. Beran; Page 427 (**top left**): Martyn F. Chillmaid/Photo Researchers, Inc.; Page 427 (**top right**): Andrew Lambert Photography/Photo Researchers, Inc.; Page 427 (**bottom**): OPC, Inc.; Page 435: Courtesy Professor Jo A. Beran; Page 443: Kathy Bendo; Page 446: Courtesy Fisher Scientific; Page 447: Michael Watson

Contents

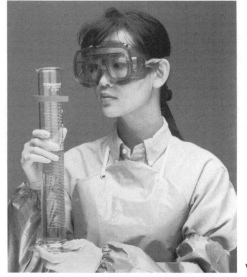

Laboratory Safety and Guidelines

Wearing proper laboratory attire protects against chemical burns and irritations.

The chemistry laboratory is one of the safest environments in an academic or industrial facility. Every chemist, trained to be aware of the potential dangers of chemicals, is additionally careful in handling, storing, and disposing of chemicals. Laboratory safety should be a constant concern and practice for everyone in the laboratory.

Be sure that you and your partners practice laboratory safety and follow basic laboratory rules. It is your responsibility, *not* the instructor's, to **play it safe.** A little extra effort on your part will assure others that the chemistry laboratory continues to be safe. Accidents do and will occur, but most often they are caused by carelessness, thoughtlessness, or neglect.

On the inside front cover of this manual, there is space to list the location of important safety equipment and other valuable reference information that are useful in the laboratory. You will be asked to complete this at your earliest laboratory meeting.

This section of the manual has guidelines for making laboratory work a safe and meaningful venture. Depending on the specific laboratory setting or experiment, other guidelines for a safe laboratory may be enforced. Study the following guidelines carefully before answering the questions on the Report Sheet of Dry Lab 1.

A. Self-Protection

1. Approved safety goggles or eye shields *must be worn* at all times to guard against the laboratory accidents of others as well as your own. Contact lenses should be replaced with prescription glasses. Where contact lenses must be worn, eye protection (safety goggles) is *absolutely necessary*. A person wearing prescription glasses must also wear safety goggles or an eye shield. Discuss any interpretations of this with your laboratory instructor.

2. Shoes *must* be worn. Wear only shoes that shed liquids. High-heeled shoes, open-toed shoes, sandals, or shoe tops of canvas, leather or fabric straps, or other woven material are *not* permitted.

3. Clothing should be only nonsynthetic (cotton). Shirts and blouses should not be torn, frilled, frayed, or flared. Sleeves should be close-fit. Clothing should cover the skin from "neck to below the knee (preferable to the ankle) and *at least* to the wrist." Long pants that cover the tops of the shoes are preferred.

    Discuss any interpretations of this with your laboratory instructor. See opening photo.

4. Laboratory aprons or coats (nonflammable, nonporous, and with snap fasteners) are highly recommended to protect outer clothing.

5. Gloves are to be worn to protect the hand when transferring corrosive liquids. If you are known to be allergic to latex gloves, consult with your instructor.

6. Jewelry should be removed. Chemicals can cause a severe irritation, if concentrated, under a ring, wristwatch, or bracelet; chemicals on fingers or gloves can cause irritation around earrings, necklaces, etc. It is just a good practice of laboratory safety to remove jewelry.

Laboratory gloves protect the skin from chemicals.

7. Secure long hair and remove (or secure) neckties and scarves.

8. Cosmetics, antibiotics, or moisturizers are *not* to be applied in the laboratory.

9. *Never* taste, smell, or touch a chemical or solution unless specifically directed to do so (see B.4 below). Individual allergic or sensitivity responses to chemicals cannot be anticipated. Poisonous substances are not always labeled.

10. Technique 3, page 14, provides an extensive overview of the proper handling of chemicals, from the dispensing of chemicals to the safety advisories for chemicals (NFPA standards). Additionally, online access to the MSDS collection of chemicals provides further specifics for all chemicals that are used in this manual.

All other techniques in the **Laboratory Techniques** section describe procedures for safely conducting an experiment. Be sure to read each technique carefully before the laboratory session for completing a safe and successful experiment.

11. Wash your hands often during the laboratory, but *always* wash your hands with soap and water before leaving the laboratory! Thereafter, wash your hands and face in the washroom. Toxic or otherwise dangerous chemicals may be inadvertently transferred to the skin and from the skin to the mouth.

For further information on personal laboratory safety see *http://membership.acs/c/ccs/publications.htm*

B. Laboratory Accidents

An eye wash can quickly remove chemicals from the eyes.

A safety shower can quickly remove chemicals from the body.

1. Locate the **laboratory safety equipment** such as eye wash fountains, safety showers, fire extinguishers, and fume hoods. Identify their locations on the inside front cover of this manual.

2. **Report all accidents** or injuries, even if considered minor, *immediately* to your instructor. A written report of any/all accidents that occur in the laboratory may be required. Consult with your laboratory instructor.

3. If an **accident occurs**, *do not panic!* The most important first action after an accident is the care of the individual. *Alert your laboratory instructor immediately!* If a person is injured, provide or seek aid *immediately;* clothing and books can be replaced and experiments can be performed again later. Second, take the appropriate action regarding the accident: clean up the chemical (see B.8, page 3), use the fire extinguisher (see B.6 below), and so on.

4. Whenever your skin (hands, arms, face, etc.) comes into contact with chemicals, quickly flush the affected area for several minutes with tap water followed by thorough washing with soap and water. Use the eyewash fountain to flush chemicals from the eyes and face. *Get help immediately*. Do *not* rub the affected area, especially the face or eyes, with your hands before washing.

5. Chemical spills over a large part of the body require immediate action. Using the safety shower, flood the affected area for at least 5 minutes. Remove all contaminated clothing if necessary. Use a mild detergent and water only (no salves, creams, lotions, etc.). Get medical attention as directed by your instructor.

6. In case of fire, discharge a fire extinguisher at the base of the flames and move it from one side to the other. Small flames can be smothered with a watchglass (do *not* use a towel, it may catch on fire). Do *not* discharge a fire extinguisher when a person's clothing is on fire—use the safety shower. Once the fire appears to be out of control, *immediately* evacuate the laboratory.

7. For abrasions or cuts, flush the affected area with water. Any further treatment should be given only after consulting with the laboratory instructor.

For burns, the affected area should be rubbed with ice, submerged in an ice/water bath, and/or placed under running water for several minutes to withdraw heat from the burned area. More serious burns require immediate medical attention. Consult with your laboratory instructor.

8. Treat chemical spills in the laboratory as follows:

   • Alert your neighbors and the laboratory instructor.
   • Clean up the spill as directed by the laboratory instructor.
   • If the substance is volatile, flammable, or toxic, warn everyone of the accident.

9. Technique 4, page 15, provides information for the proper disposal of chemicals after being used in the experiment. Improper disposal can result in serious laboratory accidents. Read that section carefully—it may prevent an "undesirable" laboratory accident. If you are uncertain of the proper procedure for the disposing of a chemical, *ask!*

C. Laboratory Rules

In addition to the guidelines for self-protection (Part A), the following rules must be followed.

1. *Smoking, drinking, eating, and chewing* (including gum and tobacco) are not permitted at any time because chemicals may inadvertently enter the mouth or lungs. Your hands may be contaminated with an "unsafe" chemical. Do not place any objects, including pens or pencils, in your mouth during or after the laboratory period. These objects may have picked up a contaminant from the laboratory bench.

2. Do *not* work in the laboratory alone. The laboratory instructor must be present.

3. Assemble your laboratory apparatus away from the edge of the lab bench ($\geq$ 8 inches or $\geq$ 20 cm) to avoid accidents.

4. Do *not* leave your experiment unattended during the laboratory period . . . this is often a time in which accidents happened.

5. Inquisitiveness and creativeness in the laboratory are encouraged. However, variations or alterations of the Experimental Procedure are forbidden without prior approval of the laboratory instructor. If your chemical intuition suggests further experimentation, first consult with your laboratory instructor.

6. Maintain an orderly, clean laboratory desk and drawer. Immediately clean up all chemical spills, paper scraps, and glassware. Discard wastes as directed by your laboratory instructor.

7. Keep drawers or cabinets closed and the aisles free of any obstructions. Do *not* place book bags, athletic equipment, or other items on the floor near any lab bench.

8. At the end of the laboratory period, completely clear the lab bench of equipment, clean it with a damp sponge or paper towel (and properly discard), and clean the sinks of all debris. Also clean all glassware used in the experiment (see Technique 2, page 13).

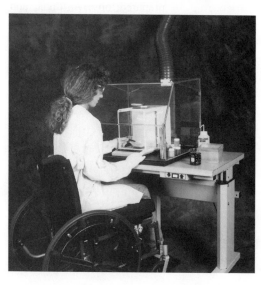

Laboratory facilities must be designed for safety.

9. Be aware of your neighbors' activities; you may be a victim of their mistakes. Advise them of improper techniques or unsafe practices. If necessary, tell the instructor.

10. For all other rules, **listen to your instructor!** Additional laboratory rules and guidelines can be added to this list at the bottom of this page.

D. Working in the Laboratory

1. Maintain a wholesome, professional attitude. Horseplay and other careless acts are prohibited. No personal audio or other "entertainment" equipment is allowed in the laboratory.

2. Do *not* entertain guests in the laboratory. Your total concentration on the experiment is required for a safe, meaningful laboratory experience. You may socialize with others in the lab, but do not have a party! You are expected to maintain a learning, scientific environment.

3. Scientists learn much by discussion with one another. Likewise, you may profit by discussion with your laboratory instructor or classmates, but *not* by copying from them.

4. *Prepare for each experiment.* Review the **Objectives** and **Introduction** to determine the "chemistry" of the experiment, the chemical system, the stoichiometry of the reactions, the color changes to anticipate, and the calculations that will be required. A thorough knowledge of the experiment will make the laboratory experience more time-efficient and scientifically more meaningful (and result in a better grade!). Complete the **Prelaboratory Assignment.**

5. Review the **Experimental Procedure**.

   - Try to understand the purpose of each step.
   - Determine if any extra equipment is needed and be ready to obtain it all at once from the stockroom.
   - Determine what data are to be collected and how it is to be analyzed (calculations, graphs, etc.).
   - Review the **Laboratory Techniques** and the **Cautions,** as they are important for conducting a safe and rewarding experiment.

6. Review the **Report Sheet.** Complete any calculations required before data collection can begin during the laboratory period. Determine the data to be collected, the number of suggested trials, and the data analysis required (e.g., calculations, graphs).

7. Review the **Laboratory Questions** at the conclusion of the Report Sheet before *and* as you perform the experiment. These questions are intended to enhance your understanding of the chemical principles on which the experiment is based.

8. Above all, *enjoy* the laboratory experience . . . be prepared, observe, think, and anticipate during the course of the experiment. Ultimately you will be rewarded.

NOTES ON LABORATORY SAFETY AND GUIDELINES

Laboratory data should be carefully recorded.

Laboratory Data

The lifeblood of a good scientist depends on the collection of reliable and reproducible data from experimental observations and on the analysis of that data. The data must be presented in a logical and credible format, that is, the data must appear such that other scientists will believe in and rely on the data that you have collected.

Believe in your data, and others should have confidence in it also. A scientist's most priceless possession is integrity. **Be a scientist.** A scientist is conscientious in his/her efforts to observe, collect, record, and interpret the experimental data as best possible. Only honest scientific work is acceptable.

You may be asked to present your data on the Report Sheet that appears at the end of each experiment, or you may be asked to keep a laboratory notebook (see Part D for guidelines). For either method, a customary procedure for collecting, recording, and presenting data is to be followed. A thorough preview of the experiment will assist in your collection and presentation of data.

A. Recording Data

1. Record all data entries *as they are being collected* on the Report Sheet or in your laboratory notebook. Be sure to include appropriate units after numerical entries. Data on scraps of paper (such as mass measurements in the balance room) will be confiscated.

2. Record the data *in permanent ink* as you perform the experiment.

3. If a mistake is made in recording data, cross out the incorrect data entry with a *single* line (do *not* erase, white-out, overwrite, or obliterate) and clearly enter the corrected data nearby (see Figure A.1). If a large section of data is deemed incorrect, write a short notation as to why the data is in error, place a single diagonal line across the data, and note where the correct data is recorded.

4. For clarity, record data entries of values <1 with a zero in the "one" position of the number; for example, record a mass measurement as 0.218 g rather than .218 g (see Figure A.1).

Trial 1

Mass of CaCO₃ sample, initial   0.218 g

Mass of CaCO₃, after heating   0.164
                                      ~~0.184~~ g

Mass of CO₂ in sample   0.054 g

Figure A.1 Procedures for recording and correcting data.

5. Data collected from an instrument and/or computer printout should be securely attached to the Report Sheet.

B. Reporting Data

The quantitative data that are collected must reflect the reliability of the instruments and equipment used to make the measurements. For example, most bathroom scales in the United States weigh to the nearest pound (± 1 lb); therefore, reporting a person's weight should reflect the precision of the measurement—a person's weight should be expressed as, e.g., 145 ± 1 pounds and *not* 145.000 . . . pounds! Conversely, if the mass of a substance is measured on a balance that has a precision of ± 0.001 g, the mass of the object should be expressed as, e.g., 0.218 g and *not* as 0.2 g.

Scientists use *significant figures to clearly express the precision of measurements.* The number of significant figures used to express the measurement is determined by the specific instrument used to make the measurement.

The number of significant figures in a measurement equals the number of figures that are certain in the measurement plus one additional figure that expresses uncertainty. The first uncertain figure in a measurement is the last significant figure of the measurement. The above mass measurement (0.218 g) has three significant figures— the first uncertain figure is the "8" which means that the confidence of the measurement is between 0.219 g and 0.217 g or 0.218 ± 0.001 g.

Rules for expressing the significant figures of a measurement and manipulating data with significant figures can be found in most general chemistry texts.

A simplified overview of the "Rules for Significant Figures" is as follows:

- Significant figures are used to express measurements, dependent on the precision of the measuring instrument.
- All definitions (e.g., 12 inches = 1 foot) have an infinite number of significant figures.
- For the addition and subtraction of data with significant figures, the answer is rounded off to the number of decimal places equal to the *fewest* number of decimal places in any one of the measurements.
- For the multiplication and division of data with significant figures, the answer is expressed with the number of significant figures equal to the *fewest* number of significant figures for any one of the measurements.

Expressing measurements in scientific notation often simplifies the recording of measurements with the correct number of significant figures. For example, the mass measurement of 0.218 g, expressed as 2.18×10^{-1} g, clearly indicates three significant figures in the measurement. Zeros at the front end of a measurement are not significant.

Zeros at the end of a measurement of data may or may not be significant. However, again that dilemma is clarified when the measurement is expressed in scientific notation. For example, the volume of a sample written as 200 mL may have one, two, or three significant figures. Expressing the measurement as 2×10^3 mL, 2.0×10^3 mL, or 2.00×10^3 mL clarifies the precision of the measurement as having one, two, or three significant figures respectively. Zeros at the end of a number *and* to the right of a decimal point are always significant.

In reporting data for your observations in this laboratory manual, follow closely the guidelines for using significant figures for correctly expressing the precision of your measurements and the reliability of your calculations.

C. Accessing Supplementary Data

You will also profit by frequent references to your textbook or, for tabular data on the properties of chemicals, the *CRC Handbook of Chemistry and Physics,* published by the Chemical Rubber Publishing Company of Cleveland, Ohio, or the *Merck Index,* published by Merck & Co., Inc., of Rahway, New Jersey. Books are generally more reliable and more complete sources of technical information than are classmates.

The World Wide Web has a wealth of information available at your fingertips. Search the web for additional insights into each experiment. In your search, keep in mind that many websites are not peer-reviewed and therefore must be judged for accuracy and truth before being used.

(Suggested only) websites that have been reviewed by the author and may enhance your appreciation of the laboratory experience are listed here:

- *http://webbook.nist.gov/chemistry* (database of technical data)
- *http://www.ilpi.com/msds* (MSDS information of chemicals)
- *http://physics.nist.gov/cuu* (database of technical data)
- *http://www.cas.org* (>31 million compounds)
- *http://en.wikipedia.org/wiki/category:chemistry*
- *http://chemfinder.camsoft.com* (information on compounds)
- *http://www.chemcenter.org* (chemistry news)
- *http://cen-online.org*
- *http://webelements.com*
- *http://www.chemdex.org*
- *http://chemistry.about.com*
- *http://www.chemtutor.com*
- *http://chem-courses.ucsd.edu* (search undergraduate labs)
- *http://www.chemistrycoach.com*
- *http://JChemEd.chem.wisc.edu*
- *http://www.siraze.net/chemistry*
- *http://www.yahoo.com* (search Science and Chemistry)
- *http://www.chemlin.net/chemistry*
- *http://antoine.frostburg.edu/chem/senese/101*

Scientific data can be obtained from the internet or analyzed with appropriate software.

D. Laboratory Notebook

The laboratory notebook is a personal, permanent record, i.e., a journal, of the activities associated with the experiment or laboratory activity. The first 3–4 pages of the notebook should be reserved for a table of contents. The laboratory notebook should have a sewn binding and the pages numbered in sequence.

Each new experiment in the laboratory notebook should begin on the right-hand side of a new page in the laboratory notebook, and should include the following sections with clear, distinct headings:

- the title of the experiment
- beginning date of the experiment
- bibliographic source of the experiment
- co-workers for the experiment
- the purpose and/or objective(s) of the experiment
- a brief, but clearly written experimental procedure that includes the appropriate balanced equations for the chemical reactions and/or any modifications of the procedure
- a list of cautions and safety concerns
- a brief description or sketch of the apparatus
- a section for the data that is recorded (see Parts A and B, Recording and Reporting) as the experiment is in progress, i.e., the Report Sheet. This data section must be planned and organized carefully. The quantitative data is to be organized, neat, and recorded with the appropriate significant figures and units; any observed, qualitative data must be written legibly, briefly, and with proper grammar. All data must be recorded in *permanent ink*. Allow plenty of room for recording observations, comments, notes, etc.

Laboratory Notebook.

- a section for data analysis that includes representative calculations, an error analysis (see Appendix B), instrument and computer printouts, graphical analyses (see Appendix C), and organized tables. Where calculations using data are involved, be orderly with the first set of data. Do *not* clutter the data analysis section with arithmetic details. All computer printouts must be securely attached.
- a section for results and discussion

At the completion of each day's laboratory activities, the laboratory activity should be dated and signed by the chemist, any co-worker, *and* the laboratory instructor at the bottom of of each page.

The laboratory instructor will outline any specific instructions that are unique to your laboratory program.

Marking pens help to organize samples.

Test tubes are a chemist's companion.

Erlenmeyer flasks are convenient for containing solutions.

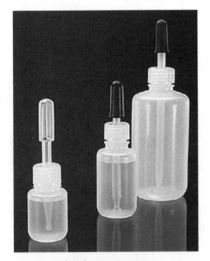

Dropping bottles assist in transferring small volumes of solutions.

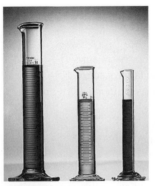

Graduated cylinders measure quantitative volumes of solutions.

A wash bottle containing deionized water must always be handy.

Common Laboratory Desk Equipment Check List

No.	Quantity	Size	Item	First Term In	First Term Out	Second Term In	Second Term Out	Third Term In	Third Term Out
1	1	10-mL	graduated cylinder	_____	_____	_____	_____	_____	_____
2	1	50-mL	graduated cylinder	_____	_____	_____	_____	_____	_____
3	5	—	beakers	_____	_____	_____	_____	_____	_____
4	2	—	stirring rods	_____	_____	_____	_____	_____	_____
5	1	500-mL	wash bottle	_____	_____	_____	_____	_____	_____
6	1	75-mm, 60°	funnel	_____	_____	_____	_____	_____	_____
7	1	125-mL	Erlenmeyer flask	_____	_____	_____	_____	_____	_____
8	1	250-mL	Erlenmeyer flask	_____	_____	_____	_____	_____	_____
9	2	25 × 200-mm	test tubes	_____	_____	_____	_____	_____	_____
10	6	18 × 150-mm	test tubes	_____	_____	_____	_____	_____	_____
11	8	10 × 75-mm	test tubes	_____	_____	_____	_____	_____	_____
12	1	large	test tube rack	_____	_____	_____	_____	_____	_____
13	1	small	test tube rack	_____	_____	_____	_____	_____	_____
14	1	—	glass plate	_____	_____	_____	_____	_____	_____
15	1	—	wire gauze	_____	_____	_____	_____	_____	_____
16	1	—	crucible tongs	_____	_____	_____	_____	_____	_____
17	1	—	spatula	_____	_____	_____	_____	_____	_____
18	2	—	litmus, red and blue	_____	_____	_____	_____	_____	_____
19	2	90-mm	watch glasses	_____	_____	_____	_____	_____	_____
20	1	75-mm	evaporating dish	_____	_____	_____	_____	_____	_____
21	4	—	dropping pipets	_____	_____	_____	_____	_____	_____
22	1	—	test tube holder	_____	_____	_____	_____	_____	_____
23	1	large	test tube brush	_____	_____	_____	_____	_____	_____
24	1	small	test tube brush	_____	_____	_____	_____	_____	_____
	1	—	marking pen	_____	_____	_____	_____	_____	_____

Special Laboratory Equipment

Number	Item	Number	Item
1	reagent bottles	16	porcelain crucible and cover
2	condenser	17	mortar and pestle
3	500-mL Erlenmeyer flask	18	glass bottle
4	1000-mL beaker	19	pipets
5	Petri dish	20	ring and buret stands
6	Büchner funnel	21	clamp
7	Büchner (filter) flask	22	double buret clamp
8	volumetric flasks	23	Bunsen burner
9	500-mL Florence flask	24	buret brush
10	$-10°C-110°C$ thermometer	25	clay pipe-stem triangle
11	100-mL graduated cylinder	26	rubber stoppers
12	50-mL buret	27	wire loop for flame test
13	glass tubing	28	pneumatic trough
14	U-tube	29	rubber pipet bulb
15	porous ceramic cup	30	iron support ring

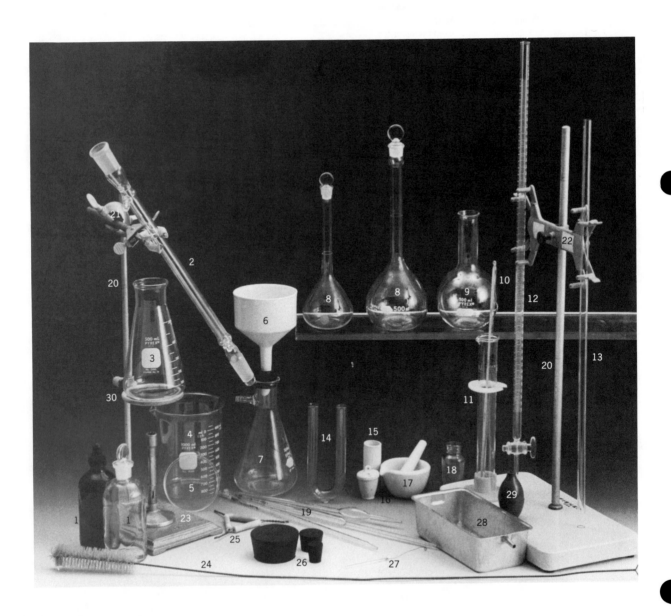

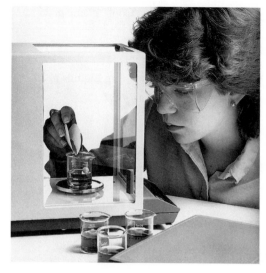

Laboratory Techniques

The application of proper laboratory techniques improves data reliability.

Scientific data that are used to analyze the characteristics of a chemical or physical change must be collected with care and patience. The data must be precise; that is, it must be reproducible to within an "acceptable" margin of error. Reproducible data implies that the data collected from an observed chemical or physical change can be again collected at a later date by the same scientist or another scientist in another laboratory (see Appendix B).

A scientist who has good laboratory skills and techniques generally collects good, reproducible data (called **quantitative data**). For that reason, careful attention as to the method (or methods) and procedures by which the data are collected is extremely important. This section of the laboratory manual describes a number of techniques that you will need to develop for collecting quantitative data in the chemistry laboratory. You do not need to know the details for all of the techniques at this time (that will come with each successive experiment that you encounter), but you should be aware of their importance, features, and location in the laboratory manual. Become *very* familiar with this section of the laboratory manual! Consult with your laboratory instructor about the completion of the Laboratory Assignment at the end of this section.

In the Experimental Procedure of each experiment, icons are placed in the margin at a position where the corresponding laboratory technique is to be applied for the collection of "better" data. The following index of icons identifies the laboratory techniques and page numbers on which they appear:

 Technique 1. Inserting Glass Tubing through a Rubber Stopper *p. 13*

 Technique 2. Cleaning Glassware *p. 13*

 Technique 3. Handling Chemicals *p. 14*

 Technique 4. Disposing of Chemicals *p. 15*

 Technique 5. Preparing Solutions *p. 15*

 Technique 6. Measuring Mass *p. 16*

Technique 7. Handling Small Volumes *p. 17*
 A. Test Tubes for Small Volumes *p. 17*

 B. Well Plates for Small Volumes *p. 17*

Caution: *Perhaps more accidents occur in the general chemistry laboratory as a result of neglect in this simple operation than all other accidents combined. Please review and practice this technique correctly when working with glass tubing. Serious injury can occur to the hand if this technique is performed incorrectly.*

Moisten the glass tubing and the hole in the rubber stopper with glycerol or water (**note:** glycerol works best). Place your hand on the tubing 2–3 cm (1 in.) from the stopper. Protect your hand with a cloth towel (Figure T.1). Simultaneously *twist* and *push* the tubing slowly and carefully through the hole. Wash off any excess glycerol on the glass or stopper with water and dry.

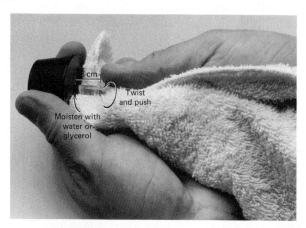

Figure T.1 Inserting glass tubing through a rubber stopper.

TECHNIQUE 2. CLEANING GLASSWARE

A chemist is very concerned about contaminants causing errors in experimental data. Cleanliness is extremely important in minimizing errors in the precision and accuracy of data. *Glassware should be clean before you begin an experiment and should be cleaned again immediately after completing the experiment.*

Clean all glassware with a soap or detergent solution using *tap* water. Use a laboratory sponge or a test tube, pipet, or buret brush as appropriate. Once the glassware is thoroughly cleaned, first rinse several times with tap water and then once or twice with *small amounts* of deionized water. Roll each rinse around the entire inner surface of the glass wall for a complete rinse. Discard each rinse through the delivery point of the vessel (i.e., buret tip, pipet tip, beaker spout). For conservation purposes, deionized water should never be used for washing glassware, but should be reserved for final rinsing only.

Invert the clean glassware on a paper towel or rubber mat to dry (Figure T.2a); do *not* wipe or blow-dry because of possible contamination. Do *not* dry heavy glassware (graduated cylinders, volumetric flasks, or bottles), or for that matter any glassware, over a direct flame.

The glassware is clean if, following the final rinse with deionized water, no water droplets adhere to the clean part of the glassware (Figure T.2b).

A laboratory detergent.

Figure T.2a Invert clean glassware on a paper towel or rubber mat to air-dry.

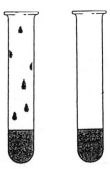

Figure T.2b Water droplets (left) do *not* adhere to the wall of clean glassware (right).

TECHNIQUE 3. HANDLING CHEMICALS

Laboratory Safety and Guidelines also include the handling of chemicals. Chemicals are safe to handle when only a few precautionary guidelines are followed.

- *Read the label* on a reagent bottle at least *twice* before removing any chemicals (Figure T.3a). The wrong chemical may lead to serious accidents or "unexplainable" results in your experiments (see Dry Lab 2 for an understanding of the rules of chemical nomenclature). Techniques 9 and 10 illustrate the correct procedures for transferring solids and liquid reagents.
- Avoid using excessive amounts of reagents. *Never* dispense more than the experiment calls for. *Do not return excess chemicals to the reagent bottle!*
- *Never* touch, taste, or smell chemicals unless specifically directed to do so. Skin, nasal, and/or eye irritations may result. If inadvertent contact with a chemical does occur, wash the affected area immediately with copious amounts of water and inform your laboratory instructor (see Laboratory Safety B.4, 5).
- Properly dispose of chemicals. See Technique 4.

Chemicals are often labeled according to National Fire Protection Association (NFPA) standards which describe the four possible hazards of a chemical and a numerical rating from 0 to 4. The four hazards are health hazard (blue), fire hazard (red), reactivity (yellow), and specific hazard (white). A label is shown in Figure T.3b.

If you wish to know more about the properties and hazards of the chemicals with which you will be working in the laboratory, safety information about the chemicals is available in a bound collection of the Material Safety Data Sheets (MSDS). The MSDS collection is also accessible on various websites (see Laboratory Data, Part C), for example at *www.ilpi.com/msds*.

In this manual, the international caution sign (shown at left) is used to identify a potential danger in the handling of a solid chemical or reagent solution or hazardous equipment.

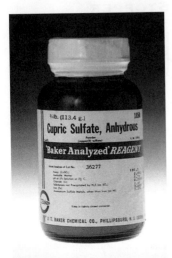

Figure T.3a Chemicals are labeled with systematic names (see Dry Lab 2).

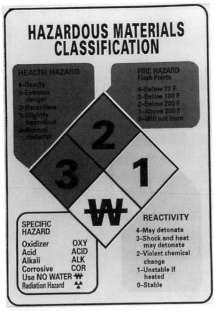

Figure T.3b Hazardous materials classification system.

Figure T.4 Waste disposal containers are available in the laboratory.

Most all chemicals used in the experiments of this manual are considered "safe" but must be properly disposed after use for safety and environmental concerns.

- Assume *nothing* (besides soap and water) is to be discarded in the sink.
- Discard waste chemicals as directed in the Experimental Procedure or by the laboratory instructor. **Read the label** on the waste container *at least twice* (Figure T.4) before discarding the chemical. Carelessness that may result in improper mixing of waste chemicals can cause serious laboratory accidents. *"When in doubt, ask your instructor, it's the safe thing to do!"*
- Note the position of each disposal icon in the Experimental Procedure as the point at which disposal is to occur.

The final disposal of chemicals is the responsibility of the stockroom personnel. Information for the proper disposal of chemicals is also available from the MSDS collection or at various websites.

The preparation of an aqueous solution is often required for an experimental procedure. The preparation begins with either a solid reagent or a solution more concentrated than the one needed for the experiment. At either starting point, the number of *moles* of compound required for the experiment is calculated: (1) from a solid, the mass and the molar mass of the compound are needed to calculate the number of moles of compound required for the preparation of the solution; (2) from a more concentrated solution, the concentration and volume (or mass) of the diluted solution must be known in order to calculate the number of moles of compound needed for the preparation of the aqueous solution. In both cases, the calculated (and then also measured) moles of compound are diluted to final volume. Knowledge of moles and mole calculations is absolutely necessary.

$$moles\ solute = \frac{grams\ solute}{molar\ mass\ solute}$$

$$V_{concentrated} = \frac{V_{dilute} \times M_{dilute}}{M_{concentrated}}$$

In the laboratory preparation, *never* insert a pipet, spatula, or dropping pipet into the reagent used for the solution preparation. Always transfer the calculated amount from the reagent bottle as described in Techniques 9 and 10.

Solutions are commonly prepared in volumetric flasks (Figure T.5) according to the following procedure:

- Place water (or the less concentrated solution) into the volumetric flask until it is one-third to one-half full.

|(a)|(b)|(c)|(d)|

Figure T.5 Place water (or the less concentrated solution) into the flask before slowly adding the solid or more concentrated solution. Dilute the solution to the "mark" with water; stopper and invert the flask 10–15 times.

- Add the solid (or add the more concentrated reagent) *slowly, while swirling,* to the volumetric flask. (**Caution:** *Never dump it in!*)
- Once the solid compound has dissolved or the more concentrated solution has been diluted, add water (dropwise if necessary) until the calibrated "mark" etched on the volumetric flask is reached (see Technique 16A for reading the meniscus). While securely holding the stopper, invert the flask slowly 10–15 times to ensure that the solution is homogeneous.

TECHNIQUE 6. MEASURING MASS

The laboratory balance is perhaps the most used *and abused* piece of equipment in the chemistry laboratory. Therefore, because of its extensive use, you and others must follow several guidelines to maintain the longevity and accuracy of the balance:

- Handle with care; balances are expensive.
- If the balance is not leveled, see your laboratory instructor.
- Use weighing paper, a watchglass, a beaker, or some other container to measure the mass of chemicals; do *not* place chemicals directly on the balance pan.
- Do *not* drop anything on the balance pan.
- If the balance is not operating correctly, see your laboratory instructor. Do *not* attempt to fix it yourself.
- After completing a mass measurement, return the mass settings to the zero position.
- Clean the balance and balance area of any spilled chemicals.

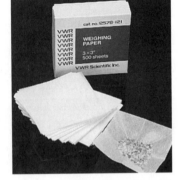

Creased weighing paper is used for measuring the mass of solids.

Tared mass: mass of sample without regard to its container

Plastic (or aluminum) weighing dishes are used for measuring the mass of solids

The mass measurement of a sample can be completed in two ways. In the traditional method, the mass of weighing paper or a clean, dry container (such as a beaker, watchglass, or weighing boat) is first measured and recorded. The sample is then placed on the weighing paper or in the container and this combined mass is measured. The mass of the weighing paper or container is then subtracted from the combined mass to record the mass of the sample.

On modern electronic balances, the mass of the weighing paper or container can be tared out—that is, the balance can be zeroed *after* placing the weighing paper or container on the balance, in effect subtracting its mass immediately (and automatically). The sample is then placed on the weighing paper or in the container, and the balance reading *is* the mass of the sample.

For either method the resultant mass of the sample is the same and is called the **tared mass** of the sample.

Different electronic balances, having varying degrees of sensitivity, are available for use in the laboratory. It is important to know (by reading the Experimental Procedure) the precision required to make a mass measurement and then to select the appropriate balance. It may save you time during the data analysis. Record mass measurements that reflect the precision of the balance, i.e., the correct number of significant figures (see Laboratory Data, Part B). These balances are shown in Figures T.6a through T.6c.

Balance	Sensitivity (g)
Top-loading (Figure T.6a)	±0.01 or ±0.001
Top-loading (Figure T.6b)	±0.0001
Analytical (Figure T.6c)	±0.00001

TECHNIQUE 7. HANDLING SMALL VOLUMES

The use of smaller quantities of chemicals for synthesis and testing in the laboratory offers many safety advantages and presents fewer chemical disposal problems. Many of the experimental procedures in this manual were designed with this in mind. Handling small volumes requires special apparatus and technique.

Figure T.6a Top-loading balance, sensitivity of ±0.01 g and/or ±0.001 g.

Figure T.6b Analytical balance, sensitivity of ±0.0001 g.

Figure T.6c Analytical balance, sensitivity of ±0.00001 g.

A. Test Tubes for Small Volumes

Small test tubes are the chemist's choice for handling small volumes. Common laboratory test tubes are generally of three sizes: the 75-mm (or 3-inch) test tube, the 150-mm (or 6-inch) test tube, and the 200-mm (or 8-inch) test tube (Figure T.7a). The approximate volumes of the three test tubes are:

75-mm (3-inch) test tube	~3 mL
150-mm (6-inch) test tube	~25 mL
200-mm (8-inch) test tube	~75 mL

The 75-mm test tube is often recommended for "small volume" experiments.

B. Well Plates for Small Volumes

Alternatively, a "well plate" can be used for a number/series of reaction vessels (Figure T.7b). The well plate is especially suited for experiments that require observations from repeated or comparative reactions. The well plate most often recommended is the 24-well plate in which each well has an approximate volume of 3.5 mL (compared to a 3 mL for a small test tube).

For either technique the Beral pipet, a plastic, disposable pipet, or a dropping pipet is often used to transfer small volumes of solutions to and from the test tubes or well plate. The Beral pipet has a capacity of about 2 mL, and some have volume graduation marks on the stem.

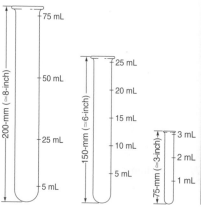

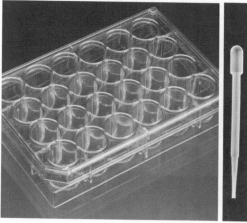

Figure T.7a The three common-sized test tubes for containing reagent solutions.

Figure T.7b A 24-well plate and Beral pipet are used for containing and transferring small quantities of reagent solutions.

TECHNIQUE 8. COLLECTING GASES

The solubility and density of a gas determine the apparatus used for its collection. Water-soluble gases should not be collected over water, but rather by air displacement.

A. Water-Soluble Gases, Air Displacement

Water-soluble gases *more* dense than air are collected by air displacement (Figure T.8a). The more dense gas pushes the less dense air up and out of the gas bottle. Water-soluble gases *less* dense than air are also collected by air displacement (Figure T.8b) except that, in this case, the less dense gas pushes the more dense air down and out of the gas bottle. Note that the gas outlet tube should extend to within 1 cm of the bottom (or top) of the gas-collecting bottle.

Figure T.8a Collection of water-soluble gases more dense than air.

Figure T.8b Collection of water-soluble gases less dense than air.

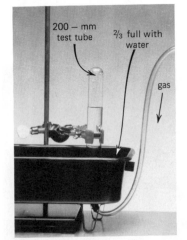

Figure T.8c Collection of water-insoluble gas by the displacement of water.

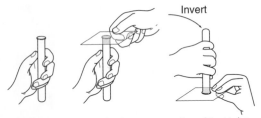

Slide glass plate or plastic wrap over top of test tube. Do not allow an air bubble to enter the test tube.

Figure T.8d Inverting a water-filled test tube.

Gases that are relatively insoluble in water are collected by water displacement. The gas pushes the water down and out of the water-filled gas-collecting vessel (Figure T.8c). The gas-collecting vessel (generally a flask or test tube) is first filled with water, covered with a glass plate or plastic wrap (no air bubbles must enter the vessel, Figure T.8d), and then inverted into a deep pan or tray half-filled with water. The glass plate or plastic wrap is removed, and the tubing from the gas generator is inserted into the mouth of the gas-collecting vessel.

Read the label on the bottle *twice* to be sure it is the correct chemical. For example, is the chemical iron(II) acetate or iron(III) acetate? Is it the anhydrous, trihydrate, or pentahydrate form of copper(II) sulfate?

Safety, again, is of primary importance. Always be aware of the importance of Technique 3. If the reagent bottle has a hollow glass stopper or if it has a screw cap, place the stopper (or cap) top side down on the bench (Figure T.9a). To dispense solid from the bottle, hold the label against your hand, tilt, and roll the solid reagent back and forth.

TECHNIQUE 9. TRANSFERRING SOLIDS

- For *larger quantities* of solid reagent, dispense the solid into a beaker (Figure T.9b) until the estimated amount has been transferred. Try not to dispense any more reagent than is necessary for the experiment. Do not return any excess reagent to the reagent bottle—share the excess with another chemist.
- For *smaller quantities* of solid reagent, first dispense the solid into the inverted hollow glass stopper or screw cap. And then transfer the estimated amount of reagent needed for the experiment from the stopper/screw cap to an appropriate vessel. Return the excess reagent in the glass stopper or screw cap to the reagent bottle—in effect, the solid reagent has never left the reagent bottle.

Figure T.9a Transferring a solid chemical from a glass ground reagent bottle. Place the glass stopper top side down.

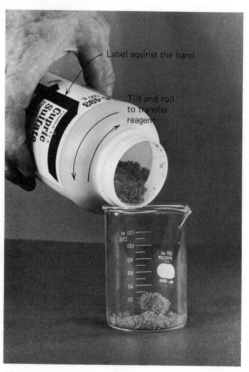

Figure T.9b Tilt and roll the reagent bottle back and forth until the desired amount of solid chemical has been dispensed.

For either situation, *never* use a spatula or any other object to break up or transfer the reagent to the appropriate container unless your laboratory instructor specifically instructs you do to so.

When you have finished dispensing the solid chemical, *recap* and return the reagent bottle to the reagent shelf.

TECHNIQUE 10. TRANSFERRING LIQUIDS AND SOLUTIONS

Read the label. When a liquid or solution is to be transferred from a reagent bottle, remove the glass stopper and hold it between the fingers of the hand used to grasp the reagent bottle (Figures T.10a, b). Never lay the glass stopper on the laboratory bench; impurities may be picked up and thus contaminate the liquid when the stopper is returned.

To transfer a liquid from one vessel to another, hold a stirring rod against the lip of the vessel containing the liquid and pour the liquid down the stirring rod, which, in turn, should touch the inner wall of the receiving vessel (Figures T.10b, c). Return the glass stopper to the reagent bottle.

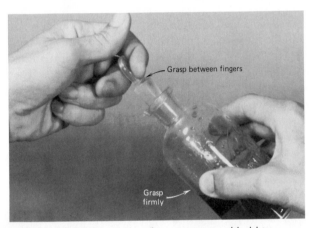

Grasp between fingers

Grasp firmly

Figure T.10a Remove the glass stopper and hold it between the fingers of the hand that grasps the reagent bottle.

Grasp stopper

Touch stirring rod to lip of bottle

Pour down stirring rod

Touch to side of beaker

Figure T.10b Transfer the liquid from the reagent bottle with the aid of the stirring rod.

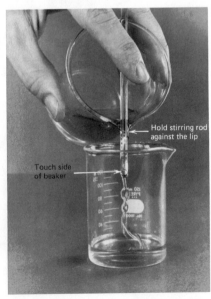

Hold stirring rod against the lip

Touch side of beaker

Figure T.10c The stirring rod should touch the lip of the transfer vessel and the inner wall of the receiving vessel.

Do *not* transfer more liquid than is needed for the experiment; do *not* return any excess or unused liquid to the original reagent bottle.

A liquid can be decanted (poured off the top) from a solid if the solid clearly separates from the liquid in a reasonably short period of time. Allow the solid to settle to the bottom of the beaker (Figure T.11a) or test tube. Transfer the liquid (called the **supernatant**) with the aid of a clean stirring rod (Figure T.11b). Do this slowly so as not to disturb the solid. Review Technique 10 for the transfer of a liquid from one vessel to another.

TECHNIQUE 11. SEPARATING A LIQUID OR SOLUTION FROM A SOLID

A. Decanting a Liquid or Solution from a Solid

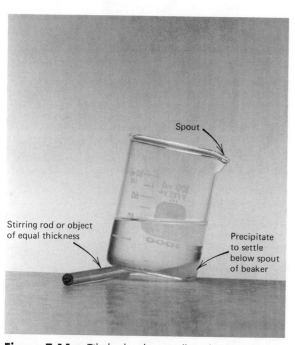

Figure T.11a Tilt the beaker to allow the precipitate to settle at the side. Use a stirring rod or a similar object to tilt the beaker.

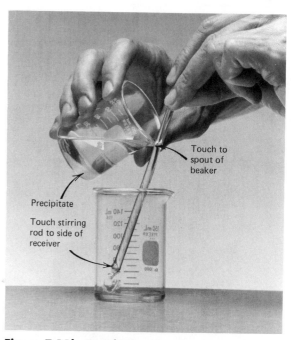

Figure T.11b Transfer the supernatant to a receiving vessel with the aid of a stirring rod.

If a solid is to be separated from the liquid using a filtering process, then the filter paper must be properly prepared. For a **gravity filtration** procedure, first fold the filter paper in half (Figure T.11c), again fold the filter paper to within about 10° of a 90° fold, tear off the corner (a *small* tear) of the outer fold unequally, and open. The tear enables a close seal to be made across the paper's folded portion when placed in a funnel.

Place the folded filter paper snugly into the funnel. Moisten the filter paper with the solvent of the liquid/solid mixture being filtered (most likely this will be deionized water) and press the filter paper against the top wall of the funnel to form a seal. Support the funnel with a clamp or in a funnel rack.

B. Preparing Filter Paper for a Filter Funnel

Transfer the liquid as described in Technique 10 (Figure T.11d). The tip of the funnel should touch the wall of the receiving beaker to reduce any splashing of the **filtrate.** Fill the bowl of the funnel until it is *less than* two-thirds full with the mixture. Always keep the funnel stem full with the filtrate; the weight of the filtrate in the funnel stem creates a slight suction on the filter in the funnel, and this hastens the filtration process.

C. Gravity Filtration

Filtrate: the solution that passes through the filter in a filtration procedure

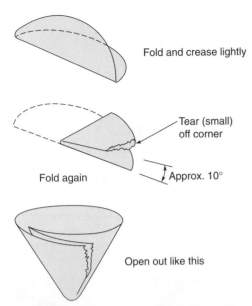

Fold and crease lightly

Tear (small) off corner

Fold again

Approx. 10°

Open out like this

Figure T.11c The sequence of folding filter paper for a filter funnel in a gravity filtration procedure.

Less than two-thirds full

Touch side of funnel

Funnel support (or iron support ring)

Touch wall of receiving flask

Figure T.11d The tip of the funnel should touch the wall of the receiving flask, and the bowl of the funnel should be one-half to two-thirds full.

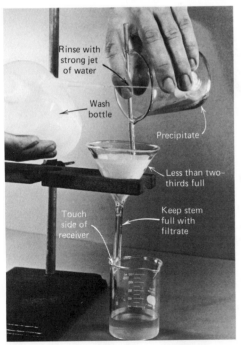

Rinse with strong jet of water

Wash bottle

Precipitate

Less than two-thirds full

Touch side of receiver

Keep stem full with filtrate

Figure T.11e Flushing the precipitate from a beaker with the aid of a "wash" bottle.

Büchner funnel

Support flask with clamp

To aspirator

To trap

Filtering flask

Support ring

Trap

Figure T.11f The aspirator should be fully open during the vacuum filtering operation.

Flush the precipitate from a beaker using a wash bottle containing the mixture's solvent (usually deionized water). Hold the beaker over the funnel or receiving vessel (Figure T.11e) at an angle such that the solvent will flow out and down the stirring rod into the funnel.

Set up the vacuum filtration apparatus as shown in Figure T.11f. A Büchner funnel (a disk of filter paper fits over the flat, perforated bottom of the funnel) set into a filter flask connected to a water aspirator is the apparatus normally used for vacuum filtration. Seal the disk of filter paper onto the bottom of the funnel by applying a light suction to the filter paper while adding a small amount of solvent.

E. Vacuum Filtration

Once the filter paper is sealed, turn the water faucet attached to the aspirator *completely* open to create a full suction. Transfer the mixture to the filter (Technique 10) and wash the precipitate with an appropriate liquid. To remove the suction, *first* disconnect the hose from the filter flask, and then turn off the water.

A centrifuge (Figure T.11g) spins at velocities of 5000 to 25,000 revolutions per minute! A solid/liquid mixture in a small test tube or centrifuge tube is placed into a sleeve of the rotor of the centrifuge. By centrifugal force the solid is forced to the bottom of the test tube or centrifuge tube and compacted. The clear liquid, called the **supernatant,** is then easily decanted without any loss of solid (Figure T.11h). This quick separation of liquid from solid requires 20–40 seconds.

F. Centrifugation

Supernatant: the clear liquid covering a precipitate

Observe these precautions in operating a centrifuge:

- Never fill the centrifuge tubes to a height more than 1 cm from the top.
- Label the centrifuge tubes to avoid confusion of samples.
- *Always* operate the centrifuge with an *even* number of centrifuge tubes containing equal volumes of liquid placed opposite one another in the centrifuge. This *balances* the centrifuge and eliminates excessive vibration and wear. If only one tube needs to be centrifuged, balance the centrifuge with a tube containing the same volume of solvent (Figure T.11i).
- *Never* attempt to manually stop a centrifuge. When the centrifuge is turned off, let the rotor come to rest on its own.

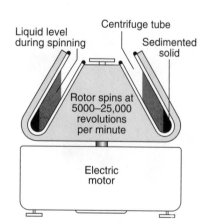

Figure T.11g A laboratory centrifuge forces the precipitate to the bottom of the centrifuge tube.

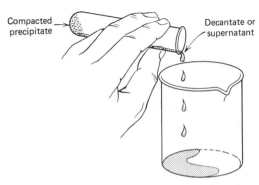

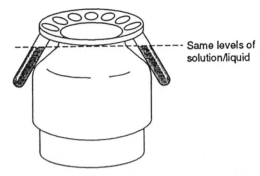

Figure T.11h Decant the supernatant from the compacted precipitate.

Figure T.11i Balance the centrifuge by placing tubes with equal volumes of liquid opposite each other inside the metal sleeves of the rotor.

TECHNIQUE 12. VENTING GASES

Fume hoods (Figure T.12a) are used for removing "undesirable" gases evolved from a reagent such as concentrated hydrochloric acid or from a chemical reaction. These gases may be toxic, corrosive, irritating, or flammable. If there is a question about the use of a fume hood, hedge on the side of safety and/or consult with your instructor.

When using a fume hood:

- turn on the hood air flow before beginning the experiment
- never place your face inside of the fume hood
- set the equipment and chemicals at least 6 inches back from the hood door
- do not crowd experimental apparatus when sharing the use of a fume hood

On occasion the space in the fume hoods is not adequate for an entire class to perform the experiment in a timely manner. With the *approval of your laboratory instructor,* an improvised hood (Figure T.12b) can be assembled. For the operation of an improvised hood, a water aspirator draws the gaseous product from above the reaction

Figure T.12a A modern laboratory fume hood.

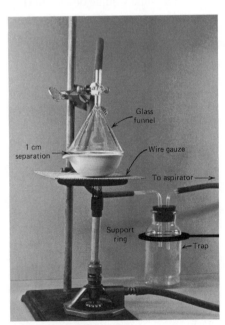

Figure T.12b Position a funnel, connected to a water aspirator, over the escaping gases. A hot plate is often substituted for the Bunsen flame.

vessel—the gas dissolves in the water. To operate the "hood," completely open the faucet that is connected to the aspirator in order to provide the best suction for the removal of the gases. But, as a reminder, *never* substitute an improvised hood for a fume hood if space is available in the fume hood.

Liquids and solutions are often heated, for example, to promote the rate of a chemical reaction to or hasten a dissolution or precipitation, in a number of different vessels. **Caution:** *Flammable liquids should **never** be heated (directly or indirectly) with a flame. Always use a hot plate—refer to Techniques 13A and 13B where hot plates are used.*

Hot liquids and solutions can be cooled by placing the glass vessel either under flowing tap water or in an ice bath.

Nonflammable liquids in beakers or flasks that are more than one-fourth full can be *slowly* heated directly with a hot plate (Figure T.13a). (**Caution:** *hot plates are hot! Do not touch!*) Caution must be taken *not* to heat the liquid too rapidly as "bumping" (the sudden formation of bubbles from the superheated liquid) may occur. To avoid/minimize bumping, place a stirring rod followed by constant stirring or **boiling chips** into the liquid. If a stirring hot plate is used, place the stir bar into the liquid and turn on the stirrer (Figure T.13b).

A direct flame may also be used to heat the liquid in a beaker or flask. Support the beaker or flask on a wire gauze that is centered over an iron ring; use a second iron ring placed around the top of the beaker or flask to prevent it from being knocked off. Position the flame directly beneath the tip of the stirring rod (Figure T.13c) or add boiling chips to avoid/minimize bumping.

TECHNIQUE 13. HEATING LIQUIDS AND SOLUTIONS

A. Beaker or flask

Boiling chips (also called boiling stones): small, porous ceramic pieces—when heated, the air contained within the porous structure is released, gently agitating the liquid and minimizing boiling. Boiling chips also provide nucleation sites on which bubbles can form.

Boiling chips.

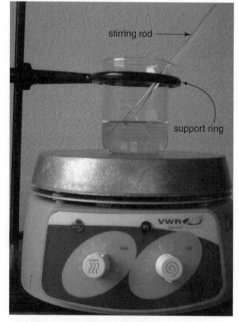

Figure T.13a A hot plate may be used to maintain solutions in a beaker or flask at a constant, elevated temperature for an extended time period.

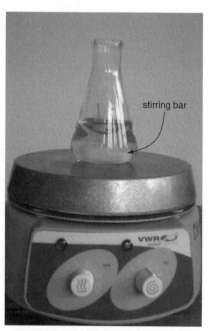

Figure T.13b A stirring hot plate may be used to heat a liquid and minimize "bumping."

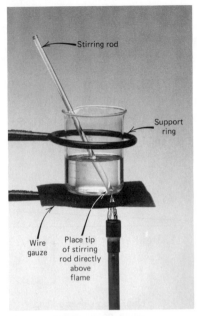

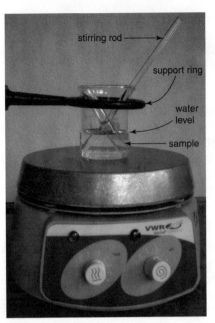

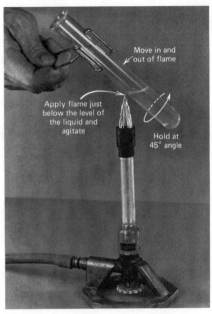

Figure T.13c Place the flame directly beneath the tip of the stirring rod in the beaker. Boiling chips may also be placed in the beaker to avoid "bumping."

Figure T.13d A hot water bath may be used to maintain solution in test tubes at a constant, elevated temperature for an extended time period.

Figure T.13e Move the test tube circularly in and out of the *cool* flame, heating the liquid or solution from top to bottom.

B. Test Tubes

Small quantities of liquids in test tubes that need to be maintained at a constant, elevated temperature over a period of time can be placed in a hot water bath (Figure T.13d). The heat source may be a hot plate or direct flame, depending on the chemicals being used. The setup is the same as that for heating a liquid in a beaker (Technique 13A).

C. Test Tube over a "Cool" Flame

Safety first should be followed when using this technique for heating liquids in test tubes.

A **cool flame** is a nonluminous flame supplied with a reduced supply of fuel. In practice, the rule of thumb for creating a cool flame for heating a liquid in a test tube is as follows: *if you can feel the heat of the flame with the hand that is holding the test tube clamp,* ***the flame is too hot!***

For heating a liquid in a test tube, the test tube should be less than one-third full of liquid. Hold the test tube with a test tube holder at an angle of about 45° with the flame. Move the test tube circularly and continuously in and out of the cool flame, heating from top to bottom, mostly near the top of the liquid (Figure T.13e). **Caution:** *Never fix the position of the flame at the base of the test tube, and never point the test tube at anyone; the contents may be ejected violently if the test tube is not heated properly.*

See Technique 13B for heating a solution in a test tube to a specified elevated temperature; the hot water bath in Technique 13B is a safer, but slower, procedure.

TECHNIQUE 14. EVAPORATING LIQUIDS

To remove a liquid from a vessel by evaporation, the flammability of the liquid must be considered. This is a safety precaution.

Use a fume hood or an improvised hood (Technique 12) as recommended to remove irritating or toxic vapors.

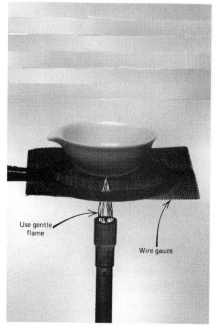

Figure T.14a Evaporation of a nonflammable liquid over a low, direct flame. A hot plate may be substituted for the flame.

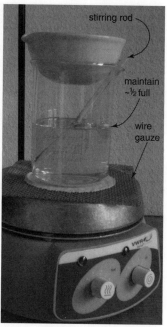

Figure T.14b Evaporation of a flammable liquid over a steam bath using a hot plate for the heat source.

A nonflammable liquid can be evaporated with a direct flame (Figure T.14a). Place the liquid in an evaporating dish centered on a wire gauze and iron ring. Use a gentle, "cool" flame to slowly evaporate the liquid.

A. Use of Direct Heat

Flammable *or* nonflammable liquids can be evaporated using a hot plate as the heat source. Place the liquid in an evaporating dish on top of a beaker according to Figure T.14b. Gentle boiling of the water in the beaker is more efficient than rapid boiling for evaporating the liquid. Avoid breathing the vapors. The use of a fume hood (Technique 12) is strongly recommended if large amounts of liquid are to be evaporated into the laboratory. Consult with your laboratory instructor.

For removing the final dampness from a solid that has formed as a result of the evaporation, consider using a drying oven as described in Technique 15A.

B. Use of Indirect Heat

Solids are heated to dry them or to test their thermal stability. A drying oven is often used for low temperature heating, and porcelain crucibles are used for high temperature heating. Beakers and test tubes can be used for moderately high temperature heating.

TECHNIQUE 15. HEATING SOLIDS

When solid chemicals are left exposed to the atmosphere they often absorb moisture. If an exact mass of a solid chemical is required for a solution preparation or for a reaction, the absorbed water must be removed before the mass measurement is made on the balance. The chemical is often placed in an open container (usually a Petri dish or beaker) in a drying oven (Figure T.15a) set at a temperature well above room temperature (most often at ~110°C) for several hours to remove the adsorbed water. The container is then removed from the drying oven and placed in a desiccator (Technique 15B) for cooling to room temperature. **Caution:** *Hot glass and cold glass look the same—the container from the drying oven is hot and should be handled accordingly.* See your laboratory instructor.

A. Heating in a Drying Oven

Figure T.15a A modern laboratory drying oven.

Figure T.15b A simple laboratory desicooler (left) or a glass desiccator (right) contains a desiccant (usually anhydrous $CaCl_2$) to provide a dry atmosphere.

B. Cooling in a Desiccator

When a Petri dish or beaker containing a solid chemical is cooled in the laboratory, moisture tends to condense on the outer surface, adding to the total mass. To minimize this mass error, and for quantitative work, substances and mixtures that may tend to be hygroscopic are placed into a desiccator (Figure T.15b) until they have reached ambient temperature.

A desiccator is a laboratory apparatus that provides a dry atmosphere. A desiccant, typically anhydrous calcium chloride, absorbs the water vapor from within the enclosure of the desiccator. The anhydrous calcium chloride forms $CaCl_2 \cdot 2H_2O$; the hydrated water molecules can be easily removed with heat (modified Technique 14A), and the calcium chloride can be recycled for subsequent use in the desiccator.

C. Using a Crucible

For high temperature combustion or decomposition of a chemical, porcelain crucibles are commonly used. To avoid contamination of the solid sample, thoroughly clean the crucible (so it is void of volatile impurities) prior to use. Often-used crucibles tend to form stress fractures or fissures. Check the crucible for flaws; if any are found return the crucible to the stockroom and check out and examine a second crucible.

1. **Drying and/or Firing the Crucible.** Support the crucible and lid on a clay triangle (Figure T.15c) and heat in a hot flame until the bottom of the crucible glows a dull red. Rotate the crucible with crucible tongs to ensure complete "firing" of the crucible, i.e., the combustion and volatilization of any impurities in the crucible. Allow the crucible and lid to cool to room temperature while on the clay triangle or *after* several minutes in a desiccator (Technique 15B). If the crucible still contains detectable impurities, add 1–2 mL of 6 *M* HNO_3 (**Caution:** *avoid skin contact, flush immediately with water*) and evaporate *slowly* to dryness in the fume hood.

2. **Igniting Contents in the Absence of Air.** To heat a solid sample to a high temperature but *not* allow it to react with the oxygen of the air, set the crucible upright in the clay triangle with the lid covering the crucible (Figure T.15d). Use the crucible tongs to adjust the lid.

3. **Igniting Contents for Combustion.** To heat a solid sample to a high temperature and allow it to react with the oxygen of the air, slightly tilt the crucible on the clay triangle and adjust the lid so that about two-thirds of the crucible remains covered (Figure T.15e). Use the crucible tongs to adjust the lid.

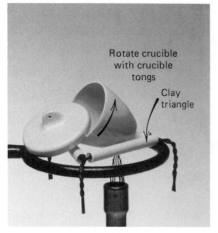

Figure T.15c Drying and/or firing a crucible and cover.

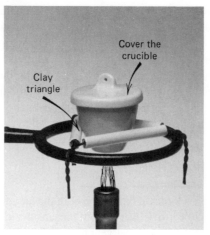

Figure T.15d Ignition of a solid sample in the absence of air.

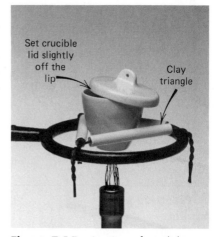

Figure T.15e Ignition of a solid sample in the presence of air for complete combustion.

The careful measurement and recording of volumes of liquids are necessary to obtain quantitative data for a large number of chemical reactions that occur in solutions. Volumes must be read and recorded as accurately as possible to minimize errors in the data.

TECHNIQUE 16. MEASURING VOLUME

1. **Reading a Meniscus.** For measurements of liquids in graduated cylinders, pipets, burets, and volumetric flasks, the volume of a liquid is read at the *bottom of its meniscus.* Position the eye horizontally at the bottom of the meniscus (Figure T.16a) to read the level of the liquid. A clear or transparent liquid is read more easily, especially in a buret, by positioning a black mark (made on a white card) behind or just below the level portion of the liquid. The black background reflects off the bottom of the meniscus and better defines the level of the liquid (Figure T.16b). Substituting a finger for the black mark on the white card also helps in detecting the bottom of the meniscus but is not as effective.

A. Reading and Recording a Meniscus

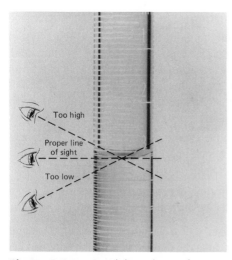

Figure T.16a Read the volume of a liquid with the eye horizontal to the bottom of the meniscus.

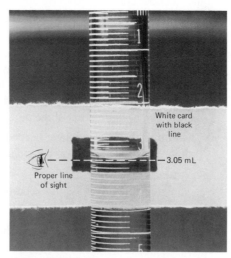

Figure T.16b Use a black line drawn on a white card to assist in pinpointing the location of the bottom of the meniscus.

Volumetric glassware: glassware that has a calibration mark(s) that indicates a calibrated volume, as determined by the manufacturer

2. **Recording a Volume.** Record the volume of a liquid in **volumetric glassware** using all certain digits (from the labeled calibration marks on the glassware) *plus* one uncertain digit (the last digit which is the best estimate between the calibration marks). This reading provides the correct number of significant figures for the measurement. See **Laboratory Data,** Part B. In Figure T.16b, the volume of solution in the buret is between the calibration marks of 3.0 and 3.1; the "3" and the "0" are certain, the "5" is the estimate between 3.0 and 3.1. The reading is 3.05 mL. Be aware that all volume readings do *not* end in "0" or "5!"

This guideline for reading volumes also applies to reading and recording temperatures on a thermometer.

B. Pipetting a Liquid

Dust/lint-free tissue.

The most common type of pipet in the laboratory is labeled TD at 20°C. A pipet labeled TD at 20°C (to deliver at 20°C) means that the volume of the pipet is calibrated according to the volume it delivers from gravity flow only.

A clean pipet in conjunction with the proper technique for dispensing a liquid from a pipet are important in any quantitative determination.

1. **Preparation of the Pipet.** See Technique 2 for cleaning glassware. A clean pipet should have no water droplets adhering to its inner wall. Inspect the pipet to ensure it is free of chips or cracks. Transfer the liquid that you intend to pipet from the reagent bottle into a clean, dry beaker; do *not* insert the pipet tip directly into the reagent bottle (Technique 5). Dry the outside of the pipet tip with a clean, dust-free towel or tissue (e.g., Kimwipe). Using the suction from a collapsed rubber (pipet) bulb, draw a 2- to 3-mL portion into the pipet as a rinse. Roll the rinse around in the pipet to make certain that the liquid washes the entire surface of the inner wall. Deliver the rinse through the pipet tip into a waste beaker and discard as directed in the experiment. Repeat the rinse 2–3 times.

2. **Filling of the Pipet.** Place the pipet tip well below the surface of the liquid in the beaker. Using the collapsed pipet bulb (or a pipet pump—*never* use your mouth!), draw the liquid into the pipet until the level is 2–3 cm above the "mark" on the pipet (Figure T.16c). Do not "jam" the pipet bulb onto the pipet! Remove the bulb and quickly cover the top of the pipet with your index finger (*not* your thumb!). Remove the tip from the liquid and wipe off the pipet tip with a clean, dust-free towel or tissue. Holding the pipet in a *vertical* position over a waste beaker, control the delivery of the excess liquid until the level is "at the mark" in the pipet (Figure T.16d). Read the meniscus correctly. Remove any drops suspended from the pipet tip by touching it to the wall of the waste beaker. This is a technique you will need to practice.

3. **Delivery of the Liquid.** Deliver the liquid to the receiving vessel (Figure T.16e) by releasing the index finger from the top of the pipet. Dispense the liquid along the wall of the receiving vessel to avoid splashing. To remove a hanging drop from the pipet tip, touch the side of the receiving flask for its removal. Do *not* blow or shake out the last bit of liquid that remains in the tip; this liquid has been included in the calibration of the pipet . . . remember this is a TD at 20°C pipet!

4. **Cleanup.** Once it is no longer needed in the experiment, rinse the pipet with several portions of deionized water and drain each rinse through the tip.

C. Titrating a Liquid (Solution)

Titrant: the reagent solution in the buret to be used for the experiment

A clean buret in conjunction with the proper technique for measuring and dispensing a liquid from a buret is important in any quantitative analysis determination.

1. **Preparation of the Buret.** See Technique 2 for cleaning glassware. If a buret brush is needed, be careful to avoid scratching the buret wall with the wire handle. Once the buret is judged to be "clean," close the stopcock. Rinse the buret with several 3- to 5-mL portions of water and then **titrant.** Tilt and roll the barrel of the buret so that the rinse comes into contact with the entire inner wall. Drain each

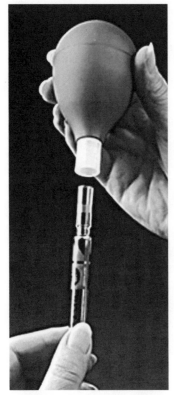

Figure T.16c Draw the liquid into the pipet with the aid of a rubber pipet bulb (*not* the mouth!).

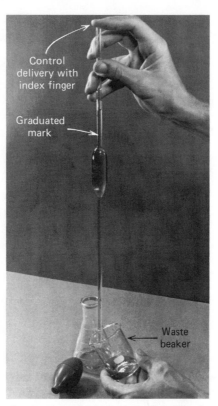

Figure T.16d Control the delivery of the liquid from the pipet with the forefinger (*not* the thumb!).

Control delivery with index finger

Graduated mark

Waste beaker

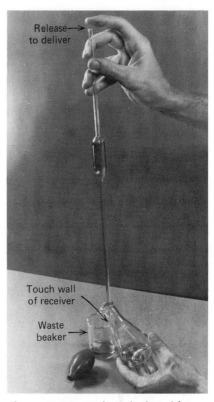

Figure T.16e Deliver the liquid from the vertically positioned pipet with the tip touching the wall of the receiving flask.

Release to deliver

Touch wall of receiver

Waste beaker

rinse through the buret tip into the waste beaker. Dispose of the rinse as advised in the experiment. Support the buret with a buret clamp (Figure T.16f).

2. **Preparation of the Titrant.** Close the stopcock. With the aid of a *clean* funnel, fill the buret with the titrant to just above the zero mark. Open the stopcock briefly to release any air bubbles in the tip *and* allow the meniscus of the titrant to go below the uppermost graduation on the buret. Allow 10–15 seconds for the titrant to drain from the wall; **record the volume** (± 0.02 mL, Technique 16A.2) of titrant in the buret. Note that the graduations on a buret *increase* in value from the top down (Figure T.16g).

3. **Operation of the Buret.** During the addition of the titrant from the buret, operate the stopcock with your left hand (if right-handed) and swirl the Erlenmeyer flask with your right hand (Figure T.16h). This prevents the stopcock from sliding out of its barrel and allows you to maintain a normal, constant swirling motion of the reaction mixture in the receiving flask as the titrant is added. The opposite procedure, of course, is applicable if you are left-handed (Figure T.16i). Use an Erlenmeyer flask as a receiving flask in order to minimize the loss of solution due to splashing.

   If a magnetic stirrer is used for mixing the titrant with the analyte, operate at a low speed to minimize any splashing of the reaction mixture and to minimize oxygen of the air from mixing and reacting with either the titrant or analyte. Occasionally, a beaker is used as a receiving flask, especially if a temperature probe (or thermometer) or a pH probe is required during the titration.

4. **Addition of Titrant to Receiving Flask.** Have a white background (a piece of white paper) beneath the receiving flask to better see the endpoint for the

Record the volume: read the volume in the buret using all certain digits (from the labeled calibration marks on the buret) plus one uncertain digit (the last digit which is the best estimate between the calibration marks)

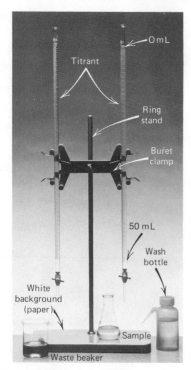

Figure T.16f Setup for a titration analysis.

Figure T.16g A 50-mL buret is marked from top to bottom, 0 to 50 mL, with 1-mL gradations divided into 0.1-mL increments.

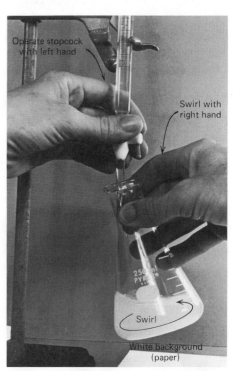

Figure T.16h Titration technique for right-handers.

Figure T.16i Titration technique for left-handers.

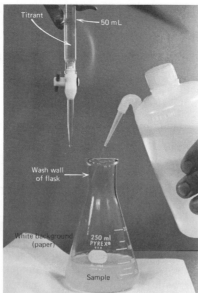

Figure T.16j Place a white background beneath the receiving flask and wash the wall of the receiving flask periodically during the titration.

Figure T.16k A slow color fade of the indicator occurs near the endpoint in the titration.

titration (the point at which the indicator turns color). If the endpoint is a change from colorless to white, a black background is preferred. Add the titrant to the Erlenmeyer flask as described above; periodically stop its addition and wash the inner wall of the flask with the solvent (generally deionized water) from a wash bottle (Figure T.16j). Near the endpoint (slower color fade of the indicator, Figure T.16k), slow the rate of titrant addition until a drop (or less) makes the color change of the indicator persist for 30 seconds. **Stop,** allow 10–15 seconds for the titrant to drain from the buret wall, read, and record the volume in the buret (Technique 16A.2).

To add less than a drop of titrant (commonly referred to as a "half-drop") to the receiving flask, suspend a drop from the buret tip, touch it to the side of the receiving flask, and wash the wall of the receiving flask (with deionized water).

5. **Cleanup.** After completing a series of titrations, drain the titrant from the buret, rinse the buret with several portions of deionized water, and drain each rinse through the tip. Discard the excess titrant and the rinses as advised in the experiment. Store the buret as advised by your laboratory instructor.

An educated nose is an important and very useful asset to the chemist. Use it with caution, however, because some vapors induce nausea and/or are toxic. *Never* hold your nose directly over a vessel. Fan some vapor toward your nose (Figure T.17a). *Always* consult your laboratory instructor before testing the odor of any chemical.

TECHNIQUE 17.
QUICK TESTS
A. Testing for Odor

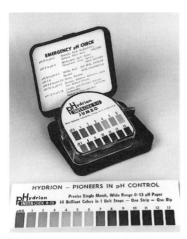

Figure T.17a Fan the vapors gently toward the nose.

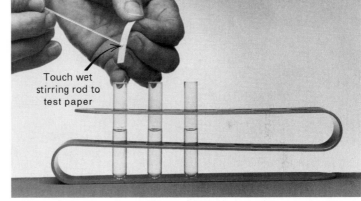

Touch wet stirring rod to test paper

Figure T.17b Test for acidity/basicity.

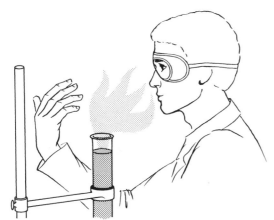

Figure T.17c Test papers impregnated with a mixture of acid-base indicators can be used to measure the approximate pH of a solution.

B. Testing for Acidity/Basicity

To test the acidity/basicity of a solution with test paper, insert a *clean* stirring rod into the solution, withdraw it, and touch it to the pH test paper (Figure T.17b). For litmus paper acidic solutions turn blue litmus red; basic solutions turn red litmus blue. *Never* place the test paper directly into the solution.

Other paper-type indictors, such as pHydrion paper (Figure T.17c), are also used to gauge the acidity/basicity of a solution.

Disclaimer: The material contained in the Laboratory Safety and Laboratory Techniques sections of this manual has been compiled from sources believed to be reliable and to represent the best opinions of safety in the laboratory. This manual is intended to provide basic guidelines for safe practices in the undergraduate chemistry laboratory. It cannot be assumed that all necessary warning and precautionary measures are contained in this manual, or that other or additional information or measures may not be required.

Further discussions of these and other laboratory techniques can be found on the World Wide Web. Refer to Laboratory Data, Part C.

Notes on Laboratory Techniques

Laboratory Techniques

Date _____ Lab Sec. _____ Name _____ Desk No. _____

Identify the Technique Icon, at left, for each of the following techniques.

Technique Icon   *Description of technique*

1. _____ Glassware is clean when no water droplets cling to the inner wall of the vessel.

2. _____ Transfer liquids/solutions from a reagent bottle or beaker with the aid of a stirring rod.

3. _____ A buret should be rinsed with several 3- to 5-mL portions of titrant before being filled.

4. _____ A "small" test tube has a volume of ~3 mL.

5. _____ Information on the properties and disposal of chemicals can be found in the stockroom or online from the MSDS collection.

6. _____ The bowl of the funnel for gravity filtration should be always less than two-thirds full.

7. _____ The color change of the indicator at the endpoint of a titration should persist for 30 seconds.

8. _____ The volume of a liquid/solution in calibrated volumetric glassware should be read using all-certain digits (from the labeled calibration marks on the glassware) *plus* one uncertain digit (the last digit which is the best estimate *between* the calibration marks).

9. _____ For heating a liquid in a test tube, the test tube should be less than one-third full, moved continuously in and out of the "cool" flame at a 45° angle, mostly near the top of the liquid.

10. _____ A centrifuge should be balanced with an even number of centrifuge/test tubes, placed across the rotor from one another, with equal volumes of liquid.

11. _____ The volume of a liquid should be read at the bottom of the meniscus.

12. _____ Water or glycerol should be applied to the glass tubing and the hole in the rubber stopper before inserting the glass tubing.

13. _____ Small quantities of liquids in test tubes are best be heated (safely) in a hot water bath with the heat source being either a hot plate or a direct flame.

14. _____ To dispense chemicals, read the label at least twice before removing any chemical from the reagent bottle.

15. _____ All chemicals must be properly disposed—either according to the Experimental Procedure or the laboratory instructor.

16. _____ To prepare a solution, always add the solid reagent or more concentrated solution to a volumetric flask that is already approximately one-third full with solvent.

17. _____ Never place reagents directly onto the weighing pan of a balance—always use weighing paper, a beaker, or some other container.

18. _____ The suction created for the vacuum filtration of a precipitate is applied with an aspirator in which the water faucet is fully open.

19. _____ A hot plate (*not* an open flame) is the heat source for heating or evaporating flammable liquids.

20. _____ To dispense a liquid from a pipet, first draw the liquid into the pipet 2–3 cm above the calibration mark with a pipet bulb and then use the index finger to control its flow.

21. _____ A litmus paper test for the acidity or basicity of a solution requires the use of a stirring rod to remove a portion of the solution to then be touched to the litmus paper.

22. _____ While dispensing titrant from a buret, the stopcock should be operated with the left hand (if right-handed) and the Erlenmeyer flask should be swirled with the right hand.

23. _____ After drying a hygroscopic solid in a drying oven, the solid should be cooled in a desiccator.

24. _____ To heat a solid to high temperatures in the absence of air, place the solid in a crucible and fully cover the crucible with the crucible lid.

25. _____ The tared mass of a sample is its mass without regard to its container.

True or False

Ask your instructor to identify the questions you are to complete.

_____ **1.** All clean glassware should be air-dried naturally.

_____ **2.** While cleaning glassware, discard all washes and rinses from the delivery point of the glass vessel.

_____ **3.** To avoid waste in the use of chemicals, share the unused portion with other chemists before discarding.

_____ **4.** Never touch, taste, or smell a chemical unless specifically told to do so.

_____ **5.** Most all chemicals used in experiments can be discarded into the sink.

_____ **6.** A dried solid should be stored in a desiccator.

_____ **7.** A 3-inch test tube has a volume of 3 mL; an 8-inch test tube must have a volume of 8 mL.

_____ **8.** To transfer a solution, a stirring rod touches the delivery point of the reagent vessel and the wall of the receiving vessel.

_____ **9.** Only one sample of a liquid mixture should be present when a centrifuge is in operation.

_____ **10.** A test tube should be more than one-third full when heating with a direct flame.

_____ **11.** If the heat of a flame is not felt with the hand holding the test tube clamp (holding the test tube), the flame is considered a cool flame.

_____ **12.** Do not blow out the solution remaining in the pipet tip after the solution has drained from the pipet.

_____ **13.** A buret must *always* be filled to the top (the zero mark) before every titration procedure.

_____ **14.** The volume of solution in a buret should be read and recorded 10–15 seconds after completing the titration.

_____ **15.** It is possible to add a half-drop of solution from a buret.

_____ **16.** To test the acidity of a solution with pH paper, place the pH paper directly in the solution.

_____ **17.** The odor of a chemical should not be tested unless specifically instructed to do so. The vapors of the chemical should be fanned toward the nose.

Summarize the "Disclaimer" in your own words.

A set of standard SI mass units.

The Laboratory and SI

- To check into the laboratory
- To become familiar with the laboratory and the laboratory manual
- To learn the rules of Laboratory Safety and the necessity of practicing these rules in the laboratory
- To learn how to properly organize and record Laboratory Data
- To develop skills in the use of *Le Système International d' Unités* (SI Units)

All chemical principles, tools, and techniques are developed in the laboratory. The experience of observing a chemical phenomenon and then explaining its behavior is one that simply cannot be gained by reading a textbook, listening to a lecturer, or viewing a video. It is in the laboratory where chemistry comes alive, where chemical principles are learned and applied to the vast natural "chemistry laboratory" that we call our everyday environment. The objectives of a laboratory experience are to design and build apparatus, develop techniques, observe, record and interpret data, and deduce rational theories so that the real world of science is better explained and understood.

In the laboratory, you will use common equipment and safe chemicals to perform experiments. You record your experimental observations and interpret the data on the basis of sound chemical principles. A good scientist is a thinking scientist trying to account for the observed data and rationalize any contradictory data. Cultivate self-reliance and confidence in your data, even if "the data do not look right." This is how many breakthroughs in science occur.

In the first few laboratory sessions you will be introduced to some basic rules, equipment, and techniques and some situations where you use them. These include laboratory safety rules, *Le Système International d' Unités* (SI Units), the Bunsen burner, and the analytical balance. Additional laboratory techniques are illustrated under Laboratory Techniques pages 11–34; others are introduced as the need arises.

Procedure Overview: Laboratory procedures are introduced. A familiarity with laboratory apparatus, the policies regarding laboratory safety, the procedures for presenting laboratory data, and an encounter with the SI units are emphasized.

At the beginning of the first laboratory period you are assigned a lab station containing laboratory equipment. Place the laboratory equipment on the laboratory bench and, with the check-in list provided on page 9, check off each piece of chemical apparatus as you return it to the drawer. If you are unsure of a name for the apparatus, refer to the list of chemical "kitchenware" on pages 8–9. Ask your instructor about any items on the check-in list that are not at your lab station.

The icon refers to Laboratory Technique 2 on page 13. Read through Laboratory Technique 2 for the proper technique of cleaning laboratory glassware.

A good scientist is always neat and well organized; keep your equipment clean and arranged in an orderly manner so that it is ready for immediate use. Have at your lab station dishwashing soap (or detergent) for cleaning glassware and paper towels or a rubber lab mat on which to set the clean glassware for drying.

Obtain your laboratory instructor's approval for the completion of the check-in procedure. Refer to Part A of the Report Sheet.

B. Laboratory Safety and Guidelines

Your laboratory instructor will discuss laboratory safety and other basic laboratory procedures with you. Remember, however, that your laboratory instructor cannot practice laboratory safety for you or for those who work with you—it is your responsibility to *play it safe!*

On the inside front cover of this manual is space to list the location of some important safety equipment and important information for reference in the laboratory. Fill this out. Obtain your laboratory instructor's approval for its completion.

Read and study the **Laboratory Safety and Guidelines** section on pages 1–4. Complete any other laboratory safety sessions or requirements that are requested by your laboratory instructor. Answer the laboratory safety questions on the Report Sheet.

C. Laboratory Data

Each of the experiments in this manual will require you to observe, record and report data, perform calculations on the data, and then analyze and interpret the data. Scientists are careful in the procedures that are used for handling data so that the reliability and credibility of the experimental data are upheld.

It is important that these procedures are followed at the outset of your laboratory experience. Read and study the **Laboratory Data** section on pages 5–8. Answer the laboratory data questions on the Report Sheet.

D. *Le Système International d'Unités* (SI Units)

A microliter syringe.

The SI, a modern version of the metric system, provides a logical and interconnected framework for all basic measurements. The SI, and some of its slight modifications, is used throughout the world by scientists and engineers as the international system for scientific measurements and, in most countries, for everyday measurements. For laboratory measurements, the SI base unit of mass is the **kilogram (kg)** [chemists are most familiar with the **gram (g)**, where 10^3 g = 1 kg], the SI base unit for length is the **meter (m)** [chemists are most familiar with several subdivisions of the meter], and the SI derived unit for volume is the **cubic meter (m^3)** [chemists are most familiar with the **liter (L)**, where 1 L = 10^{-3} m^3 = 1 dm^3]. Subdivisions and multiples of each unit are related to these units by a power of ten. The prefixes used to denote these subdivisions and multiples are shown in Table D1.1. Memorize these prefixes and their meanings.

Table D1.1 Prefixes in *Le Système International d'Unités*

Prefix	Abbreviation	Meaning (power of ten)	Example Using "grams"
femto-	f	10^{-15}	fg = 10^{-15} g
pico-	p	10^{-12}	pg = 10^{-12} g
nano-	n	10^{-9}	ng = 10^{-9} g
micro-	μ	10^{-6}	μg = 10^{-6} g
milli-	m	10^{-3}	mg = 10^{-3} g
centi-	c	10^{-2}	cg = 10^{-2} g
deci-	d	10^{-1}	dg = 10^{-1} g
kilo-	k	10^3	kg = 10^3 g
mega-	M	10^6	Mg = 10^6 g
giga-	G	10^9	Gg = 10^9 g
tera-	T	10^{12}	Tg = 10^{12} g

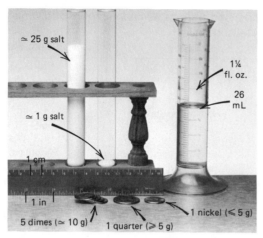

Figure D1.1 Comparisons of SI and English measurements.

In Table D1.1, *the prefix indicates the power of ten.* For example, 4.3 *milli*grams means 4.3×10^{-3} grams; *milli* has the same meaning as "$\times 10^{-3}$." Figure D1.1 shows representative SI measurements for mass, length, and volume.

Conversions of measurements within SI are quite simple if the definitions for the prefixes are known and unit conversion factors for problem solving are used. To illustrate the use of unit conversion factors, consider the following example.

Example D1.1 Convert 4.3 nanograms to micrograms.

Solution. From Table D1.1, note that nano- and 10^{-9} are equivalent and that micro- and 10^{-6} are equivalent. Considering mass, it also means that $ng = 10^{-9}$ g *and* $\mu g = 10^{-6}$ g. This produces two equivalent unit conversion factors for each equality:

$$\frac{10^{-9} \text{ g}}{ng}, \quad \frac{ng}{10^{-9} \text{ g}} \quad \text{and} \quad \frac{10^{-6} \text{ g}}{\mu g}, \quad \frac{\mu g}{10^{-6} \text{g}}$$

Beginning our solution to the problem with the measured quantity (i.e., 4.3 ng), we need to convert ng to g and g to μg. We multiply by the appropriate conversion factors to obtain proper unit cancellation:

$$4.3 \text{ ng} \times \frac{10^{-9} \text{ g}}{ng} \times \frac{\mu g}{10^{-6} \text{ g}} = 4.3 \times 10^{-3} \mu g$$

ng cancel g cancel

$$ng \quad \rightarrow \quad g \quad \rightarrow \quad \mu g \quad = \quad \mu g$$

The conversion factors in Example D1.1 have no effect on the magnitude of the mass measurement (the conversion factors = 1), only the units by which it is expressed.

The SI is compared with the English system in Table D1.2. SI units of measurement that chemists commonly use in the laboratory are listed in brackets. Appendix A has a more comprehensive table of conversion factors. Conversions between the SI and the English system are quite valuable, especially to Americans because international science and trade communications are in SI or metric units.

Appendix A

Table D1.2 Comparison of *Le Système International d'Unités* and English System of Measurements[†]

Physical Quantity	SI Unit	Conversion Factor
Length	meter (m)	1 km = 0.6214 mi 1 m = 39.37 in. 1 in. = 0.0254 m = 2.54 cm
Volume	cubic meter (m^3) [liter (L)][a]	1 L = 10^{-3} m^3 = 1 dm^3 = 10^3 mL 1 mL = 1 cm^3 1 L = 1.057 qt 1 oz (fluid) 29.57 mL
Mass	kilogram (kg) [gram (g)]	1 lb = 453.6 g 1 kg = 2.205 lb
Pressure	pascal (Pa) [atmosphere (atm)]	1 Pa = 1 N/m^2 1 atm = 101.325 kPa = 760 torr 1 atm = 14.70 lb/in^2 (psi)
Temperature	kelvin (K) [degrees Celsius (°C)]	K = 273 + °C $°C = \dfrac{°F - 32}{1.8}$
Energy	joule (J)	1 cal = 4.184 J 1 Btu = 1054 J

[a]The SI units enclosed in brackets are commonly used in chemical measurements and calculations.
[†]For additional conversion factors, go online to *www.onlineconversion.com*.

Graduated cylinders of different volumes.

Example D1.2 Using Tables D1.1 and D1.2, determine the volume of 1.00 quart of water in terms of cubic centimeters.
Solution. As 1 cm = 10^{-2} m, then 1 cm^3 = $(10^{-2}$ m$)^3$ = $(10^{-2})^3$ m^3.
From Table D1.2, we need conversion factors for quarts → liters, liters → m^3, and finally m^3 → cm^3. Starting with 1.00 qt (our measured value in the problem), we have:

$$1.00 \text{ qt} \times \frac{1 \text{ L}}{1.057 \text{ qt}} \times \frac{10^{-3} \text{ m}^3}{1 \text{ L}} \times \frac{cm^3}{(10^{-2})^3 \text{ m}^3} = 946 \text{ cm}^3$$

qt cancel   L cancel m^3 cancel

qt ⟶ L ⟶ m^3 ⟶ cm^3 = cm^3

Note again that the conversion factors in Example D1.2 do not change the magnitude of the measurement, only its form of expression.
Since 1 cm^3 = 1 mL, 1.00 quart is equivalent to 946 mL.

The Report Sheet, Part D, further acquaints you with conversions within the SI and between SI and the English system. Ask your laboratory instructor which of the questions from Part D you are to complete on the Report Sheet.

The Laboratory and SI

Date _____ Lab Sec. _____ Name _____ Desk No. _____

A. Laboratory Check-in

Instructor's approval _____

B. Laboratory Safety and Guidelines

Instructor's approval for completion of inside front cover _____

Read the **Laboratory Safety and Guidelines** section on pages 1–4 and answer the following as **true** or **false**.

_____ 1. Prescription glasses, which are required by law to be "safety glasses," can be worn in place of safety goggles in the laboratory.

_____ 2. Sleeveless blouses and tank tops are *not* appropriate attire for the laboratory.

_____ 3. Only shoes that shed liquids are permitted in the laboratory.

_____ 4. "I just finished my tennis class. I can wear my tennis shorts to lab just this one time, right?"

_____ 5. Your laboratory has an eye wash fountain.

_____ 6. "Oops! I broke a beaker containing deionized water in the sink." An accident as simple as that does not need to be reported to the laboratory instructor.

_____ 7. A beaker containing an acidic solution has broken on the bench top and spilled onto your clothes from the waist down and it burns. Ouch! You should immediately proceed to the safety shower and flood the affected area.

_____ 8. You received a "paper cut" on your finger and it is bleeding. Immediately go to the medicine cabinet to apply a disinfectant.

_____ 9. It is good laboratory protocol to inform other students when they are not practicing good laboratory safety procedures. If they continue to not follow the safety procedures, you should "rat" on them . . . tell the laboratory instructor.

_____ 10. You work more efficiently with the accompaniment of your favorite CD. Therefore, your music played aloud or with your MP3 player in the laboratory is an acceptable laboratory procedure.

_____ 11. Your friend is a senior chemistry major and thoroughly understands the difficult experiment that you are performing. Therefore, it is advisable (even recommended) that you invite him/her into the laboratory for direct assistance.

_____ 12. You missed lunch but brought a sandwich to the laboratory. Since you can't eat in the lab, it is "ok" to leave the sandwich in the hallway and then go in/out to take bites while the laboratory experiment is ongoing.

Write a short response for the following questions.

1. What does the phrase "neck to knee to wrist" mean with regard to laboratory safety?

2. The first action after an accident occurs is:

3. You want to try a variation of the Experimental Procedure because of your chemical curiosity. What is the proper procedure for performing the experiment?

4. A chemical spill has occurred. What should be your first and second action in treating the chemical spill?

5. Describe how you will be dressed when you are about to begin an experiment in the laboratory.

Instructor's approval of your knowledge of laboratory safety. _____

C. Laboratory Data

Read the Laboratory Data section on pages 5–8 and answer the following as **true** or **false**.

_____ **1.** "Quick data," such as that of a mass measurement on a balance located at the far side of the laboratory, can be recorded on a paper scrap and then transferred to the Report Sheet at your lab station.

_____ **2.** Data that has been mistakenly recorded on the Report Sheet can be erased and replaced with the correct data. This is to maintain a neat Report Sheet.

_____ **3.** All data should be recorded in permanent ink!

_____ **4.** The laboratory equipment and instrumentation determine the number of significant figures used to record quantitative data.

_____ **5.** Zeros recorded in a measurement are *never* significant figures.

D. *Le Système International d'Unités* (SI Units)

Circle the questions that have been assigned.

1. Complete the following table.

	SI Expression	Power of Ten Expression	SI Expression	Power of Ten Expression
Example	1.2 mg	1.2×10^{-3} g		
a.	3.3 gigabytes	_____	c. _____	6.72×10^{-3} ampere
b.	_____	7.6×10^{-6} L	d. 2.16 kilowatts	_____

2. Convert each of the following using the definitions in Table D1.1 and unit conversion factors. Show the cancellation of units.

a. 4.76 pm $\times$ $\dfrac{\text{m}}{\text{pm}}$ $\times$ $\dfrac{\mu\text{m}}{\text{m}}$ $=$ _____μm

b. 250 mL $\times$ _____ $\times$ _____ $=$ _____cL

3. Determine the volume (in mL) of 1.0 teaspoon. 1 tablespoon = 3 teaspoons; 1 tablespoon = ½ fluid ounce

4. A basketball is inflated to 9.0 psi (pounds per square inch) above atmospheric pressure of 14.7 psi (a total pressure of 23.7 psi). The diameter of the basketball is 10 inches.

a. What is the pressure of the air in the basketball, expressed in atmospheres?

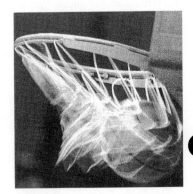

b. What is the volume (in liters) of the basketball? The volume of a sphere $= \frac{4}{3}\pi r^3$.

5. Hurricane Katrina, which hit the Louisiana/Mississippi Gulf Coast on August 29, 2005, had the second lowest ever recorded barometric pressure at 920 mb. Convert this pressure to units of atmospheres, kilopascals, and inches of Hg (the latter of which you would see on the evening weather report).

6. Measure the inside diameter and length of a test tube in centimeters:

diameter = _____ length = _____
Using the equation $V = \pi r^2 l$, calculate the volume of the test tube.

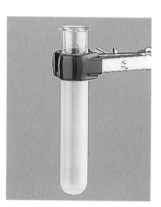

7. A family used 392 kilowatt·hours, kwh, of electricity during the month of October. Express this amount of energy in joules and Btu. See Appendix A.

8. The standard width between rails on North American and most European railroads is 4 ft 8 in. Calculate this distance in meters and centimeters.

9. The heat required to raise the temperature of a large cup of water (for coffee) from room temperature to boiling is approximately 100 kJ. Express this quantity of heat in kilocalories and British thermal units. Show your calculation

10. An aspirin tablet has a mass of about 325 mg (5 grains). Calculate the total mass of aspirin tablets in a 250-tablet bottle in grams and ounces.

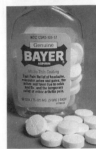

11. A concrete pile from a waterfront pier was pulled from the harbor water at Port Hueneme, California. Its dimensions were 14 inches × 14 inches × 15 feet (1 in = 2.54 cm).

    a. What is the surface area, expressed in square feet, of a single face of the pile? Exclude the ends of the pile.

    b. Determine the total surface area, expressed in square meters, of the pile, including the ends of the pile.

    c. How many cubic meters of concrete were used to make the pile?

Experiment 1

Basic Laboratory Operations

A properly adjusted Bunsen flame is a blue, nonluminous flame.

- To light and properly adjust the flame of a Bunsen burner
- To develop the skill for properly operating a balance
- To develop the technique of using a pipet
- To determine the density of an unknown substance

The following techniques are used in the Experimental Procedure

You will use a number of techniques repeatedly throughout your laboratory experience. Seventeen principal techniques are fully described under Laboratory Techniques in this manual (pps. 11–34). The application of a technique in the Experimental Procedure is noted in the page margin by the corresponding technique icon. Become familiar with each technique and icon before using it in the experiment.

In this experiment you will learn several common techniques that are used repeatedly throughout this laboratory manual—you will learn to light and adjust a Bunsen burner, to use a laboratory balance, and to use a pipet. With the skills developed in using a balance and pipet, you will determine the density of a metal and a liquid.

Bunsen Burner

Laboratory burners come in many shapes and sizes, but all accomplish one main purpose: a combustible gas–air mixture yields a hot, efficient flame. Because Robert Bunsen (1811–1899) was the first to design and perfect this burner, his name is given to most burners of this type used in the general chemistry laboratory (Figure 1.1).

The combustible gas used to supply the fuel for the Bunsen burner in most laboratories is natural gas. Natural gas is a mixture of gaseous **hydrocarbons,** but primarily the hydrocarbon methane, CH_4. If sufficient oxygen is supplied, methane burns with a blue, **nonluminous** flame, producing carbon dioxide and water as combustion products.

$$CH_4(g) + 2\ O_2(g) \rightarrow CO_2(g) + 2\ H_2O(g) \tag{1.1}$$

With an insufficient supply of oxygen, small carbon particles are produced which, when heated to **incandescence,** produce a yellow, luminous flame. The combustion products may, in addition to carbon dioxide and water, include carbon monoxide.

Hydrocarbon: a molecule consisting of only the elements carbon and hydrogen

Nonluminous: nonglowing or nonilluminating

Incandescence: glowing with intense heat

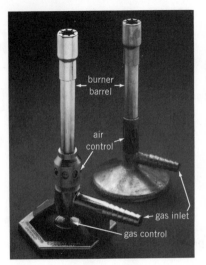

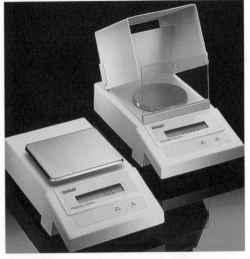

Figure 1.1 Bunsen-type burners.

Figure 1.2 Two balances with different sensitivities.

Laboratory Balances

The laboratory balance is perhaps the most often used piece of laboratory apparatus. Balances, *not* scales, are of different makes, models, and sensitivities (Figure 1.2). The selection of the appropriate balance for a mass measurement depends upon the degree of precision required for the analysis. The top-loading balances are the most common. Read Technique 6 under Laboratory Techniques (p. 16) for instructions on the proper operation of the laboratory balance.

Density

Intensive property: property independent of sample size

$$density = \frac{mass}{volume}$$

Each pure substance exhibits its own set of **intensive properties.** One such property is **density,** the mass of a substance per unit volume. In the English system the density of water at 4°C is 8.34 lb/gal or 62.2 lb/ft^3, whereas in SI the density of water at 4°C is 1.00 g/cm^3 or 1.00 g/mL. By measuring the mass and volume of a substance, its density can be calculated.

In this experiment, the data for the mass and the volume of water displaced (see Figure 1.5) are used to calculate the density of the water-insoluble solid; the density of an unknown liquid is calculated from separate mass and volume measurements of the liquid.

Chemists conventionally express the units for density in the SI as g/cm^3 for solids, g/mL for liquids, and g/L for gases at specified temperatures and pressures.

EXPERIMENTAL PROCEDURE

Procedure Overview: A Bunsen flame is ignited, adjusted, and analyzed. Various laboratory balances are operated and used. Mass and volume data are collected and used to determine the density of a solid and of a liquid.

Perform the experiment with a partner. At each circled superscript $^{1-10}$ in the procedure, *stop,* and record your observation on the Report Sheet. Discuss your observations with your lab partner and your instructor.

A. Bunsen Burner

1. **Lighting the Burner.** Properly light a burner using the following sequence of steps:

    a. Attach the tubing from the burner to the gas outlet on the lab bench. Close the gas control valve on the burner (see Figure 1.1) and fully open the gas valve at the outlet.

    b. Close the air control holes at the base of the burner and slightly open the gas control valve.

c. Bring a lighted match or striker up the outside of the burner barrel until the escaping gas at the top ignites.

d. After the gas ignites, adjust the gas control valve until the flame is pale blue and has two or more distinct cones.

e. Slowly open the air control valve until you hear a slight buzzing. This sound is characteristic of the hottest flame from the burner. Too much air may blow the flame out. When the *best* adjustment is reached, three distinct cones are visible (Figure 1.3).

f. *If the flame goes out,* immediately close the gas valve at the outlet and repeat the procedure for lighting the burner.①

2. **Observing Flame Temperatures Using a Wire Gauze.** Temperatures within the second (inner) cone of a nonluminous blue flame approach 1500°C.

   a. Using crucible tongs (or forceps), hold a wire gauze parallel to the burner barrel about 1 cm above the burner top (Figure 1.4). Observe the relative heat zones of the flame. Sketch a diagram of your observations on the Report Sheet.②

   b. *Close* the air control valve and repeat the observation with a luminous flame.③

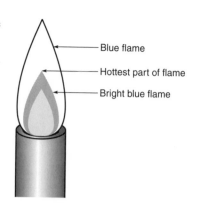

Figure 1.3 Flame of a properly adjusted Bunsen burner.

1. **Practice Using the Balances.** Refer to the Report Sheet. As suggested, measure the mass of several objects. Use the top-loading balance only after the instructor explains its operation. Be sure to record the mass of the objects according to the sensitivity of the balance.④Refer to Technique 6.

2. **Precision of Instrument.** a. Obtain a 10-mL graduated cylinder and measure and tare its mass (±0.001 g) on your assigned balance. Add 7 mL (±0.1 mL) of water and measure the combined mass.⑤ Calculate the mass of 7 mL of water. refer to Technique 16A (p. 29) for reading and recording a volume.

   b. Discard the water and again fill the graduated cylinder to the 7-mL mark. Record the mass measurement on the Report Sheet. Repeat this procedure at least five times.

B. Laboratory Balances

Ask your instructor which balance you are to use to determine the densities of your unknowns. Write the balance number on the Report Sheet.⑥

1. **Water-Insoluble Solid.** a. Obtain an unknown solid and record its number.⑦ Using the assigned balance tare the mass of a piece of weighing paper[1], place the solid on the weighing paper, and measure its mass. Record the mass according to the sensitivity of the balance.

   b. Half-fill a 10-mL graduated cylinder with water and record its volume (Figure 1.5a).

   c. Gently slide the known mass of solid into the graduated cylinder held at a 45° angle. Roll the solid around in the cylinder, removing any air bubbles that are trapped or that adhere to the solid. Record the new water level (Figure 1.5b). The volume of the solid is the difference between the two water levels.

   d. Remove the solid, dry it, and measure its volume a second time.

C. Density

Figure 1.4 Hold the wire gauze parallel to the burner barrel.

Disposal: Check with your instructor for the procedure of properly returning the solid sample.

---

[1]The balance is reset to zero *after* the weighing paper is placed on the balance pan.

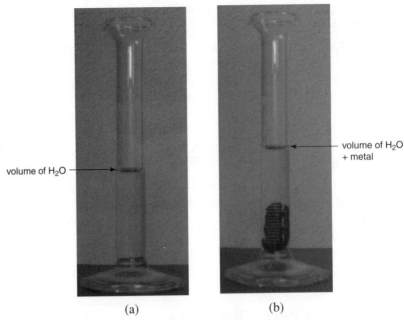

volume of H₂O

volume of H₂O + metal

(a)   (b)

Figure 1.5 Apparatus for measuring the density of a water-insoluble solid.

Refer to Appendix B for calculating average values.

Flammable: capable of igniting in air (generally initiated with a flame or spark)

2. **Liquid, Water.** a. Clean and dry your smallest laboratory beaker. Using your assigned balance, measure and record its mass. Pipet 5 mL of water into the beaker.

   b. Measure and record the mass of the beaker and water. Calculate the density of water from the available data. Repeat the density determination for Trial 2.[8]

   c. Collect and record the density value of water at room temperature from five additional laboratory measurements from classmates.[9] Calculate the average density of water at room temperature.

3. **Liquid, Unknown.** a. Dry the beaker and pipet. Ask the instructor for a liquid unknown and record its number.[10] (**Caution:** *The unknown liquid may be **flammable.** Do not inhale the fumes of the liquid; extinguish all flames.*)

   b. Rinse the pipet with two 1-mL quantities of the unknown liquid and discard. Repeat the measurements of Part C.2, substituting the unknown liquid for the water. Repeat this experiment for Trial 2. Calculate the average density of the liquid.

Disposal: Check with your instructor. Dispose of the unknown liquid and the rinses in the "Waste Liquids" container.

The Next Step

How might the density of an object less dense than water be determined, e.g., packing peanuts, a slice of bread, a feather? How might the density of a water-soluble solid be determined?

Basic Laboratory Operations

Date _____ Lab Sec. _____ Name _____ Desk No. _____

1. a. What is the dominant color of a nonluminous flame from a Bunsen burner? Explain.

   b. Is the temperature of a luminous flame greater or less than that of a nonluminous flame? Explain.

2. Diagram the cross section of a graduated cylinder, illustrating *how* to read the meniscus.

3. Experimental Procedure, Part B. What is the sensitivity of the *least* sensitive balance most likely to be in your laboratory?

4. A fire in the "pits" at the Indianapolis 500 Motor Speedway is especially dangerous because the flame from the fuel used in the race cars is nearly colorless and nonluminous, unlike that of a gasoline fire. How might the fuel used in the Indianapolis 500 race cars differ from that of gasoline?

5. Refer to Technique 16B.

   a. Remove the drop suspended from a pipet tip by . . .

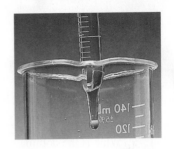

   b. The finger used to control the delivery of liquid from a pipet is the . . .

   c. A pipet is filled with the aid of a . . .

   d. Most pipets are calibrated as "TD 20°C". Define "TD" and what is its meaning regarding the volume of liquid a pipet delivers?

6. Experimental Procedure, Part C.1. The density of aluminum is 2.70 g/cm^3 and the density of chromium is 7.19 g/cm^3. If equal masses of aluminum and chromium are transferred to equal volumes of water in separate graduated cylinders, which graduated cylinder would have the greatest volume change? Explain.

7. Experimental Procedure, Part C.3. The mass of a beaker is 5.333 g. After 5.00 mL of spearmint oil is pipetted into the beaker, the combined mass of the beaker and the spearmint oil sample is 9.962 g. From the data, what is the measured density of spearmint oil?

Basic Laboratory Operations

Date _____ Lab Sec. _____ Name _____ Desk No. _____

A. Bunsen Burner

1. ①Instructor's Approval of a well-adjusted Bunsen flame. _____

2. a. At right, draw a sketch of the heat zones for a nonluminous flame as directed with the wire gauze positioned parallel to the burner barrel. Label the "cool" and "hot" zones.②

   b. What happens to the "cool" and "hot" zones of the flame when the air control valve is closed? Explain.③

B. Laboratory Balances

1. ④Determine the mass of the following objects on your assigned balance. Express your results with the correct sensitivity.

   Balance No. _____ Sensitivity _____

Object	Description of Object (size or volume)	Mass (g)
Test tube	_____	_____
Beaker	_____	_____
Spatula	_____	_____
Graduated cylinder	_____	_____
Other (see instructor)	_____	_____

2. ⑤**Precision of Instrument.** Balance No. _____

   Mass of graduated cylinder (if not tared) _____

	Trial 1	Trial 2	Trial 3	Trial 4	Trial 5	Trial 6
a. Mass of Cylinder +H_2O (*l*) (g)	_____	_____	_____	_____	_____	_____
b. Mass of 7 mL H_2O(*l*) (g)						
c. Average mass of 7 mL H_2O(*l*) (g)						

Comment on the precision of the mass measurements for the water samples.

C. Density

⑥Balance No. _____

1. ⑦**Solid Unknown Number** _____

	Trial 1	Trial 2
a. Tared mass of solid (g)	_____	_____
b. Volume of water (cm^3)	_____	_____
c. Volume of water and solid (cm^3)	_____	_____
d. Volume of solid (cm^3)		
e. Density of solid (g/cm^3)		
f. Average density of solid (g/cm^3)		

2. **Liquid**

	Water		Liquid Unknown No. _____ ⑩	
	Trial 1	Trial 2	Trial 1	Trial 2
a. Mass of beaker (g)	_____	_____	_____	_____
b. Mass of beaker + liquid (g)	_____	_____	_____	_____
c. Mass of liquid (g)				
d. Volume of liquid (mL)	_____	_____	_____	_____
e. ⑧Density of liquid (g/mL)				
f. Average density of liquid (g/mL)				

⑨**Class Data/Group**	1	2	3	4	5	6
Density of Water	_____	_____	_____	_____	_____	_____

Average density of water at room temperature:

Laboratory Questions

Circle the questions that have been assigned.

1. What are the products of the combustion of candle wax? Of a Bunsen flame?

2. Why is it possible to extinguish the flame of a candle or match by *very gently* blowing on it?

3. What substance is "burning" (the fuel) and what substance is supporting the burning of candle wax? Compare that to the flame of a Bunsen burner.

4. Part C.1c. The solid is not completely submerged in the water. Explain how this technique error affects the reported density of the solid.

5. Part C.2. Suppose that after delivery several drops of the water cling to the inner wall of the pipet (because the pipet wall is dirty). How does this technique error affect the reported density of water? See drawing at right.

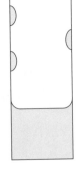

6. Part C.3. The unknown liquid is volatile. If some of the liquid evaporates between the time that the liquid is delivered to the beaker and the time that its mass is measured, will the reported density of the liquid be too high, too low, or unaffected? Explain.

Experiment 2

Identification of a Compound: Chemical Properties

A potassium chromate solution added to a silver nitrate solution results in the formation of insoluble silver chromate.

OBJECTIVES

- To identify a compound on the basis of its **chemical properties**
- To design a systematic procedure for determining the presence of a particular compound in aqueous solution

Chemical property: characteristic of a substance that is dependent on its chemical environment

TECHNIQUES

The following techniques are used in the Experimental Procedure

INTRODUCTION

Chemists, and scientists in general, develop and design experiments in an attempt to understand, explain, and predict various chemical phenomena. Carefully controlled (laboratory) conditions are needed to minimize the many parameters that affect the observations. Chemists organize and categorize their data, and then systematically analyze the data to reach some conclusion; often, the conclusion may be to carefully plan more experiments!

It is presumptuous to believe that a chemist must know the result of an experiment before it is ever attempted; most often an experiment is designed to determine the presence or absence of a **substance** or to determine or measure a parameter. A goal of the environmental or synthesis research chemist is, for example, to separate the substances of a reaction mixture (either one generated in the laboratory or one found in nature) and then identify each substance through a systematic, or sometimes a **trial-and-error**, study of their chemical and physical properties. As you will experience later, Experiments 37–39 are designed to identify a specific ion (by taking advantage of its unique chemical properties) in a mixture of ions through a systematic sequence of analyses.

On occasion an experiment reveals an observation and data that are uncharacteristic of any other set of properties for a known substance, in which case either a new compound has been synthesized/discovered or experimental errors and/or interpretations have infiltrated the study.

In this experiment, you will observe chemical reactions that are characteristic of various compounds under controlled conditions. After collecting and organizing your data, you will be given an unknown compound; a compound that you have previously investigated. The interpretations of the collected data will assist you in identifying your compound.

What observations will you be looking for? Chemical changes are generally accompanied by one or more of the following:

Substance: a pure element or compound having a unique set of chemical and physical properties

Trial-and-error study: a method that is often used to seek a pattern in the accumulated data

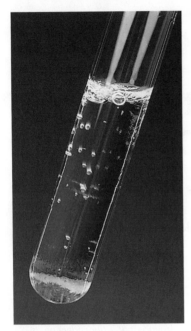

Figure 2.1 A reaction mixture of NaHCO₃(aq) and HCl(aq) produces CO₂ gas.

Reagent: a solid chemical or a solution having a known concentration of solute

- A *gas* is evolved. This evolution may be quite rapid or it may be a "fizzing" sound (Figure 2.1).
- A *precipitate* appears (or disappears). The nature of the precipitate is important; it may be crystalline, it may have color, it may merely cloud the solution.
- *Heat* may be evolved or absorbed. The reaction vessel becomes warm if the reaction is exothermic or cools if the reaction is endothermic.
- A *color change* occurs. A substance added to the system may cause a color change.
- A *change in odor* is detected. The odor of a substance may appear, disappear, or become more intense during the course of a chemical reaction.

The chemical properties of the following compounds, dissolved in water, are investigated in Part A of this experiment:

Sodium chloride	$NaCl(aq)$
Sodium carbonate	$Na_2CO_3(aq)$
Magnesium sulfate	$MgSO_4(aq)$
Ammonium chloride	$NH_4Cl(aq)$
Water	$H_2O(l)$

The following test **reagents** are used to identify and characterize these compounds:

Silver nitrate	$AgNO_3(aq)$
Sodium hydroxide	$NaOH(aq)$
Hydrochloric acid	$HCl(aq)$

In Part B of this experiment the chemical properties of five compounds in aqueous solutions, labeled 1 through 5, are investigated with three reagents, labeled A, B, and C. Chemical tests will be performed with these eight solutions. An unknown will then be issued and matched with one of the solutions, labeled 1 through 5.

EXPERIMENTAL PROCEDURE

Procedure Overview: In Part A a series of tests for the chemical properties of known compounds in aqueous solutions are conducted. A similar series of tests are conducted on an unknown set of compounds in Part B. In each case, an unknown compound is identified on the basis of the chemical properties observed.

You should discuss and interpret your observations on the known chemical tests with a partner, but each of you should analyze your own unknown compound. At each circled superscript ⌐⊃in the procedure, *stop*, and record your observation on the Report Sheet.

To organize your work, you will conduct a test on each known compound in the five aqueous solutions and the unknown compound with a single test reagent. It is therefore extremely important that you are organized and that you describe your observations as completely and clearly as possible. The Report Sheet provides a "reaction matrix" for you to describe your observations; because the space is limited, you may want to devise a code, e.g.,

- p—precipitate + color
- c—cloudy + color
- nr—no reaction
- g—gas, no odor
- go—gas, odor

A. Chemical Properties of Known Compounds

1. **Observations with Silver Nitrate Test Reagent. a.** Use a permanent marker to label five small, clean test tubes (Figure 2.2a) or set up a clean 24-well plate (Figure 2.2b). Ask your instructor which setup you should use. Place 5–10 drops of each of the five "known" solutions into the labeled test tubes (or wells A1–A5).

    b. Use a dropper pipet (or a dropper bottle) for the delivery of the silver nitrate solution. (**Caution:** *AgNO₃ forms black stains on the skin. The stain, caused by*

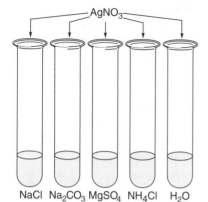

Figure 2.2a Arrangement of test tubes for testing with the silver nitrate reagent.

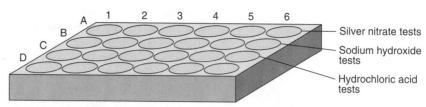

Figure 2.2b Arrangement of test solutions in the 24-well plate for testing salts.

silver metal, causes no harm.) If after adding several drops you observe a chemical change, add 5–10 drops to see if there are additional changes. Record your observations in the matrix on the Report Sheet.[1] Save your test solutions for Part A.4. Write the formulas for each of the precipitates that formed. Ask your lab instructor for assistance.

Appendix G

2. **Observations with Sodium Hydroxide Test Reagent.** a. Use a permanent marker to label five additional small, clean test tubes (Figure 2.3). Place 5–10 drops of each of the five "known" solutions into this second set of labeled test tubes (or wells B1–B5, Figure 2.2b).

b. To each of these solutions slowly add 5–10 drops of the sodium hydroxide solution; make observations as you add the solution. Check to see if a gas evolves in any of the tests. Check for odor. What is the nature of any precipitates that form? Observe *closely.*[2] Save your test solutions for reference in Part A.4. Write the formulas for each of the precipitates that formed.

Appendix G

3. **Observations with Hydrochloric Acid Test Reagent.** a. Use a permanent marker to label five additional small, clean test tubes (Figure 2.4). Place 5–10 drops of each of the five "known" solutions into this third set of labeled test tubes (or wells C1–C5, Figure 2.2b).

b. Add the hydrochloric test reagent to the solutions and record your observations. Check to see if any gas is evolved. Check for odor. Observe closely.[3] Save your test solutions for reference in Part A.4. Write the formulas for any compound that forms.

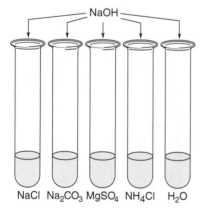

Figure 2.3 Arrangement of test tubes for testing with the sodium hydroxide reagent.

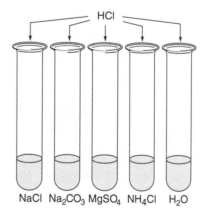

Figure 2.4 Arrangement of test tubes for testing with the hydrochloric acid reagent.

4. **Identification of Unknown.** Obtain an unknown for Part A from your laboratory instructor. Repeat the three tests with the reagents in Parts A.1, 2, and 3 on your unknown. On the basis of the data from the "known" solutions (collected and summarized in the Report Sheet matrix) and that of your unknown solution, identify the compound in your unknown solution.[4]

Disposal: Discard the test solutions in the "Waste Salts" container.

CLEANUP: Rinse the test tubes or well plate twice with tap water and twice with deionized water. Discard each rinse in the "Waste Salts" container.

B. Chemical Properties of Unknown Compounds

A dropper pipet. 20 drops is ~1 mL of solution.

The design of the experiment in Part B is similar to that of Part A. Therefore, fifteen clean test tubes or a clean 24-well plate is necessary.

1. **Preparation of Solutions.** On the reagent shelf are five solutions, labeled 1 through 5, each containing a different compound. Use small clean test tubes or the well plate as your testing laboratory. About 1 mL of each test solution is necessary for analysis.

2. **Preparation of Reagents.** Also on the reagent shelf are three reagents labeled A, B, and C. Use a dropper pipet (or dropper bottle) or a Beral pipet to deliver Reagents A through C to the solutions.

3. **Testing the Solutions.** Test each of the five solutions with drops (and then excess drops) of Reagent A. If, after adding several drops, you observe a chemical change, add 5–10 drops more to see if there are additional changes. Observe closely and describe any evidence of chemical change; record your observations.[5] With a fresh set of solutions 1–5 in clean test tubes (or wells), test each with Reagent B.[6] Repeat with Reagent C.[7]

4. **Identification of Unknown.** An unknown solution will be issued that is one of the five solutions from Part B.1. On the basis of the data in your reaction matrix and the data you have collected, identify which of the five solutions is your unknown.

Disposal: Discard the test solutions in the "Waste Salts" container.

CLEANUP: Rinse the test tubes or well plate twice with tap water and twice with deionized water. Discard each rinse in the "Waste Salts" container.

The Next Step

This experiment will enable you to better understand the importance of "separation and identification," a theme that appears throughout this manual. For example, refer to Experiments 3, 4, 37, 38, and 39. These experiments require good experimental techniques that support an understanding of the chemical principles involved in the separation and identification of the various compounds/ions. Additionally, the amounts of a substance of interest are also determined in other experiments.

Obtain a small (~50 cm³) sample of soil, add water, and filter. Test the filtrate with the silver nitrate test reagent. Test a second soil sample *directly* with the hydrochloric acid test reagent. What are your conclusions?

Experiment 2 *Prelaboratory Assignment*

Identification of a Compound: Chemical Properties

Date __________ Lab Sec. ______ Name ______________________________ Desk No. __________

1. Experimental Procedure, Part A. a. What is the criterion for clean glassware?

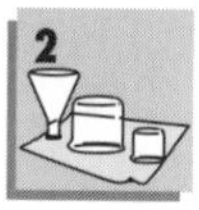

   b. What is the size and volume of a "small, clean test tube?"

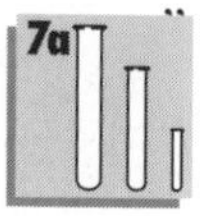

2. Experimental Procedure, Part A.2. Describe the technique for testing the odor of a chemical.

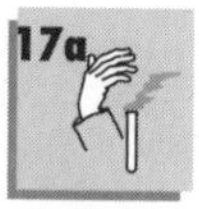

3. a. Depending upon the tip of a dropper pipet, there are approximately 20 drops per milliliter of water. What is the approximate volume (in mL) of one drop of an aqueous solution?

   b. A micropipet delivers 153 drops of alcohol for each milliliter. Calculate the volume (in mL) of alcohol in each drop.

4. Write a balanced equation for the following observed reactions:

   a. Aqueous solutions of sodium hydroxide, $NaOH$, and sulfuric acid, H_2SO_4, are mixed. The neutralization products are water and one other compound with the evolution of heat.

   b. Aqueous solutions of copper(II) nitrate, $Cu(NO_3)_2$, and sodium carbonate, Na_2CO_3, are mixed. A blue precipitate of copper(II) carbonate forms in addition to one other compound.

5. Experimental Procedure, Part A. The substances, NaCl, Na₂CO₃, MgSO₄, and NH₄Cl used for test solutions, are all soluble ionic compounds. For each substance indicate the ions present in its respective test solution.

NaCl: _____

Na₂CO₃: _____

MgSO₄: _____

NH₄Cl: _____

6. Three colorless solutions in test tubes, with no labels, are in a test tube rack on the laboratory bench. Lying beside the test tubes are three labels: potassium iodide, KI, silver nitrate, AgNO₃, and sodium sulfide, Na₂S. You are to place the labels on the test tubes using only the three solutions present. Here are your tests:

- A portion of test tube #1 added to a portion of test tube #3 produces a yellow, silver iodide precipitate.
- A portion of test tube #1 added to a portion of test tube #2 produces a black, silver sulfide precipitate.

a. Your conclusions are:
Test Tube 1_____

Test Tube 2_____

Test Tube 3_____

b. Write the balanced equation for the formation of silver iodide, AgI.

c. Write the balanced equation for the formation of sulfide, Ag₂S.

Identification of a Compound: Chemical Properties

Date _____ Lab Sec. _____ Name _____ Desk No. _____

A. Chemical Properties of Known Compounds

Test	NaCl(aq)	Na$_2$CO$_3$(aq)	MgSO$_4$(aq)	NH$_4$Cl(aq)	H$_2$O(l)	Unknown
① AgNO$_3$(aq)	_____	_____	_____	_____	_____	_____
② NaOH(aq)	_____	_____	_____	_____	_____	_____
③ HCl(aq)	_____	_____	_____	_____	_____	_____

Write formulas for the precipitates that formed in Part A. (See Appendix G)

Part A.1	_____	_____	_____	_____	_____	_____
Part A.2	_____	_____	_____	_____	_____	_____
Part A.3	_____	_____	_____	_____	_____	_____

Sample No. of unknown for Part A.4 _____

④ Compound in unknown solution _____

B. Chemical Properties of Unknown Compounds

Solution No.	1	2	3	4	5	Unknown
⑤ Reagent A	_____	_____	_____	_____	_____	_____
⑥ Reagent B	_____	_____	_____	_____	_____	_____
⑦ Reagent C	_____	_____	_____	_____	_____	_____

Sample No. of unknown for Part B.4 _____

Compound of unknown is the same as Solution No. _____

Laboratory Questions

Circle the questions that have been assigned.

1. What *chemical* test will distinguish calcium chloride, $CaCl_2$, from calcium carbonate, $CaCO_3$? Explain.

2. What test reagent will distinguish a soluble Cl^- salt from a soluble SO_4^{2-} salt? Explain.

3. Predict what would be observed (and why) from an aqueous mixture for each of the following:

    a. potassium carbonate and hydrochloric acid
    b. zinc chloride and silver nitrate
    c. magnesium chloride and sodium hydroxide
    d. ammonium nitrate and sodium hydroxide

4. Three colorless solutions in test tubes, with no labels, are in a test tube rack on the laboratory bench. Lying beside the test tubes are three labels: silver nitrate, $AgNO_3$, hydrochloric acid, HCl, and sodium carbonate, Na_2CO_3. You are to place the labels on the test tubes using only the three solutions present. Here is your analysis procedure:

    - A portion of test tube 1 added to a portion of test tube 2 produces carbon dioxide gas, CO_2.
    - A portion of test tube 2 added to a portion of test tube 3 produces a white, silver carbonate precipitate.

    a. On the basis of your observations how would you label the three test tubes?
    b. What would you expect to happen if a portion of test tube 1 is added to a portion of test tube 3?

5. For individual solutions of the cations Ag^+, Ba^{2+}, Mg^{2+}, or Cu^{2+}, the following experimental observations were collected:

	$NH_3(aq)$	$HCl(aq)$	$H_2SO_4(aq)$
Ag^+	No change	White ppt[a]	No change
Ba^{2+}	No change	No change	White ppt
Mg^{2+}	White ppt	No change	No change
Cu^{2+}	Blue ppt/deep blue soln with excess	No change	No change

[a]Example: when an aqueous solution of hydrochloric acid is added to a solution containing Ag^+, a white precipitate (ppt) forms.

    a. Identify a reagent that distinguishes the chemical properties of Ag^+ and Mg^{2+}. Explain.
    b. Identify a reagent that distinguishes the chemical properties of HCl and H_2SO_4. Explain.
    c. Identify a reagent that distinguishes the chemical properties of Ba^{2+} and Cu^{2+}. Explain.
    *d. Identify a reagent that distinguishes the chemical properties of Cu^{2+} and Mg^{2+}. Explain.

6. Three colorless solutions in test tubes, with no labels, are in a test tube rack on the laboratory bench. Lying beside the tests tubes are three labels: 0.10 M Na_2CO_3, 0.10 M HCl, and 0.10 M KOH. You are to place the labels on the test tubes using only the three solutions present. Here are your tests:

    - A few drops of the solution from test tube #1 added to a similar volume of the solution in test tube #2 produces no visible reaction but the solution becomes warm.
    - A few drops of the solution from test tube #1 added to a similar volume of the solution in test tube #3 produces carbon dioxide gas.

    Identify the labels for test tube #1, #2, and #3.

Experiment 3

Water Analysis: Solids

"Clear" water from streams contains small quantities of dissolved and suspended solids.

- To determine the total, dissolved, and suspended solids in a water sample
- To determine the ions present in the solids of a water sample.

OBJECTIVES

The following techniques are used in the Experimental Procedure:

TECHNIQUES

INTRODUCTION

Surface water is used as the primary drinking water source for many large municipalities. The water is piped into a water treatment facility where impurities are removed and bacteria are killed before the water is placed into the distribution lines. The contents of the surface water must be known and predictable so that the treatment facility can properly and adequately remove these impurities. Tests are used to determine the contents of the surface water.

Water in the environment has a large number of impurities with an extensive range of concentrations. **Dissolved solids** are water-soluble substances, most often salts although some dissolved solids may come from organic sources. Naturally occurring dissolved salts generally result from the movement of water over or through mineral deposits, such as limestone. These dissolved solids, characteristic of the **watershed,** generally consist of the sodium, calcium, magnesium, and potassium cations and the chloride, sulfate, bicarbonate, carbonate, bromide, and fluoride anions. Anthropogenic (human-related) dissolved solids include nitrates from fertilizer runoff and human wastes, phosphates from detergents and fertilizers, and organic compounds from pesticides, sewage runoff, and industrial wastes.

Watershed: the land area from which the natural drainage of water occurs

Dissolved salts can be problematic in potable water. Typical dissolved salt concentrations range from 20 to 1000 mg/L, although most waters are less than 500 mg/L. High concentrations also indicate "hard" water (Experiment 21) which can clog pipes and industrial cooling systems. Also, high concentrations of dissolved salts can cause diarrhea or constipation in some people.

Salinity, a measure of the total salt content in a water sample, is expressed as the grams of dissolved salts per kilogram of water or as **parts per thousand** (ppt). The average ocean salinity is 35 ppt whereas fresh water salinity is usually less than 0.5 ppt. Brackish water, where fresh river water meets salty ocean water, varies from 0.5 ppt to 17 ppt. For water samples with low organic levels, the salinity of a water sample approximates that of total dissolved solids (TDS) content.

Parts per thousand (ppt): 1g of substance per 1000g (1 kg) of sample

More specifically, the anions that account for the salinity of the water are generally the carbonates and bicarbonates, CO_3^{2-} and HCO_3^-, the halides (Cl^-, Br^-, and I^-), the phosphates, PO_4^{3-}, and the sulfates, SO_4^{2-}. A qualitative testing of a water sample can determine the presence of these various ions.

To test for the presence of carbonates and bicarbonates in the water, an acid, generally nitric acid, HNO_3, is added to the sample resulting in the evolution of carbon dioxide gas. The **ionic equation** for the reaction is

$$CO_3^{2-}(aq) + 2\,H^+(aq) + NO_3^-(aq) \rightarrow CO_2(g) + H_2O(l) + NO_3^-(aq) \quad (3.1)$$

A qualitative test for the halides is the addition of silver ion, resulting in a silver halide precipitate:

$$Ag^+(aq) + Cl^-, Br^-, I^-(aq) \rightarrow$$
$$AgCl\,(s, \text{white}) + AgBr\,(s, \text{light brown}) + AgI(s, \text{dark brown}) \quad (3.2)$$

A "dirty" appearance of the precipitate is evidence for a mixture of the halides in the water. Additionally, if the phosphates are present, a white silver phosphate precipitate forms:

$$3\,Ag^+(aq) + PO_4^{3-}(aq) \rightarrow Ag_3PO_4(s) \quad (3.3)$$

The presence of phosphate ion (or HPO_4^{2-}) can also be determined independently by the addition of an ammonium molybdate solution to the sample, resulting in the formation of a yellow precipitate:

$$HPO_4^{2-}(aq) + 12(NH_4)_2MoO_4(aq) + 23\,H^+(aq) \rightarrow$$
$$(NH_4)_3PO_4(MoO_3)_{12}(s) + 21\,NH_4^+(aq) + 12\,H_2O(l) \quad (3.4)$$

While these tests may have interferences as performed in this experiment, a more detailed experimental and systematic procedure is presented in Experiment 37.

The most common cation present in "natural" waters is calcium. Calcium is the principal ion responsible for water hardness. See Experiment 21. A qualitative test for its presence is its formation of insoluble calcium oxalate.

$$Ca^{2+}(aq) + C_2O_4^{2-}(aq) \rightarrow CaC_2O_4(s) \quad (3.5)$$

Suspended solids are very finely divided particles, kept in suspension by the turbulent action of the moving water, are *insoluble* in water but are filterable. Total suspended solids (TSS) is a measure of the turbidity and the clarity of the water.

High concentrations of suspended solids, such as decayed organic matter, sand, silt, and clay, can settle to cover (and suffocate) the existing ecosystem at the bottom of a lake, can make disinfectants for water treatment less effective, and can absorb/adsorb various organic and inorganic pollutants resulting in an increase of their residence in the water sample.

Total solids (TS) are the sum of the dissolved and suspended solids in the water sample. In this experiment the total solids and the dissolved solids are determined directly; the suspended solids are assumed to be the difference, since

total solids (TS) = total dissolved solids (TDS) + total suspended solids (TSS) (3.6)

The U.S. Public Health Service recommends that drinking water not exceed 500 mg total solids/kg water, or 500 **ppm.** However, in some localities, the total solids content may range up to 1000 mg/kg of potable water; that's 1 g/L!! An amount over 500 mg/kg water does not mean the water is unfit for drinking; an excess of 500 mg/kg is merely not recommended.

Ionic equation: ionic equations, though appearing somewhat premature in this manual, are written to better illustrate the ions in solution that are involved in the chemical reactions and observations. Experiment 6 will further illustrate the use of ionic equations for chemical reactions.

Suspended solids: solids that exhibit colloidal properties or solids that remain in the water because of turbulence

ppm: 1 mg substance per kilogram sample = 1 part per million (ppm)

EXPERIMENTAL PROCEDURE

Filtrate: the solution that passes through the filter into a receiving flask.

Procedure Overview: The amounts of total, dissolved, and suspended solids in a water sample are determined in this experiment. The water sample is filtered to remove the suspended solids and the **filtrate** is evaporated to dryness to determine the dissolved solids. The water sample may be from the ocean, a lake, a stream, or from an underground aquifer.

rinse splattered material
from watch glass

wire
gauze

wash
bottle

hot plate

Figure 3.1 Wash the spattered material from the convex side of the watch glass.

Obtain 100 mL of a water sample from your instructor. Preferably the water sample is high in **turbidity.** Record the sample number and write a short description of the sample on the Report Sheet. With approval, bring your own "environmental" water sample to the laboratory for analysis. Ask your instructor whether evaporating dishes or 250 mL beakers are to be used for the analysis.

Turbid sample: a cloudy suspension due to stirred sediment.

If time allows, the experiment should be repeated twice. A basis for water quality is *not* determined from a single analysis—a minimum of three trials is necessary for reputable analytical data. Ask your instructor for additional information.

Assume the density of your water sample to be 1.02 g/m L.

1. **Filter the Water Sample.** Gravity filter about 50 mL of a *thoroughly stirred or shaken* water sample into a clean, dry 100-mL beaker. While waiting for the filtration to be completed, proceed to Part B.

2. **Evaporate the Filtrate to Dryness.** a. Clean, dry, and measure the mass (± 0.001 g) of an evaporating dish (or 250-mL beaker).

   b. Pipet a 25-mL aliquot (portion) of the filtrate into the evaporating dish (250-mL beaker). Determine the combined mass of the sample and evaporating dish (beaker).

   c. Use a hot plate or direct flame (Figure T.14a or T.14b) to *slowly* heat—do not boil—the mixture to dryness.

   d. As the mixture nears dryness, cover the evaporating dish (beaker) with a watch glass and reduce the intensity of the heat.[1] If spattering occurs, allow the dish to cool to room temperature, rinse the adhered solids from the watch glass (see Figure 3.1) and return the rinse to the dish.

3. **A Final Heating to Dryness.** Again heat slowly, being careful to avoid further spattering. After all of the water has evaporated, reduce the heat of the hot plate, or maintain a **"cool" flame** beneath the dish for 3 minutes. Allow the dish to cool to room temperature and determine its final mass. Cool the evaporating dish and sample in a desiccator, if available.

A. Total Dissolved Solids (TDS)

Cool flame: a Bunsen flame of low intensity—a slow rate of natural gas is flowing through the burner barrel.

[1]This reduces the spattering of the remaining solid and its subsequent loss in analysis.

B. Total Solids (TS) and Total Suspended Solids (TSS)

1. **Evaporate an Original Water Sample to Dryness.** a. Clean, dry, and measure the mass (± 0.001 g) of a second evaporating dish (or 250-mL beaker).

   b. Thoroughly stir or agitate 100 mL of the original water sample; pipet[2] a 25-mL aliquot of this sample into the evaporating dish (250-mL beaker). Record the combined mass of the water sample and evaporating dish (beaker).

   c. Evaporate *slowly* the sample to dryness as described in Part A.2. Record the mass of the solids remaining in the evaporating dish.

2. **Total Suspended Solids.** Collect the appropriate data to determine the total suspended solids in the water sample.

C. Analysis of Data

Appendix B

1. **Precision of Data?** Compare your TDS and TSS data with three other chemists in your laboratory who have analyzed the *same* water sample. Record their results on the Report Sheet. Calculate the average value for the TSS in the water sample.

D. Chemical Tests[3]

Appendix G

1. **Test for Carbonates and Bicarbonates.** With your spatula loosen small amounts of the dried samples from Part A and Part B and transfer each to *separately* marked 75-mm test tubes. Add several drops of 6 M HNO_3 (**Caution:** *HNO_3 is corrosive and a severe skin irritant*) and observe. What can you conclude from your observation?

2. **Test for Chlorides (Halides).** To each of the test tubes add 10 drops of water, agitate the solution, and add several drops of 0.01 M $AgNO_3$ (**Caution:** *$AgNO_3$ is a skin irritant*) and observe. What can you conclude from your observation?

3. **Test for Phosphates.** With your spatula loosen small amounts of the dried sample from Parts A and B and transfer each to *separately* marked 75-mm test tubes. To each test tube add ~ 10 drops of water and several drops of 6 M HNO_3 (**Caution:** *Avoid skin contact. Do not inhale fumes.*) followed by 1 mL of 0.5M $(NH_4)_2MoO_4$. Shake and warm the resulting solutions in a warm water bath ($\sim 60°C$, Figure T.13b) and let stand for 10–15 minutes. The yellow color can be slow to form *if* PO_4^{3-} is present in the sample.

4. **Test for Calcium Ion.** With your spatula loosen small amounts of the dried samples from Parts A and B and transfer each to *separately* marked 75-mm test tubes. Add about 10 drops of water, agitate the solution, and add several drops of 1 M $K_2C_2O_4$ and observe. What can you conclude from your observation?

Disposal: Discard the dried salts from Part A and B and the test solutions from Part D in the "Waste Salts" container.

CLEANUP: Rinse the test tubes and evaporating dishes (250-mL beakers) with tap water and twice with deionized water.

The Next Step

Devise a plan to determine the changes in TSS and TDS at various points along a water source (river, stream, lake, drinking water, etc.). Explain why the values change as a result of location, rainfall, season, time of day, etc. Test to determine the ions that are primary contributors to the TDS of the sample.

---

[2]If the solution appears to be so turbid that it may plug the pipet tip, use a 25-mL graduated cylinder to measure the water sample as accurately as possible.
[3]For each of the tests in Part D there are other ions that may show a positive test. However the ions being tested are those most common in environmental water samples.

Water Analysis: Solids

Date _____ Lab Sec. _____ Name _____ Desk No. _____

1. List several anions, by formula, that contribute to the salinity of a water sample.

2. Distinguish between and characterize the "total dissolved solids" and "total suspended solids" in a water sample.

3. Experimental Procedure, Part A.2c. Explain why a "cool flame" is important to use in heatng a solution to dryness.

4. a. What is an **aliquot** of a sample?

   b. What is the **filtrate** in a gravity filtration procedure?

   c. How full (the maximum level) should a funnel be filled with solution in a filtration procedure?

5. Experimental Procedure, Part D. What observation is "expected" when
   a. an acid (nitric acid, HNO_3) is added to a solution containing carbonate or bicarbonate ions? See Experiment 2, Experimental Procedure, Part A.3.

b. silver ion is added to a solution containing chloride (or bromide or iodide) ions? See Appendix G.

c. a solution of 0.5 M $(NH_4)_2MoO_4$ is added to a water sample containing phosphate ion (and heated)? Explain.

6. The following data were collected for determining the concentration of suspended solids in a water sample (density = 1.02 g/mL).

	Trial 1	Trial 2	Trial 3	Trial 4	Trial 5	Trial 6
Volume of sample (mL)	25.0	20.0	50.0	25.0	20.0	25.0
Mass of sample (g)						
Mass of dry solid (g)	10.767	8.436	21.770	10.826	8.671	10.942
Mass of solid/mass of sample (g/g)						

a. What is the *average* total suspended solids (TSS) in the water sample? Express this measurement in ppt (parts per thousand, g/kg). See Appendix B.

*b. Calculate the standard deviation and the relative standard deviation (% RSD) for the analyses.

7. A 25.0-mL aliquot of a well-shaken sample of river water is pipetted into a 25.414-g evaporating dish. After the mixture is evaporated to dryness, the dish and dried sample has a mass of 36.147 g. Determine the total solids in the sample; express total solids in units of g/kg sample (parts per thousand, ppt). Assume the density of the river water to be 1.01 g/mL.

Water Analysis: Solids

Date _____ Lab Sec. _____ Name _____ Desk No. _____

Sample Number: _____ Describe the nature of your water sample, i.e., its color, turbidity, etc.

A. Total Dissolved Solids (TDS)   *Trial 1*

1. Mass of evaporating dish (beaker) (*g*)   _____

2. Mass of water sample plus evaporating dish (beaker) (*g*)   _____

3. Mass of water sample (*g*)   _____

4. Mass of *dried* sample plus evaporating dish (*g*)   _____

5. Mass of dissolved solids in 25-mL aliquot of filtered sample (*g*)   _____

6. Mass of dissolved solids per total mass of sample (*g* solids/*g* sample)   _____

7. Total dissolved solids (TDS) or salinity (*g* solids/*kg* sample, ppt)   _____

B. Total Solids (TS) and Total Suspended Solids (TSS)

1. Mass of evaporating dish (beaker) (*g*)   _____

2. Mass of water sample plus evaporating dish (beaker) (*g*)   _____

3. Mass of water sample (*g*)   _____

4. Mass of *dried* sample (*g*)   _____

5. Mass of total solids in 25-mL aliquot of unfiltered sample (*g*)   _____

6. Mass of total solids per total mass of sample (*g* solids/*g* sample)   _____

7. Total solids (*g* solids/*kg* sample, ppt)   _____

8. Total suspended solids (*g* solids/*kg* sample, ppt)   _____

C. Analysis of Data

Chemist No.	#1 (you)	#2	#3	#4
TDS (*g/kg*)				
TS (*g/kg*)				
TSS (*g/kg*)				

Average value of total suspended solids (TSS) from four chemists (*x*) = _____

D. Chemical Tests.

Test	Observation	Conclusion
1. CO_3^{2-}, HCO_3^- (TDS)	_____	_____
CO_3^{2-}, HCO_3^- (TS)	_____	_____
2. Cl^-, Br^-, I^- (TDS)	_____	_____
Cl^-, Br^-, I^- (TS)	_____	_____
3. PO_4^{3-} (TDS)	_____	_____
PO_4^{3-} (TS)	_____	_____
4. Ca^{2+} (TDS)	_____	_____
Ca^{2+} (TS)	_____	_____

Write a summary of your assessment of the quality of your water sample.

Laboratory Questions

Circle the questions that have been assigned.

1. Part A.2. The evaporating dish was not properly cleaned of a volatile material before its mass was determined. When the sample is heated to dryness the volatile material is removed. How does this error in technique affect the reported TDS for the water sample? Explain.

2. Part A.2. Some spattering of the sample onto the watchglass does occur near dryness. In a hurry to complete the analysis, the chemist chooses not to return the spattered solids to the original sample and skips the first part of Part A.3. Will the reported TDS for the water sample be too high or too low? Explain.

3. Part A.3. The sample in the evaporating dish is *not* heated to total dryness. How will this error in technique affect the reported value for TDS? Explain. TSS? Explain.

4. Part A.3. As the sample cools moisture from the atmosphere condenses on the outside of the evaporating dish (beaker) before the mass is measured. How does the presence of the condensed moisture affect the reported TDS for the water sample? Explain.

5. Part B.1. The sample in the evaporating dish (beaker) is *not* heated to total dryness. How will this error in technique affect the reported value for total solids? Explain. TSS? Explain.

6. Parts A and B. Suppose the water sample has a relatively high percent of volatile solid material. How would this have affected the reported mass of
   a. dissolved solids? Explain.
   b. total solids? Explain.
   c. suspended solids? Explain.

7. Part D.2. When several drops of 0.010 M $AgNO_3$ are added to a test sample, a white precipitate forms. What can you conclude from this observation? Explain.

8. Part D.3. a. When 1 mL of 0.5 M $(NH_4)_2MoO_4$ is added to the water sample, a yellow solution is observed but no precipitate. What can you conclude from this observation? Explain.
   b. How does this observation help in interpreting any observations from Part D.2?

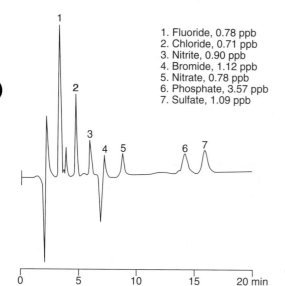

1. Fluoride, 0.78 ppb
2. Chloride, 0.71 ppb
3. Nitrite, 0.90 ppb
4. Bromide, 1.12 ppb
5. Nitrate, 0.78 ppb
6. Phosphate, 3.57 ppb
7. Sulfate, 1.09 ppb

Experiment 4

Paper Chromatography

Trace levels of anions in high purity water can be determined by chromatography.

OBJECTIVES

- To become familiar with chromatography, a technique for separating the components of a mixture
- To separate a mixture of transition metal cations by paper chromatography

TECHNIQUES

The following techniques are used in the Experimental Procedure

INTRODUCTION

Most substances found in nature, and many prepared in the laboratory, are impure; that is, they are a part of a mixture. One goal of chemical research is to devise methods to remove impurities from the chemical of interest.

A **mixture** is a physical combination of two or more pure substances wherein each substance retains its own chemical identity. For example, each component in a sodium chloride/water mixture possesses the same chemical properties as in the pure state: water consists of H_2O molecules, and sodium chloride is sodium ions, Na^+, and chloride ions, Cl^-.

The method chosen for separating a mixture is based on the differences in the chemical and/or physical properties of the components of the mixture. Some common *physical* methods for separating the components of a mixture include:

- **Filtration:** removing a solid substance from a liquid by passing the suspension through a filter (See Techniques 11B–E and Experiment 3 for details.).
- **Distillation:** vaporizing a liquid from a solid (or another liquid) and condensing the vapor (See margin photo.).
- **Crystallization:** forming a crystalline solid by decreasing its solubility by cooling the solution, evaporating the solvent, or adding a solvent in which the substance is less soluble (see Experiment 15.)
- **Extraction:** removing a substance from a solid or liquid mixture by adding a solvent in which the substance is more soluble (see Experiment 11.)

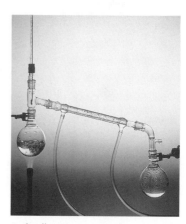

A distillation apparatus.

- **Centrifugation:** removing a substance from a mixture using a centrifuge (See Technique 11F for details.)
- **Sublimation:** vaporizing a solid and recondensing its vapor (not all solids sublime, however)
- **Chromatography:** separating the components of a mixture on the basis of their differing adsorptive tendencies on a stationary phase

Mobile phase: the phase (generally liquid) in which the components of the mixture exist

Eluent: the solvent in which the components of the mixture are moved along the stationary phase

Stationary phase: the phase (generally solid) to which the components of the mobile phase are characteristically adsorbed

In the chromatography[1] technique, two phases are required for the separation of the components (or compounds) of a mixture, the mobile phase and the stationary phase. The **mobile phase** consists of the components of the mixture and the solvent—the solvent being called the **eluent** or **eluting solution** (generally a mixture of solvents of differing polarities). The **stationary phase** is an adsorbent that has an intermolecular affinity not only for the solvent,[2] but also for the individual components of the mixture.

As the mobile phase passes over the stationary phase, the chromatogram develops. The components of the mobile phase have varying affinities for the stationary phase. The components of the mixture having a stronger affinity for the stationary phase move shorter distances while those with a lesser affinity move greater distances along the stationary phase during the time the chromatogram is being developed. Separation, and subsequent identification, of the components of the mixture is thus achieved. The leading edge of the mobile phase in the chromatogram is called the **eluent front.** When the eluent front reaches the edge of the chromatographic paper, the developing of the chromatogram is stopped.

Of the different types of chromatography including, for example, gas–solid chromatography, liquid–solid chromatography, column chromatography, thin-layer chromatography, and ion chromatography, this experiment uses paper chromatography as a separation technique.

In this experiment, chromatographic paper (similar to filter paper), a paper that consists of polar cellulose molecules, is the stationary phase. The mobile phase consists of one or more of the transition metal cations, Mn^{2+}, Fe^{3+}, Co^{2+}, Ni^{2+}, and Cu^{2+}, dissolved in an acetone–hydrochloric acid eluent.[3] In this experiment, the chromatographic paper is first "spotted" (and marked with a pencil *only*) with individual solutions for *each* of the cations and then for a solution with a mixture of cations. The paper is then dried. The spotted chromatographic paper is then placed in contact with the eluent to form the mobile

Capillary action: action by which the ions have an adhesive attraction (ion-dipole attractive forces) to the fibers of the stationary phase (paper)

Band: the identifying position of the component on the chromatography paper

phase. By **capillary action** the transition metal ions are transported along the paper. Each transition metal has its own (unique) adsorptive affinity for the polar, cellulose chromatographic paper; some are more strongly adsorbed than others. Also, each ion has its own solubility in the eluting solution. As a result of these two factors, some transition metal ions move further along the chromatographic paper than others to form **bands** at some distance from the origin, therefore indicating that the ions are separating.[4]

For a given eluting solution, stationary phase, temperature, and so on, each ion is characterized by its own R_f (ratio of fronts) factor:

$$R_{f, ion} = \frac{\text{distance from origin to final position of ion}}{\text{distance from origin to eluent front}} = \frac{D_{ion}}{D_{solvent}} \qquad (4.1)$$

---

[1]*Chromatography* means "the graphing of colors" where historically this technique was used to separate the various colored compounds from naturally occurring colored substances (e.g., dyes, plant pigments).

[2]This attraction is due to the similar type of intermolecular forces between the adsorbent (or the stationary phase) and the components of the mobile phase (e.g., dipole–dipole, ion–dipole, dipole–hydrogen bonding).

[3]The eluting solution (acetone–hydrochloric acid) actually converts the transition metal ions to their chloro ions, and it is in this form that they are separated along the paper.

[4]Consider, for example, a group of students moving through a cafeteria food line: the stationary phase is the food line; the mobile phase consists of students. You are well aware that some students go through the food line quickly because they are less attracted to the various food items. Others take forever to pass through the food line because they must stop to consider each item; they thus have a longer residence time in the food line.

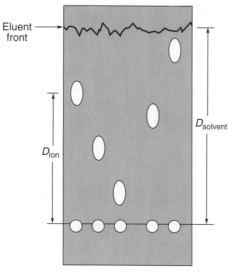

Figure 4.1 Determination of the R_f value for a transition metal ion.

Eluent front

$D_{solvent}$

D_{ion}

The origin of the **chromatogram** is defined as the point where the ion is "spotted" on the paper. The eluent front is defined as the most advanced point of movement of the mobile phase along the paper from the origin (Figure 4.1).

Chromatogram: the "picture" of the separated components on the chromatography paper

The next step is to identify the exact position of these bands. Some transition metal ions are already colored; for others, a characteristic reagent for each metal ion is used to enhance the metal ion's appearance and location on the stationary phase.

In this experiment you will prepare a chromatogram for each transition metal ion, locate its characteristic band, and determine its R_f value. A chromatogram of a test solution(s) will also be analyzed to identify the transition metal ions present in a solution mixture.

Procedure Overview: Chromatographic paper is used to separate an aqueous mixture of transition metal cations. An eluent is used to move the ions along the paper; the relative solubility of the cations in the solution versus the relative adsorptivity of the cations for the paper results in their separation along the paper. An enhancement reagent is used to intensify the appearance of the metal cation band on the paper.

Obtain about 2 mL of three unknown cation mixtures from your instructor in carefully marked 75-mm test tubes. Record the number of each unknown on the second page of Report Sheet.

EXPERIMENTAL PROCEDURE

1. **Developing Chamber.** Obtain a 600-mL beaker and enough plastic wrap (e.g., Saran Wrap®) for a cover (Figure 4.2). In the fume hood, prepare 10 mL of an eluting solution that consists of 9 mL of acetone and 1 mL of 6 *M* HCl. (**Caution:** *Acetone is flammable; extinguish all flames; HCl is corrosive.*) Pour this eluent into the middle of the beaker using a stirring rod; be careful not to wet the sides of the beaker. The depth of the eluent in the beaker should be 0.75–1.25 cm but *less than* 1.5 cm. Cover the beaker with the plastic wrap for about 10 minutes.[5] Hereafter, this apparatus is called the **developing chamber.**

   Figure 4.3 shows a commercial developing chamber.

2. **Ammonia Chamber.** The ammonia chamber may already be assembled in the hood. *Ask your instructor.* Obtain a dry, 1000-mL beaker and place it in the *fume hood.* Pour 5 mL of conc NH_3 into a 30-mL beaker and position it at the center of the 1000-mL beaker (Figure 4.4). (**Caution:** *Do not inhale the ammonia fumes.*)

A. Preparation of Chromatography Apparatus

Figure 4.2 Use plastic wrap to cover the developing chamber.

[5]The "atmosphere" in the beaker should become saturated with the vapor of the acetone/HCl eluent.

Figure 4.3 A commercial developing chamber.

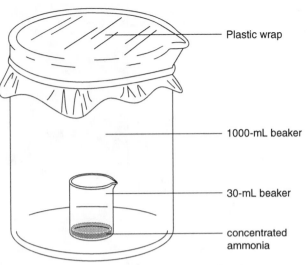

Plastic wrap

1000-mL beaker

30-mL beaker

concentrated ammonia

Figure 4.4 Apparatus for the ammonia chamber.

Cover the top of the 1000-mL beaker with plastic wrap. This apparatus will be used in Part C.3.

3. **Capillary Tube.** Obtain eight capillary tubes from the stockroom. When a glass capillary tube touches a solution, the solution should be drawn into the tip. When the tip is then touched to a piece of filter paper it should deliver a microdrop of solution. Try "spotting" a piece of filter paper with a known solution or water until the diameter of the drop is only 2–4 mm.

4. **Stationary Phase.** Obtain one piece of chromatographic paper (approximately 10 $\times$ 20 cm). Handle the paper only along its top 20-cm edge (by your designation) and lay it flat on a clean piece of paper, *not* directly on the lab bench. Draw a *pencil* line 1.5 cm from the *bottom* 20-cm edge of the paper (Figure 4.5). Starting 2 cm from the 10-cm edge and along the 1.5-cm line, make eight X's with a 2-cm separation. Use a pencil to label each X *below* the 1.5-cm line with the five cations being investigated and unknowns U1, U2, and U3.

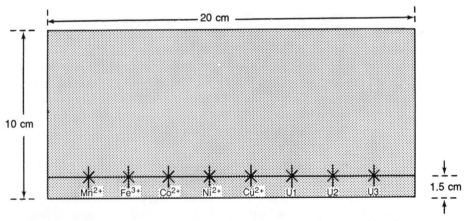

20 cm

10 cm

Mn²⁺ Fe³⁺ Co²⁺ Ni²⁺ Cu²⁺ U1 U2 U3

1.5 cm

Figure 4.5 Labeling the stationary phase (the chromatographic paper) for the chromatogram.

1. **Spot the Stationary Phase with the Knowns and Unknown(s).** Using the capillary tubes (remember you'll need *eight* of them, one for each solution) "spot" the chromatographic paper at the marked X's with the five known solutions containing the cations and the three unknown solutions. The microdrop should be 2–4 mm in diameter. Allow the "spots" to dry; a heat lamp or hair dryer may be used to hasten the drying—do *not* touch the paper along this bottom edge.

   Repeat the spotting/drying procedure two more times in order to increase the amount of metal ion at the "spot" on the chromatographic paper. Be sure to dry the sample between applications. Dry the paper with **caution:** *The heat lamp or hair dryer is hot!*

2. **Prepare the Stationary Phase for Elution.** Form the chromatographic paper into a cylinder and, near the top, attach the ends with tape, a staple, or small paper clip; do *not* allow the two ends of the paper to touch (Figure 4.6). Be sure the spots are dry and will *not* come into direct contact with the eluting solution!

3. **Develop the Chromatogram.** Place the paper cylinder into the developing chamber (Part A.1). The entire "bottom" of the cylindrical chromatographic paper must sit on the bottom of the developing chamber. Do *not* allow the paper to touch the wall. Make certain that the eluent is *below* the 1.5-cm line. Replace the plastic wrap. Do not disturb the developing chamber once the paper has been placed inside.

   When the eluent front has moved to within 1.5 cm of the top of the chromatographic paper remove the plastic wrap.

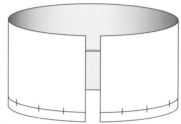

Figure 4.6 Formation of the stationary phase for placement in the developing chamber.

1. **Detection of Bands.** Remove the paper from the developing chamber and *quickly* mark (with a pencil) the position of the *eluent front.*[6] Allow the chromatogram to dry. While the chromatogram is drying, cover the developing chamber with the plastic wrap. Analyze the paper and circle (with a pencil) any colored bands, those from the solutions containing the known cations and those from the unknown solutions.[7]

2. **Enhancement of the Chromatogram.** To enhance the appearance and locations of the bands, move the chromatogram to the fume hood. Position the paper in the ammonia chamber (Part A.2) and cover the 1000-mL beaker with the plastic wrap.

   After the deep blue color of Cu^{2+} is evident, remove the chromatogram and circle any new transition metal ion bands that appear. Table 4.1 identifies the colors of other cations in the presence of ammonia. Mark the *center* of each band with a pencil. Allow the chromatogram to dry.

Table 4.1 Spot Solutions That Enhance the Band Positions of the Various Cations

			Cation		
	Mn^{2+}	Fe^{3+}	Co^{2+}	Ni^{2+}	Cu^{2+}
NH_3 test	Tan	Red-brown	Pink (brown)	Light blue	Blue
Spot solution	0.1 M $NaBiO_3$ (acidic) (purple)	0.2 M KSCN (blood red)	Satd KSCN in acetone (blue-green)	0.1 M NaHDMG (brick-red)	0.2 M $K_4[Fe(CN)_6]$ (red)

3. **Band Enhancement** (Optional, seek advice from your instructor). The "exact" band positions for the known cations and those of the mixture may be better defined using a second "spot solution." As necessary, use a capillary tube to spot the center of each band (from the known and unknown test solutions) with the corresponding spot solution identified in Table 4.1. This technique more vividly

---

[6]Do this quickly because the eluent evaporates.
[7]For the unknown solutions, there will most likely be more than one band.

locates the position of the band. Remember, the unknown test solution(s) may have more than one cation band.

4. **Analysis of Your Chromatogram.** Where is the center of each band? Mark the center of each band with a pencil. What is the color of each transition metal ion on the chromatogram?

Calculate and record the R_f values for each transition metal ion.

5. **Composition of the Unknown(s).** Look closely at the bands for your unknown(s). What is the R_f value for each band in each unknown? Which ion(s) is(are) present in your unknown(s)? Record your conclusions on the Report Sheet. Submit your chromatogram to your instructor for approval.

Disposal: Dispose of the eluting solution in the "Waste Organics" container. Allow the conc NH_3 in the ammonia chamber to evaporate in the hood.

The Next Step

Ink is a mixture of dyes. (1) Research the use of a chromatographic technique for the separation of the dyes in ink. (2) Design a project to separate the components of leaves from different trees, shrubs, and/or grasses. (3) Design a project for the separation of amino acids in fruit juices.

ATTACH OR SKETCH THE CHROMATOGRAM THAT WAS DEVELOPED IN THE EXPERIMENT

Paper Chromatography

Date _____ Lab Sec. _____ Name _____ Desk No. _____

1. a. Define the mobile phase in chromatography.

   b. What is the chemical composition of the mobile phase in this experiment?

2. a. Define the stationary phase in chromatography.

   b. What is the stationary phase in this experiment?

3. a. Define the eluent in chromatography.

   b. What is the eluent (eluting solution) in this experiment?

4. Experimental Procedure, Part A. Distinguish between the functions of the developing chamber and the ammonia chamber.

5. Experimental Procedure, Parts A.4 and B.1. Specifically, what must be done to prepare the stationary phase for the developing chamber in this experiment?

6. The chromatogram for the separation of amino acids using a butanoic acid/acetic eluent is sketched at right.

Use a ruler to determine and calculate the R_f for each amino acid.

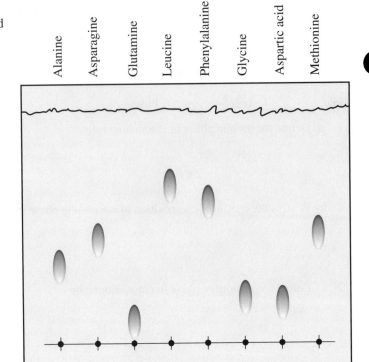

7. A student developed a chromatogram and found that the eluent front traveled 69 mm and the Zn^{2+} cation traveled 24 mm. In the development of a chromatogram of a mixture of cations, the eluent front traveled 52 mm. If Zn^{2+} is a cation of the unknown mixture, where will its band appear in the chromatogram?

Paper Chromatography

Date _____ Lab Sec. _____ Name _____ Desk No. _____

C. Analysis of the Chromatogram

Distance of eluent front from the origin: _____ mm

	Color (original)	Color (with NH_3)	Color (with spot solution)	Distance (mm) Traveled	R_f
Mn^{2+}	_____	_____	_____	_____	
Fe^{3+}	_____	_____	_____	_____	
Co^{2+}	_____	_____	_____	_____	
Ni^{2+}	_____	_____	_____	_____	
Cu^{2+}	_____	_____	_____	_____	

Instructor's approval of chromatogram _____

Show your calculations for R_f.

Unknown number(s)   U1=_____   U2=_____   U3=_____

	Band 1		Band 2		Band 3		Band 4	
Unknown	Distance (mm) Traveled	R_f	Distance (mm) Traveled	R_f	Distance (mm) Traveled	R_f	Distance (mm) Traveled	R_f
U1	_____		_____		_____		_____	
U2	_____		_____		_____		_____	
U3	_____		_____		_____		_____	

Show your calculations for R_f.

Cations present in U1   _____   _____   _____   _____

Cations present in U2   _____   _____   _____   _____

Cations present in U3   _____   _____   _____   _____

Laboratory Questions

Circle the questions that have been assigned.

1. Part A.1 and Part B.3.
   a. Why is it important to keep the developing chamber covered with plastic wrap during the development of the chromatogram?
   b. The developing chamber is *not* covered with plastic wrap during the development of the chromatogram. How does this technique error affect the D_{ion} for a cation? Explain.

2. Part A.4. Why was a pencil used to mark the chromatogram and not a ballpoint or ink pen?

3. Part B.1. Explain why the cation samples are repeatedly "spotted" and dried on the chromatographic paper.

4. Part B.3. The eluent is to be below the 1.5-cm line on the chromatographic paper. Describe the expected observation if the eluent were above the 1.5-cm line.

5. Part C.2. Explain why the *center* of the band is used to calculate the R_f value for a cation rather than the leading edge of the band.

6. Part C.2. The ammonia chamber is to be covered during the enhancement of the chromatogram. What is the consequence of the ammonia chamber being left uncovered?

7. Suppose two cations have the same R_f value. How might you resolve their presence in a mixture using paper chromatography?

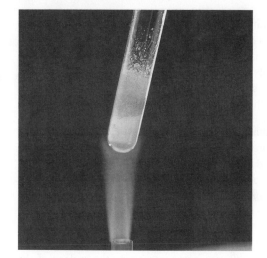

Percent Water in a Hydrated Salt

Heat readily removes the hydrated water molecules (top of sample in test tube) from copper(II) sulfate pentahydrate forming anhydrous copper(II) sulfate (bottom).

* To determine the percent by mass of water in a hydrated salt
* To learn to handle laboratory apparatus without touching it

OBJECTIVES

The following techniques are used in the Experimental Procedure

TECHNIQUES

INTRODUCTION

Many salts occurring in nature, purchased from the grocery shelf or from chemical suppliers are **hydrated;** that is, a number of water molecules are chemically bound to the ions of the salt in its crystalline structure. These water molecules are referred to as **waters of crystallization.** The number of moles of water per mole of salt is usually a constant. For example, iron(III) chloride is purchased as $FeCl_3 \cdot 6H_2O$, not as $FeCl_3$, and copper(II) sulfate as $CuSO_4 \cdot 5H_2O$, not as $CuSO_4$. For some salts, the water molecules are so weakly bound to the ions that heat removes them to form the **anhydrous** salt. Hydrated salts that spontaneously (without heat) lose water molecules to the atmosphere are **efflorescent,** whereas salts that readily absorb water are **deliquescent.**

Hydrate: water molecules are chemically bound to the ions of the salt as part of the structure of the compound

Anhydrous: without water

For Epsom salt (magnesium sulfate heptahydrate, Figure 5.1), the anhydrous salt, $MgSO_4$, forms with gentle heating.

$$MgSO_4 \cdot 7H_2O(s) \xrightarrow{\Delta(>200°C)} MgSO_4(s) + 7 H_2O(g) \tag{5.1}$$

In other salts such as $FeCl_3 \cdot 6H_2O$, the water molecules are so strongly bound to the salt that anhydrous $FeCl_3$ cannot form regardless of the intensity of the heat.

In Epsom salt 7 moles of water, or 126.1 g of H_2O, are bound to each mole of magnesium sulfate, or 120.4 g of $MgSO_4$. The percent by mass of water in the salt is

$$\frac{126.1 \text{ g } H_2O}{(126.1 \text{ g} + 120.4 \text{ g)} \text{ salt}} \times 100 = 51.16\% \text{ } H_2O \tag{5.2}$$

A **gravimetric analysis** is an analytical method that relies almost exclusively on mass measurements for the analysis. Generally the substance being analyzed must have a mass large enough to be measured easily with the balances that are available in the laboratory.

This experiment uses the gravimetric analysis method to determine the percent by mass of water in a hydrated salt. The mass of a hydrated salt is measured, the sample is heated to drive off the hydrated water molecules (the waters of crystallization), and then the mass of remaining sample is measured again. Cycles of heating and measuring of the sample's mass are continued until reproducibility of the mass measurements is attained.

Figure 5.1 Epsom salt, $MgSO_4 \cdot 7H_2O$.

EXPERIMENTAL PROCEDURE

Procedure Overview: The mass of a hydrated salt is measured before and after it is heated to a high temperature. This mass difference and the mass of the anhydrous salt are the data needed to calculate the percent water in the original hydrated salt.

You are to complete *at least* two trials in this experiment. Obtain a hydrated salt from your instructor. Record the unknown number of your hydrated salt on the Report Sheet. A 150-mm test tube may be substituted for the crucible and lid. Ask your instructor.

A. Sample Preparation

Fired: heated to a very high temperature to volatilize impurities

Figure 5.2 Handle the crucible with tongs after heating.

1. **Prepare a Clean Crucible.** Obtain a clean crucible and lid. Check the crucible for stress fractures or fissures, which are common in often-used crucibles. If none are found, support the crucible and lid on a clay triangle and heat with an intense flame for 5 minutes (see Technique 15C, Figure T.15c). Allow them to cool slowly. (**Caution:** *Do not set them on the lab bench for fear of contamination.*)[1,2]

   Caution: *hot and cool crucibles look the same—do not touch!* Determine the mass (± 0.001 g) of the **fired**, *cool* crucible and lid and record. Handle the crucible and lid with the crucible tongs for the remainder of the experiment (Figure 5.2); do *not* use your fingers—oil from the fingers can contaminate the surface of the crucible and lid. Be sure the crucible tongs are clean and dry.

2. **Determine the Mass of Sample.** Add no more than 3 g of your hydrated salt to the crucible. Measure and record the combined mass (± 0.001 g) of the crucible, lid, and hydrated salt. Calculate the mass of the hydrated salt.

3. **Adjust the Crucible Lid.** Return the crucible (use crucible tongs only) with the sample to the clay triangle; set the lid just off the lip of the crucible to allow the evolved water molecules to escape on heating (Figure T.15e).

B. Thermal Decomposition of the Sample

1. **Heat the Sample.** Initially heat the sample slowly and then gradually intensify the heat. Do *not* allow the crucible to become red hot. This may cause the anhydrous salt to decompose as well. Maintain the high temperature on the sample for 10 minutes.

   Cover the crucible with the lid; allow them to cool to room temperature (see footnote 1). Determine the combined mass of the crucible, lid, and anhydrous salt on the same balance that was used for earlier measurements.

2. **Have You Removed All of the Water?** Reheat the sample for 2 minutes (Figure T.15e), but do *not* intensify the flame—avoid the decomposition of the salt. Cool to room temperature and again measure the combined mass. If this second mass measurement of the anhydrous salt disagrees by greater than ± 0.010 g from that in Part B.1, repeat Part B.2.

3. **Repeat with a New Sample.** Repeat the experiment two more times with original hydrated salt samples.

Disposal: Dispose of all waste anhydrous salt in the "Waste Solids" container.

CLEANUP: Rinse the crucible with 2–3 milliliters of 1 *M* HCl and discard in the "Waste Acids" container. Then rinse several times with tap water and finally deionized water. Each water rinse can be discarded in the sink.

The Next Step

(1) Most any substance has water as a part of its composition, either adsorbed or bonded to the substance. Develop a procedure for a determining the percent water in a soil sample . . . remember that dissolved gases may also be present. (2) How rapidly do potato chips or crackers absorb water from the atmosphere? Develop a data plot of percent water vs. time for "soggy crackers." (3) A coal analysis involves the determination of the (a) percent moisture, (b) percent volatile combustible matter, VCM, (c) percent ash, and (d) percent fixed carbon, FC, where %FC = 100% − (%H_2O + %VCM + %ash). Develop a procedure for the analysis for a coal sample . . . the analysis indicates the quality of the coal for combustion, coking, etc.

[1]Place the crucible and lid in a desiccator (if available) for cooling.
[2]If the crucible remains dirty after heating, move the apparatus to the fume hood, add 1–2 mL of 6 *M* HNO_3, and gently evaporate to dryness. (**Caution:** *Avoid skin contact, flush immediately with water.*)

Percent Water in a Hydrated Salt

Date _____ Lab Sec. _____ Name _____ Desk No. _____

1. Calcium chloride, a deliquescent salt, is used as a desiccant in laboratory desiccators. Explain.

2. Experimental Procedure, Part A.1. What is the purpose of firing the crucible?

3. A 1.994-g sample of gypsum, a hydrated salt of calcium sulfate, $CaSO_4$, is heated at a temperature greater than 170°C in a crucible until a constant mass is reached. The mass of the anhydrous $CaSO_4$ salt is 1.577 g.

   Calculate the percent by mass of water in the hydrated calcium sulfate salt.

4. The gravimetric analysis of this experiment is meant to be quantitative; therefore, all precautions should be made to minimize errors in the analysis.
   a. The crucible and lid are handled exclusively with crucible tongs in the experiment. How does this technique maintain the integrity of the analysis?

   b. Mass measurements of the crucible, lid, and sample are performed only at room temperature. Why is this technique necessary for a gravimetric analysis?

   c. Why is the position of the crucible lid critical to the dehydration of the salt during the heating process? Explain.

5. The following data were collected from the gravimetric analysis of a hydrated salt:

Mass of crucible and lid (g)	19.437
Mass of crucible, lid, and hydrated salt (g)	21.626
Mass of crucible, lid, and anhydrous salt (g)	21.441

Determine the percent water in the hydrated salt.

6. a. What is the percent by mass of water in copper(II) sulfate pentahydrate, $CuSO_4 \cdot 5H_2O$? See opening photo.

b. What mass due to waters of crystallization is present in a 3.38-g sample of $CuSO_4 \cdot 5H_2O$?

Percent Water in a Hydrated Salt

Date _____ Lab Sec. _____ Name _____ Desk No. _____

Unknown No. _____	Trial 1	Trial 2	Trial 3
1. Mass of crucible and lid (g)	_____	_____	_____
2. Mass of crucible, lid, and hydrated salt (g)	_____	_____	_____
3. Mass of crucible, lid, and anhydrous salt	_____	_____	_____
1st mass measurement (g)	_____	_____	_____
2nd mass measurement (g)	_____	_____	_____
3rd mass measurement (g)	_____	_____	_____
4. Final mass of crucible, lid, and anhydrous salt (g)	_____	_____	_____

Calculations

	Trial 1	Trial 2	Trial 3
1. Mass of hydrated salt (g)	_____	_____	_____
2. Mass of anhydrous salt (g)	_____	_____	_____
3. Mass of water lost (g)	_____	_____	_____
4. Percent by mass of volatile water in hydrated salt (%)	_____ *	_____	_____

5. Average percent H_2O in hydrated salt ($\%H_2O$) _____

6. Standard deviation of $\%H_2O$ _____ *

7. Relative standard deviation of $\%H_2O$ in hydrated salt ($\%RSD$) _____ *

*Show calculations on next page.

*Calculations for Trial 1. Show your work.

*Calculation of Standard Deviation and %RSD. Appendix B.

Laboratory Questions

Circle the questions that have been assigned.

1. Part A.1. During the cooling of the fired crucible, water vapor condensed on the crucible wall before its mass measurement. The condensation did not occur following thermal decomposition of the hydrated salt in Part B. Will the reported percent water in the hydrated salt be reported too high or too low? Explain.

2. Part A.1. The fired crucible is handled with (oily) fingers before its mass measurement. Subsequently in Part B.1, the oil from the fingers is burned off. How does this technique error affect the reported percent water in the hydrated salt? Explain.

3. Part A.1. The crucible is handled with (oily) fingers after its mass measurement but before the ~3 g sample of the hydrated salt is measured (Part A.2). Subsequently in Part B.1, the oil from the fingers is burned off. How does this technique error affect the reported percent water in the hydrated salt? Explain.

4. Part A.1. Suppose the original sample is unknowingly contaminated with a second anhydrous salt. Will the reported percent water in the hydrated salt be too high, too low, or unaffected by its presence? Explain.

*5. After heating the crucible in Part A.1, the crucible is set on the lab bench where it is contaminated with the cleaning oil used to clean the lab bench but before its mass is measured. The analysis continues through Part B.1 where the mass of the anhydrous salt is determined. While heating, the cleaning oil is burned off the bottom of the crucible. Describe the error that has occurred; that is, is the mass of the anhydrous salt remaining in the crucible reported as being too high or too low? Explain.

6. Part B.1. The hydrated salt is overheated and the anhydrous salt thermally decomposes, one product being a gas. Will the reported percent water in the hydrated salt be reported too high, too low, or be unaffected? Explain.

7. Part B.2. Because of a lack of time, Bill decided to skip this step in the Experimental Procedure. Will his haste in reporting the "percent H_2O in the hydrated salt" likely be too high, too low, or unaffected? Explain.

Inorganic Nomenclature I. Oxidation Numbers

Sodium chloride salt crystals are a one-to-one combination of sodium cations and chloride anions. The sodium cation has an oxidation number of +1 and the chloride anion has an oxidation number of −1.

- To become familiar with the oxidation numbers of various elements
- To learn to write formulas of compounds based on oxidation numbers

INTRODUCTION

You have probably noticed that members of virtually all professions have a specialized language. Chemists are no exception. For chemists to communicate internationally, some standardization of the technical language is required. Historically, common names for many compounds evolved and are still universally understood—for example, water, sugar (sucrose), and ammonia—but with new compounds being synthesized and isolated daily, a "familiar" system is no longer viable. Two of the more recent "newsworthy" common names to appear in the literature are buckminsterfullerene (also called bucky ball) and sulflower.[1]

The Chemical Abstracts Service (CAS) of the American Chemical Society currently has over 31 million inorganic and organic compounds registered in their database[2] with over 4000 new compounds being added *daily*. Not only does each compound have a unique name, according to a systematic method of nomenclature, but also each compound has a unique molecular structure and set of chemical and physical properties. Therefore, it has become absolutely necessary for the scientific community to properly name compounds according to a universally accepted system of nomenclature.

In the three parts of Dry Lab 2 (A, B, and C) you will learn a few systematic rules, established by the International Union for Pure and Applied Chemistry (IUPAC), for naming and writing formulas for inorganic compounds. You are undoubtedly already familiar with some symbols for the elements and the names for several common compounds. For instance, NaCl is sodium chloride. Continued practice and work in writing formulas and naming compounds will make you even more knowledgeable of the chemist's vocabulary.

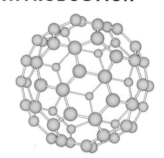

Buckminsterfullerene

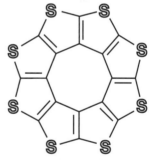

Sulflower

Elements or groups of chemically bonded elements (called **polyatomic** or **molecular groups**) that have **charge** are called **ions**—if the charge is positive they are called **cations,** if negative, they are **anions.** Oftentimes elements combine to form a compound in which the elements may not have an actual charge, but rather an "apparent" charge, called its **oxidation number** (or oxidation state). The oxidation number, the charge an atom would have if the electrons in the bond were assigned to the more electronegative element, may be either positive or negative.

Oxidation Number

Charges are conventionally written "number and charge" (e.g., 3+), but oxidation numbers are written "charge and number" (e.g., −2)

---

[1]An extensive listing of common chemical names appears in Appendix D.
[2]http://www.cas.org

While monoatomic and polyatomic ions have an actual charge, the charge of an atom/element within a compound is often uncertain. However, with the application of a few rules, the apparent charge or oxidation number of an atom/element in a compound can be easily determined. Thus, the use of oxidation numbers has a greater range of applicability. For example, knowledge of the oxidation numbers for a range of elements allows us to write the correct chemical formula for a large number of compounds. It is *not* necessary to simply memorize disconnected chemical formulas.

The following "rules" will only use the term "oxidation number" when considering the common charge *or* "apparent" charge for an element. Note in Rule #6 that polyatomic ions have an actual charge (not an oxidation number).

1. Any element in the free state (not combined with another element) has an oxidation number of zero, regardless of the complexity of the molecule in which it occurs. Each atom in Ne, O_2, P_4, and S_8 has an oxidation number of 0.

2. Monoatomic ions have an oxidation number equal to the charge of the ion. The ions Ca^{2+}, Fe^{3+}, and Cl^- have oxidation numbers of $+2$, $+3$, and -1 respectively.

3. Oxygen in compounds has an assigned oxidation number of -2 (except for a -1 in peroxides, e.g., H_2O_2, and $+2$ in OF_2). The oxidation number of oxygen is -2 in FeO, Fe_2O_3, $KMnO_4$, and KIO_3.

4. Hydrogen in compounds has an oxidation number of $+1$ (except for -1 in metal hydrides, e.g., NaH). Its oxidation number is $+1$ in HCl, $NaHCO_3$, and NH_3.

5. Some elements exhibit only one common oxidation number in certain types of compounds:

   a. Group 1A elements always have an oxidation number of $+1$ in compounds.

   b. Group 2A elements always have an oxidation number of $+2$ in compounds.

   c. Boron and aluminum always possess an oxidation number of $+3$ in compounds.

   d. In **binary compounds** *with metals,* the nonmetallic elements of Group 6A generally exhibit an oxidation number of -2.

   e. In binary compounds *with metals,* the elements of Group 7A have an oxidation number of -1.

6. **Polyatomic ions** have a charge equal to the sum of the oxidation numbers of the elements of the polyatomic group. See Example D2A.2. The polyatomic ions SO_4^{2-}, NO_3^-, and PO_4^{3-}, have charges of 2−, 1−, and 3−, respectively.

7. In assigning oxidation numbers to elements in a compound, the element closest to fluorine (the most electronegative element) in the periodic table is always assigned the negative oxidation number. In the compound, P_4O_{10}, oxygen has the negative oxidation number of -2.

8. a. For compounds the sum of oxidation numbers of all atoms in the compound must equal zero.

      Example D2A.1 For Na_2S, the sum of the oxidation numbers equals zero.

$$2 \text{ Na atoms, } +1 \text{ for each (Rule 5a)} = +2$$

$$1 \text{ S atom, } -2 \text{ for each (Rule 5d)} = -2$$

      Sum of oxidation numbers $(+2) + (-2) = 0$

   b. For polyatomic ions the sum of the oxidation numbers of the elements must equal the charge of the ion.

      Example D2A.2 For CO_3^{2-}, the sum of the oxidation numbers of the elements equals a charge of 2−.

$$3 \text{ O atoms, } -2 \text{ for each (Rule 3)} = -6$$

$$1 \text{ C atom which must be } +4 = +4$$

      so that $(-6) + (+4) = 2-$, the charge of the CO_3^{2-} ion.

Binary compounds: compounds consisting of only two elements (see Dry Lab 2B)

Polyatomic ions are also called molecular ions (see Dry Lab 2C)

9. Some chemical elements show more than one oxidation number, depending on the compound. The preceding rules may be used to determine their values. Consider the compounds $FeCl_2$ and $FeCl_3$. Since the chlorine atom has an oxidation number of -1 when combined with a metal (Rule 5e), the oxidation numbers of iron are $+2$ in $FeCl_2$ and $+3$ in $FeCl_3$.

An extensive listing of oxidation numbers for monoatomic ions and polyatomic ions are presented in Dry Lab 2C, Table D2C.1.

Writing Formulas from Charges and Oxidation Numbers

The formulas of compounds can be written from a knowledge of the oxidation numbers of the elements that make up the compound. The sum of the oxidation numbers for a compound must equal zero.

Example D2A.3   Write the formula of the compound formed from Ca^{2+} ions and Cl^- ions.

As the sum of the oxidation numbers must equal zero, two Cl^- are needed for each Ca^{2+}; therefore the formula of the compound is $CaCl_2$.

DRY LAB PROCEDURE

Procedure Overview: The oxidation number of an element in a selection of compounds and ions is determined by application of the rules in the Introduction. Additionally, a knowledge of oxidation numbers of elements and the charge on the polyatomic groups will allow you to write the correct formulas for compounds.

Your instructor will indicate the questions you are to complete. Answer them on a separate piece of paper. Be sure to indicate the date, your lab section, and your desk number on your report sheet.

1. Indicate the oxidation number of carbon and sulfur in the following compounds.

   a. CO　　　d. $Na_2C_2O_4$　g. SO_2　　j. Na_2SO_3　m. SCl_2
   b. CO_2　　e. CH_4　　h. SO_3　　k. $Na_2S_2O_3$　n. Na_2S_2
   c. Na_2CO_3　f. H_2CO　　i. Na_2SO_4　l. $Na_2S_4O_6$　o. $SOCl_2$

2. Indicate the oxidation number of phosphorus, iodine, nitrogen, tellurium, and silicon in the following ions.

   a. PO_4^{3-}　　d. $P_3O_{10}^{5-}$　g. IO^-　　j. NO_2^-　　m. $N_2O_2^{2-}$
   b. PO_3^{3-}　　e. IO_3^-　　h. NH_4^+　k. NO^+　　n. TeO_4^{2-}
   c. HPO_4^{2-}　f. IO_2^-　　i. NO_3^-　l. NO_2^+　　o. SiO_3^{2-}

3. Indicate the oxidation number of the metallic element(s) in the following compounds.

   a. Fe_2O_3　　　　g. CrO_3　　　　m. MnO_2
   b. FeO　　　　　h. K_2CrO_4　　　n. PbO_2
   c. CoS　　　　　i. $K_2Cr_2O_7$　　o. Pb_3O_4
   d. $CoSO_4$　　　　j. $KCrO_2$　　　p. ZrI_4
   e. K_3CoCl_6　　k. K_2MnO_4　　q. U_3O_8
   f. $CrCl_3$　　　　l. Mn_2O_7　　　r. UO_2Cl_2

4. Write formulas for all of the compounds resulting from matching all cations with all anions in each set.

Set 1		Set 2		Set 3	
Cations	Anions	Cations	Anions	Cations	Anions
Li^+	Cl^-	Os^{8+}	P^{3-}	Pb^{2+}	SiO_3^{2-}
Ca^{2+}	SO_4^{2-}	Fe^{2+}	N^{3-}	Zr^{4+}	S^{2-}
Na^+	NO_3^-	Al^{3+}	H^-	Co^{3+}	I^-
NH_4^+	O^{2-}	Zn^{2+}	OH^-	Mn^{3+}	ClO^-
V^{5+}	CO_3^{2-}	Cr^{3+}	IO_3^-	Hg^{2+}	CrO_4^{2-}

Inorganic Nomenclature II.* Binary Compounds

Calcium fluoride, magnesium oxide, and sodium chloride (left to right) are binary compounds of a metal and a nonmetal, called **salts.**

OBJECTIVES

- To name and write formulas for the binary compounds of a metal and a nonmetal, of two nonmetals (or metalloid and nonmetal), and of acids.
- To name and write formulas for hydrated compounds.

INTRODUCTION

The oxidation number of an element helps us to write the formula of a compound and to characterize the chemical nature of the element in the compound. Such information may tell us of the reactivity of the compound. In this dry lab, the correct writing and naming of binary compounds become the first steps in understanding the systematic nomenclature of a large array of chemical compounds.

Binary compounds are the simplest of compounds consisting of a chemical combination of two elements, *not* necessarily two atoms. Compounds such as NaCl, SO_3, Cr_2O_3, and HCl are all binary compounds. Binary compounds can be categorized into those of a metal cation and a nonmetal anion, of two nonmetals (or a metalloid and a nonmetal), and of acids.

Metal Cation and Nonmetal Anion—A Binary Salt

Compounds consisting of only two elements are named directly from the elements involved. In naming a binary salt, it is customary to name the more metallic (more electropositive) element first. The root of the second element is then named with the suffix -*ide* added to it. NaBr is sodium brom*ide;* Al_2O_3 is aluminum ox*ide.*

Two polyatomic anions, OH^- and CN^-, have names ending in -*ide,* even though their salts are not binary. Thus, KOH is potassium hydrox*ide* and NaCN is sodium cyan*ide.* One polyatomic cation, NH_4^+, is also treated as a single species in the naming of a compound; NH_4Cl is ammonium chlor*ide,* even though NH_4Cl is not a binary compound.

When a metal cation exhibits more than one oxidation number, the nomenclature must reflect that. Multiple oxidation numbers normally exist for the transition metal cations. Two systems are used for expressing different oxidation numbers of a metal cation:

1. The "old" system uses a different suffix to reflect the oxidation number of the metal cation; the suffix is usually added to the root of the Latin name for the cation. The -*ous* ending designates the lower of two oxidation numbers for the metal cation, whereas the -*ic* ending indicates the higher one.

*For more information on chemical nomenclature, go online to *http://en.wikipedia.org/wiki/Category:Chemistry.*

2. The Stock system uses a Roman numeral after the English name to indicate the oxidation number of the metal cation in the compound.

Examples for the copper, iron, and tin salts are

Formula	"Old" System	Stock System
Cu_2O	*cuprous* oxide	copper(I) oxide
CuO	*cupric* oxide	copper(II) oxide
$FeCl_2$	*ferrous* chloride	iron(II) chloride
$FeCl_3$	*ferric* chloride	iron(III) chloride
$SnCl_2$	*stannous* chloride	tin(II) chloride
$SnCl_4$	*stannic* chloride	tin(IV) chloride

In naming compounds of a metal known to have variable oxidation numbers, the Stock system is preferred in chemistry today. Following are the metal cations that are commonly named using both the "old" and the Stock system.

Fe^{3+}	ferric	iron(III)	Cu^{2+}	cupric	copper(II)
Fe^{2+}	ferrous	iron(II)	Cu^+	cuprous	copper(I)
Sn^{4+}	stannic	tin(IV)	Hg^{2+}	mercuric	mercury(II)
Sn^{2+}	stannous	tin(II)	Hg_2^{2+}	mercurous	mercury(I)
Cr^{3+}	chromic	chromium(III)	Co^{3+}	cobaltic	cobalt(III)
Cr^{2+}	chromous	chromium(II)	Co^{2+}	cobaltous	cobalt(II)
Pb^{4+}	plumbic	lead(IV)	Mn^{3+}	manganic	manganese(III)
Pb^{2+}	plumbous	lead(II)	Mn^{2+}	manganous	manganese(II)

"Old" names are still often used in chemical nomenclature.

Often more than two compounds form from the chemical combination of two non-metals or a metalloid and a nonmetal. For example, nitrogen and oxygen combine to form the compounds N_2O, NO, NO_2, N_2O_3, N_2O_4, and N_2O_5. It is obvious that these are all nitrogen ox*ides*, but each must have a unique name.

To distinguish between the nitrogen oxides and, in general, to name compounds formed between two nonmetals or a metalloid and a nonmetal, Greek prefixes are used to designate the number of atoms of each element present in the molecule. The common Greek prefixes follow:

Two Nonmetals (or a Metalloid and a Nonmetal)

Prefix	Meaning	Prefix	Meaning
mono-*	one	hepta-	seven
di-	two	octa-	eight
tri-	three	nona-	nine
tetra-	four	deca-	ten
penta-	five	dodeca-	twelve
hexa-	six		

*mono- is seldom used, as "one" is generally implied.

The "prefix" system is the *preferred* method for naming the binary compounds of two nonmetals and of a metalloid and a nonmetal. The Stock system is occasionally used, but as you can see, this system does not differentiate between the NO_2 and N_2O_4 nitrogen oxides below.

Formula	"Prefix" System	Stock System
N_2O	*di*nitrogen oxide	nitrogen(I) oxide
NO	nitrogen oxide	nitrogen(II) oxide
NO_2	nitrogen *di*oxide	nitrogen(IV) oxide
N_2O_3	*di*nitrogen *tri*oxide	nitrogen(III) oxide
N_2O_4	*di*nitrogen *tetr*oxide	nitrogen(IV) oxide
N_2O_5	*di*nitrogen *pent*oxide	nitrogen(V) oxide

dinitrogen trioxide

dinitrogen tetroxide

dinitrogen pentoxide

Note that on occasion the end vowel of the Greek prefix is omitted from the spelling for clarity in pronunciation.

Hydrates

Prefixes are also used for the naming of **hydrates,** inorganic salts in which water molecules are a part of the crystalline structure of the solid. The "waters of hydration" (also called waters of crystallization) are bound to the ions of the salt and, in most cases, can be removed with the application of heat (see Experiment 5). For example, barium chloride is often purchased as barium chloride *di*hydrate, $BaCl_2 \cdot 2H_2O$. The formula of iron(III) chloride *hexa*hydrate is $FeCl_3 \cdot 6H_2O$.

Binary Acids

A binary acid is an *aqueous* solution of a compound formed by hydrogen and a more electronegative nonmetal. To name the acid, the prefix *hydro-* and suffix *-ic* are added to the root of the nonmetal name.

Formula	Name
HCl(*aq*)	*hydro*chloric acid
HBr(*aq*)	*hydro*bromic acid
H$_2$S(*aq*)	*hydro*sulfuric acid

Writing Formulas from Names

In writing formulas for compounds it is mandatory that the sum of the oxidation numbers (see Dry Lab 2A) of the elements equals zero. The name of the compound must dictate how the formula is to be written.

Example D2B.1   Write the formula for calcium nitride.
Calcium is in Group 2A and therefore has an oxidation number of $+2$: Ca^{2+}. The nitrogen atom is always -3 when combined with a metal: N^{3-}. Therefore, for the sum of the oxidation numbers to equal zero there must be three Ca^{2+} (a total of $+6$) for every two N^{3-} (a total of -6): the formula is Ca_3N_2.

DRY LAB PROCEDURE

Procedure Overview: Given the formula of the compound, the proper names for a large number of compounds are to be written, and given the name of the compound, the formulas for a large number of compounds are to be written.

Your instructor will assign the exercises you are to complete. Answer them on a separate piece of paper. Be sure to indicate the date, your lab section, and your desk number on your report sheet. Use the rules that have been described.

1. Name the following binary salts of the representative elements.

   a. Na$_3$P   d. CaC$_2$ g. Ca$_3$P$_2$ j. K$_2$S   m. NH$_4$Br   p. AlCl$_3$
   b. Na$_2$O   e. CaI$_2$ h. KCN   k. K$_2$Te   n. (NH$_4$)$_2$S   q. Al$_2$O$_3$
   c. Na$_3$N   f. CaH$_2$ i. KOH   l. K$_2$O$_2$ o. NH$_4$CN   r. AlN

2. Name the following salts according to the "old" "*-ic, -ous*" system *and* the Stock system.

a. CrS	g. HgCl$_2$	m. CoO
b. Cr$_2$O$_3$	h. Hg$_2$Cl$_2$	n. CoBr$_3 \cdot$6H$_2$O
c. CrI$_3 \cdot$6H$_2$O	i. HgO	o. SnF$_4$
d. CuCl	j. Fe$_2$O$_3$	p. SnO$_2$
e. CuI$_2$	k. FeS	q. Cu$_2$O
f. CuBr$_2 \cdot$4H$_2$O	l. FeI$_3 \cdot$6H$_2$O	r. Fe(OH)$_3$

3. Name the following binary acids.

   a. HF(aq)   d. HBr(aq)

   b. HI(aq)   e. H$_2$Te(aq)

   c. H$_2$Se(aq)   f. HCl(aq)

4. Name the following binary compounds consisting of two nonmetals or a metalloid and a nonmetal.

   a. SO$_2$ d. SF$_6$ g. N$_2$O$_5$ j. SiCl$_4$ m. AsH$_3$ p. XeF$_4$

   b. SO$_3$ e. SCl$_4$ h. N$_2$S$_4$ k. SiO$_2$ n. AsF$_5$ q. XeF$_6$

   c. S$_4$N$_4$ f. NO$_2$ i. NF$_3$ l. AsCl$_3$ o. HCl   r. XeO$_3$

5. Write formulas for the following compounds consisting of a metal cation and a nonmetal anion.

   a. ferrous sulfide   h. nickel(III) oxide

   b. iron(III) hydroxide   i. chromium(III) oxide

   c. ferric oxide   j. titanium(IV) chloride

   d. aluminum iodide   k. cobalt(II) chloride hexahydrate

   e. copper(I) chloride   l. cobaltous oxide

   f. cupric cyanide tetrahydrate   m. mercury(I) chloride

   g. manganese(IV) oxide   n. mercuric iodide

6. Write formulas for the following compounds consisting of two nonmetals (or a metalloid and a nonmetal).

   a. hydrochloric acid   g. iodine pentafluoride

   b. hydrosulfuric acid   h. krypton difluoride

   c. hydroiodic acid   i. tetrasulfur tetranitride

   d. silicon tetrafluoride   j. dichlorine heptaoxide

   e. arsenic pentafluoride   k. phosphorus trihydride (phosphine)

   f. xenon hexafluoride   l. tetraphosphorus decoxide

7. Write formulas and name the binary compounds resulting from matching all cations with all anions in each set.

Set 1		Set 2		Set 3	
Cations	Anions	Cations	Anions	Cations	Anions
K$^+$	Cl$^-$	Co^{2+}	P^{3-}	Mn^{3+}	O^{2-}
H$^+$(aq)	S^{2-}	Co^{3+}	Br$^-$	Sn^{2+}	N^{3-}
Na$^+$	F$^-$	Pb^{2+}	O^{2-}	NH$_4$$^+$	S^{2-}
Fe^{3+}	CN$^-$	Pt^{4+}	F$^-$	Hg$_2$$^{2+}$	Se^{2-}
Cu$^+$	I$^-$	Ba^{2+}	OH$^-$	Ce^{4+}	I$^-$

*Name as acids

8. Write the formulas for the compounds shown in the photo.

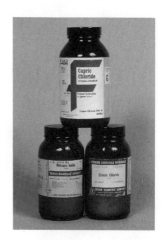

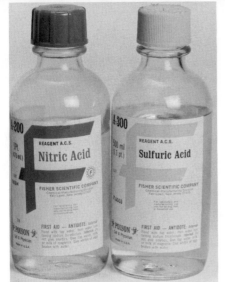

Inorganic Nomenclature III. Ternary Compounds

Common ternary acids (or oxoacids) in the laboratory are nitric acid and sulfuric acid.

OBJECTIVES

- To name and write formulas for salts and acids containing polyatomic anions
- To name and write formulas for acid salts

INTRODUCTION

In Dry Lab 2B, the naming and the writing of formulas for binary compounds, the simplest of compounds, were introduced.

Ternary compounds are generally considered as having a polyatomic anion containing oxygen. Compounds such as $KMnO_4$, $BaSO_4$, $KClO_3$, and HNO_3 are ternary compounds that have the polyatomic anions MnO_4^-, SO_4^{2-}, ClO_3^-, and NO_3^-, respectively. Ternary compounds can be categorized as salts and acids.

Metal Cation and a Polyatomic Anion Containing Oxygen—A Ternary Salt

Many polyatomic anions consist of one element (usually a nonmetal) and oxygen; the entire grouping of atoms carries a negative charge. The anion is named by using the root name of the element (not the oxygen) with the suffix *-ate*. As examples, SO_4^{2-} is the sulf*ate* ion and NO_3^- is the nit*rate* ion.

If *only two* polyatomic anions form from an element and oxygen, the polyatomic anion having the element with the higher oxidation number uses the suffix *-ate;* the polyatomic anion having the element with the lower oxidation number uses the suffix *-ite*. As examples,

sulfate ion

nitrate ion

phosphate ion

Ion	Oxidation Number	Name	Ion	Oxidation Number	Name
SO_4^{2-}	S is +6	sulf*ate*	SO_3^{2-}	S is +4	sulf*ite*
NO_3^-	N is +5	nit*rate*	NO_2^-	N is +3	nit*rite*
PO_4^{3-}	P is +5	phosph*ate*	PO_3^{3-}	P is +3	phosph*ite*

The formulas and names of common polyatomic anions are listed in Table D2C.1.

Salts that have a polyatomic anion are named in the same manner as the binary salts: the metal cation (along with the Stock system or *-ic, -ous* distinction) is named first followed by the polyatomic anion. Some examples follow:

Salt	Name
Na_2SO_4	sodium sulfate
Na_2SO_3	sodium sulfite
$Fe(NO_3)_3$	iron(III) or ferric nitrate
$FeSO_4$	iron(II) or ferrous sulfate

Table D2C.1 Names and Oxidation Numbers of Common Monoatomic and Polyatomic Ions

A. Metallic and Polyatomic Cations

Oxidation Number of +1

NH_4^+	ammonium	Hg_2^{2+}	mercury(I), mercurous
Cu^+	copper(I), cuprous	K^+	potassium
H^+	hydrogen	Ag^+	silver (I)
Li^+	lithium	Na^+	sodium

Oxidation Number of +2

Ba^{2+}	barium	Mn^{2+}	manganese(II), manganous
Cd^{2+}	cadmium	Hg^{2+}	mercury(II), mercuric
Ca^{2+}	calcium	Ni^{2+}	nickel(II)
Cr^{2+}	chromium(II), chromous	Sr^{2+}	strontium
Co^{2+}	cobalt(II), cobaltous	Sn^{2+}	tin(II), stannous
Cu^{2+}	copper(II), cupric	UO_2^{2+}	uranyl
Fe^{2+}	iron(II), ferrous	VO^{2+}	vanadyl
Pb^{2+}	lead(II), plumbous	Zn^{2+}	zinc
Mg^{2+}	magnesium		

Oxidation Number of +3

Al^{3+}	aluminum	Co^{3+}	cobalt(III), cobaltic
As^{3+}	arsenic(III)	Fe^{3+}	iron(III), ferric
Cr^{3+}	chromium(III), chromic	Mn^{3+}	manganese(III), manganic

Oxidation Number of +4

Pb^{4+}	lead(IV), plumbic	Sn^{4+}	tin(IV), stannic

Oxidation Number of +5

V^{5+}	vanadium(V)	As^{5+}	arsenic(V)

B. Nonmetallic and Polyatomic Anions

Oxidation Number of −1

$CH_3CO_2^-$	acetate *or*	H^-	hydride
$C_2H_3O_2^-$	acetate	ClO^-	hypochlorite
Br^-	bromide	I^-	iodide
ClO_3^-	chlorate	NO_3^-	nitrate
Cl^-	chloride	NO_2^-	nitrite
ClO_2^-	chlorite	ClO_4^-	perchlorate
CN^-	cyanide	IO_4^-	periodate
F^-	fluoride	MnO_4^-	permanganate
OH^-	hydroxide		

Oxidation Number of −2

CO_3^{2-}	carbonate	O_2^{2-}	peroxide
CrO_4^{2-}	chromate	SiO_3^{2-}	silicate
$Cr_2O_7^{2-}$	dichromate	SO_4^{2-}	sulfate
MnO_4^{2-}	manganate	S^{2-}	sulfide
O^{2-}	oxide	SO_3^{2-}	sulfite
$C_2O_4^{2-}$	oxalate	$S_2O_3^{2-}$	thiosulfate

Oxidation Number of −3

N^{3-}	nitride	BO_3^{3-}	borate
PO_4^{3-}	phosphate	AsO_3^{3-}	arsenite
PO_3^{3-}	phosphite	AsO_4^{3-}	arsenate
P^{3-}	phosphide		

If *more than two* polyatomic anions are formed from a given element and oxygen, the prefixes *per-* and *hypo-* are added to distinguish the additional ions. This appears most often among the polyatomic anions with a halogen as the distinguishing element. The prefixes *per-* and *hypo-* are often used to identify the extremes (the high and low) in oxidation numbers of the element in the anion. Table D2C.2 summarizes the nomenclature for these and all polyatomic anions.

For example, ClO_4^- is *per*chlor*ate* because the oxidation number of Cl is +7 (the highest oxidation number of the four polyatomic anions of chlorine), whereas ClO^- is *hypo*chlor*ite* because the oxidation number of Cl is +1. Therefore, $NaClO_4$ is sodium perchlorate and NaClO is sodium hypochlorite. The name of $Co(ClO_3)_2$ is cobalt(II)

Table D2C.2 Nomenclature of Polyatomic Anions and Ternary Acids

Name of Polyatomic Anion	Acid	Group Number 4A	5A	6A	7A
		Oxidation Number			
per___ate	per___ic acid	—	—	—	+7
___ate	___ic acid	+4	+5	+6	+5
___ite	___ous acid	—	+3	+4	+3
hypo___ite	hypo___ous acid	—	+1	+2	+1

chlorate because the chlorate ion is ClO_3^- (see Table D2C.1) and the oxidation number of Cl is +5 (see Table D2C.2). The cobalt is Co^{2+} because two chlorate, ClO_3^-, ions are required for the neutral salt.

Ternary Acids or Oxoacids

sulfuric acid

perchloric acid

The ternary acids (also referred to as **oxoacids**) are compounds of hydrogen and a polyatomic anion. In contrast to the binary acids, the naming of the ternary acids does not include any mention of hydrogen. The ternary acids are named by using the root of the element in the polyatomic anion and adding a suffix; if the polyatomic anion ends in -*ate*, the ternary acid is named an -*ic acid*. If the polyatomic anion ends in -*ite*, the ternary acid is named an -*ous acid*. To understand by example, the following are representative names for ternary acids.

Ion	Name of Ion	Acid	Name of Acid
SO_4^{2-}	sulf*ate* ion	H_2SO_4	sulfur*ic acid*
SO_3^{2-}	sulf*ite* ion	H_2SO_3	sulfur*ous acid*
CO_3^{2-}	carbon*ate* ion	H_2CO_3	carbon*ic acid*
NO_3^-	nitr*ate* ion	HNO_3	nitr*ic acid*
ClO_4^-	perchlor*ate* ion	$HClO_4$	perchlor*ic acid*
IO^-	hypoiod*ite* ion	HIO	hypoiod*ous acid*

Acid Salts

Acid salts are salts in which a metal cation replaces *less than* all of the hydrogens of an acid having more than one hydrogen (a polyprotic acid). The remaining presence of the hydrogen in the compound is indicated by inserting its name into that of the salt.

Salt	Name
$NaHSO_4$	sodium *hydrogen* sulfate (one Na^+ ion replaces one H^+ ion in H_2SO_4)
$CaHPO_4$	calcium *hydrogen* phosphate (one Ca^{2+} ion replaces two H^+ ions in H_3PO_4)
NaH_2PO_4	sodium *dihydrogen* phosphate (one Na^+ ion replaces one H^+ ion in H_3PO_4)
$NaHCO_3$	sodium *hydrogen* carbonate (one Na^+ ion replaces one H^+ ion in H_2CO_3)
$NaHS$	sodium *hydrogen* sulfide (one Na^+ ion replaces one H^+ ion in H_2S)

An older system of naming acid salts substitutes the prefix "bi" for a *single* hydrogen before naming the polyatomic anion. For example, $NaHCO_3$ is sodium *bi*carbonate, $NaHSO_4$ is sodium *bi*sulfate, and $NaHS$ is sodium *bi*sulfide.

The Next Step

The naming of organic compounds also follows a set of guidelines. The International Union of Pure and Applied Chemistry (IUPAC) regularly meets to ensure that "new" compounds have a systematic name. Review the nomenclature of the simple organic compounds, such as the alcohols, ethers, acids, amines, etc.

DRY LAB PROCEDURE

Procedure Overview: Given the formula of the compounds, the proper names for a large number of ternary compounds are to be written. Given the name of the compounds, the formulas for a large number of compounds are to be written.

Your instructor will assign the exercises you are to complete. Answer them on a separate piece of paper. Be sure to indicate the date, your lab section, and your desk number on your report sheet. Use the rules that have been described.

1. Use Table D2C.2 to name the following polyatomic anions.

   a. BrO_3^- d. $N_2O_2^{2-}$ g. IO_2^- j. TeO_4^{2-}

   b. IO_3^- e. AsO_2^- h. SO_3^{2-} k. SeO_4^{2-}

   c. PO_2^{3-} f. BrO_2^- i. SiO_3^{2-} l. NO_2^-

2. Name the following salts of the representative elements.

   a. Na_2SO_4 e. $Ca_3(PO_3)_2$ i. K_2MnO_4 m. $Li_2S_2O_3$

   b. K_3AsO_4 f. Na_2SiO_3 j. $KMnO_4$ n. $Ba(NO_2)_2$

   c. Li_2CO_3 g. K_2CrO_4 k. Li_2SO_3 o. $Ba(NO_3)_2$

   d. $Ca_3(PO_4)_2$ h. $K_2Cr_2O_7$ l. Li_2SO_4 p. KCH_3CO_2

3. Name the following salts of the transition and post-transition elements using the Stock system.

   a. $Fe(OH)_3$ e. $CuCO_3$ i. $MnSO_4$ m. $CrPO_4$

   b. $FePO_4 \cdot 6H_2O$ f. $CuSO_4 \cdot 5H_2O$ j. $Mn(CH_3CO_2)_2$ n. $CrSO_4 \cdot 6H_2O$

   c. $FeSO_4 \cdot 7H_2O$ g. $Sn(NO_3)_2$ k. $Hg_2(NO_3)_2$ o. $Co_2(CO_3)_3$

   d. $CuCN$ h. $Sn(SO_4)_2$ l. $Hg(NO_3)_2 \cdot H_2O$ p. $CoSO_4 \cdot 7H_2O$

4. Name the following ternary acids.

   a. H_2SO_4 e. $HMnO_4$ i. HNO_2 m. $HClO_4$

   b. H_2SO_3 f. H_2CrO_4 j. H_2CO_3 n. $HClO_3$

   c. $H_2S_2O_3$ g. H_3BO_3 k. $H_2C_2O_4$ o. $HClO_2$

   d. H_3PO_4 h. HNO_3 l. CH_3COOH p. $HClO$

5. Name the following acid salts. Use the "older" method wherever possible.

   a. $NaHCO_3$ d. NH_4HCO_3 g. $NaHSO_4 \cdot H_2O$ j. $MgHAsO_4$

   b. $Ca(HCO_3)_2$ e. $NaHS$ h. Li_2HPO_4 k. KH_2AsO_4

   c. KHC_2O_4 f. $KHSO_3$ i. LiH_2PO_4 l. $KHCrO_4$

6. Write formulas for the following compounds.

   a. potassium permanganate

   b. potassium manganate

   c. calcium carbonate

   d. lead(II) carbonate

   e. ferric carbonate

   f. silver thiosulfate

   g. sodium sulfite

   h. ferrous sulfate heptahydrate

   i. iron(II) oxalate

   j. sodium chromate

   k. potassium dichromate

   l. nickel(II) nitrate hexahydrate

   m. chromous nitrite

   n. vanadyl nitrate

   o. uranyl acetate

   p. barium acetate dihydrate

   q. sodium silicate

   r. calcium hypochlorite

   s. potassium chlorate

   t. ammonium oxalate

   u. sodium borate

   v. cuprous iodate

7. Write formulas for the following acids.

   a. sulfuric acid

   b. thiosulfuric acid

   c. sulfurous acid

   d. periodic acid

   e. iodic acid

   f. hypochlorous acid

   g. nitrous acid

   h. nitric acid

   i. phosphorous acid

   j. phosphoric acid

   k. carbonic acid

   l. bromous acid

   m. chromic acid

   n. permanganic acid

   o. manganic acid

   p. boric acid

   q. oxalic acid

   r. silicic acid

8. Write formulas and name the hodgepodge of compounds resulting from matching all cations with all anions for each set.

Set 1		Set 2		Set 3	
Cations	Anions	Cations	Anions	Cations	Anions
Li^+	Cl^-	Fe^{3+}	PO_4^{3-}	Pb^{2+}	SiO_3^{2-}
Cd^{2+}	SO_4^{2-}	Fe^{2+}	HPO_4^{2-}	NH_4^+	S^{2-}
Na^+	NO_3^-	Al^{3+}	HCO_3^-	$H^+(aq)*$	MnO_4^-
Cu^{2+}	O^{2-}	Zn^{2+}	CN^-	Mn^{3+}	HSO_4^-
V^{5+}	CO_3^{2-}	K^+	$CH_3CO_2^-$ or $C_2H_3O_2^-$	Hg^{2+}	$Cr_2O_7^{2-}$
Mg^{2+}	I^-	VO^{2+}	IO^-	Sr^{2+}	$C_2O_4^{2-}$

*Name as acids

9. Write correct formulas for the following hodgepodge of compounds from Dry Labs 2A, 2B, and 2C.

a. vanadium(V) fluoride
b. stannic oxide
c. silicon tetrafluoride
d. mercuric oxide
e. lithium hypochlorite
f. iodine trifluoride
g. ferrous oxalate
h. cuprous oxide
i. copper(I) chloride
j. calcium hydride
k. cadmium iodide
l. barium acetate dihydrate
m. ammonium sulfide

n. vanadium(V) oxide
o. titanium(IV) chloride
p. scandium(III) nitrate
q. nickel(II) acetate hexahydrate
r. mercurous nitrate
s. lead(II) acetate
t. ferric phosphate hexahydrate
u. ferric chromate
v. dinitrogen tetrasulfide
w. chromous acetate
x. calcium nitride
y. ammonium dichromate
z. silver acetate

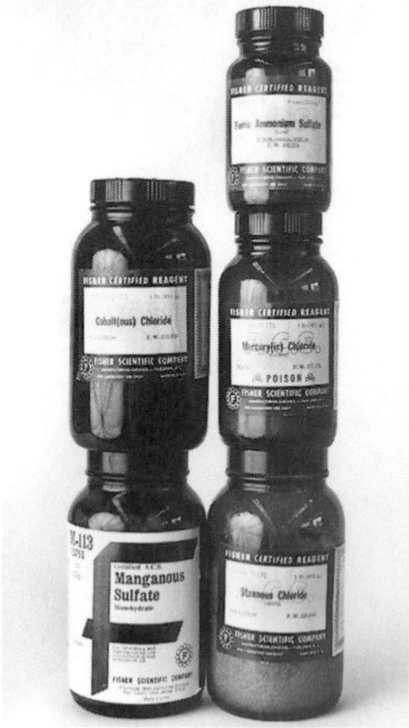

10. Write the correct formulas for the compounds shown in the photo.

11. Write formulas for the compounds having the following common names.

Common Name	Chemical Name
a. acid of sugar	oxalic acid
b. aqua fortis	nitric acid
c. barium white, fixed white	barium sulfate dihydrate
d. bitter salt, Epsom salts	magnesium sulfate heptahydrate
e. blue vitrol	copper(II) sulfate pentahydrate
f. calomel	mercurous chloride
g. caustic potash	potassium hydroxide
h. Chile saltpeter, sodium nitre	sodium nitrate
i. chrome yellow	lead(IV) chromate
j. Indian red, jeweler's rouge	ferric oxide
k. lime	calcium oxide
l. oil of vitrol	sulfuric acid
m. talc or talcum	magnesium silicate
n. Glauber's salt	sodium sulfate

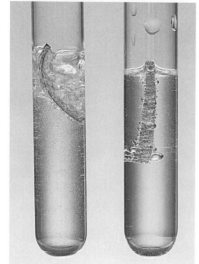

Experiment 6

Acids, Bases, and Salts

Hydrogen gas is produced at different rates from the reaction of zinc metal with hydrochloric acid (left) and phosphoric acid (right).

OBJECTIVES

- To become familiar with the chemical properties of acids, bases, and salts
- To develop the concept of pH and approximate the pH of common acids, bases, and salts
- To observe the relative solubility of common salts
- To observe the relative acidity and basicity of common salts
- To write equations that account for observations from chemical reactions

TECHNIQUES

The following techniques are used in the Experimental Procedure

INTRODUCTION

Many of the chemical compounds that you will encounter in the laboratory (and in lecture) can be classified as an acid, a base, or a salt. The nomenclature and the formulas of these compounds were derived systematically in Dry Lab 2. Using the correct formulas in a chemical equation allows chemists to express chemical reactions and chemical systems that are simply understood without writing extensive paragraphs. *Balanced* equations lead to an understanding of **stoichiometry.**

Stoichometry: a study of a chemical reaction using a balanced equation

However, names, formulas, and equations have little meaning unless there is some tangible relationship to chemicals and chemical reactions. To a chemist, sulfur is not simply an element with the symbol S that reacts with oxygen to form sulfur dioxide, but rather a yellow solid that can be held in the hand and burns in air with a blue flame, producing a choking irritant called sulfur dioxide.

In this experiment, you will observe some of the chemical properties and reactions of acids, bases, and salts. What does it mean when a substance is identified as being acidic? basic? soluble? insoluble?

Acids

Acidic solutions have a sour or tart taste, cause a pricking sensation on the skin, and turn blue **litmus** red. Nearly all of the foods and drinks we consume are acidic . . . think of lemon juice as being quite acidic to taste but milk not quite so. All acids are

Litmus: a common laboratory acid-base indicator

substances that produce **hydronium ion, H_3O^+,** in aqueous solutions. For example, the most versatile of all chemicals worldwide is sulfuric acid,[1] H_2SO_4, a **diprotic** acid producing H_3O^+ in two steps:

$$H_2SO_4(aq) + H_2O(l) \rightarrow H_3O^+(aq) + HSO_4^-(aq) \tag{6.1}$$
$$HSO_4^-(aq) + H_2O(l) \rightarrow H_3O^+(aq) + SO_4^{2-}(aq) \tag{6.2}$$

Other prominent "inorganic" acids are hydrochloric acid (also called **muriatic acid**), nitric acid, and phosphoric acid (Figure 6.1).

Some common "organic" acids are acetic acid found in vinegar, citric acid found in citrus fruits, and ascorbic acid, the vitamin C acid (Figure 6.2).

Many cations, such as ammonium and (**hydrated**) ferric ions, are also capable of producing acidic solutions as well.

$$NH_4^+(aq) + H_2O(l) \leftrightarrow NH_3(aq) + H_3O^+(aq) \tag{6.3}$$
$$Fe(H_2O)^{3+}(aq) + H_2O(l) \leftrightarrow FeOH^{2+}(aq) + H_3O^+(aq) \tag{6.4}$$

Bases

Figure 6.1 Phosphoric acid is the additive that delivers the tart taste in many soft drinks.

Basic solutions have a bitter taste, are slippery to the touch, and turn red litmus blue. Our palates are unaccustomed to the taste of bases . . . think of antacids, soaps, detergents, and household ammonia. All bases are substances that produce hydroxide ion, OH^-, in aqueous solutions. For example, the most common base is ammonia (the #2 produced chemical), the most common laboratory base is sodium hydroxide. In water,

$$NH_3(aq) + H_2O(l) \leftrightarrow NH_4^+(aq) + OH^-(aq) \tag{6.5}$$
$$NaOH(aq) \rightarrow Na^+(aq) + OH^-(aq) \tag{6.6}$$

Sodium hydroxide, found in oven and drain cleaners, is commonly called **lye** or **caustic soda;** calcium hydroxide is called **slaked lime;** potassium hydroxide is called **caustic potash;** magnesium hydroxide, called **milk of magnesia,** is an antacid and purgative (Figure 6.3).

A large number of anions, such as carbonate and phosphate ions, also are capable of producing basic solutions as well. For example, sodium carbonate, known as soda ash and **washing soda** (often added to detergents), produces OH^- in solution.

$$CO_3^{2-}(aq) + H_2O(l) \leftrightarrow HCO_3^-(aq) + OH^-(aq) \tag{6.7}$$

Salts

Salts are produced when aqueous solutions of acids and bases are mixed. The **neutralization reaction** may result in the salt being soluble or insoluble in water. For example, a mix of sulfuric acid and sodium hydroxide produces water and the water-soluble sodium sulfate salt:

$$H_2SO_4(aq) + 2\,NaOH(aq) \rightarrow 2\,H_2O(l) + Na_2SO_4(aq) \tag{6.8}$$

A **salt** therefore can be defined as any ionic compound that is a neutralization product of an acid-base reaction.

Ionic Compounds, Reactions, and Equations

Salts are ionic compounds. If the salt is **soluble,** then it exists as ions in solution; if the salt is **insoluble,** then it exists as a solid precipitate.

For example, consider two beakers, one containing silver nitrate and a second containing sodium chloride (Figure 6.4). Both silver nitrate and sodium chloride are soluble and therefore exist in solution as $Ag^+(aq)$, $NO_3^-(aq)$ and $Na^+(aq)$, $Cl^-(aq)$ in their respective beakers . . . *no* $AgNO_3$ or $NaCl$ molecules are present! When the two solu-

---

[1] Also called **oil of vitriol,** sulfuric acid production (#1 produced chemical worldwide) exceeds the #2 produced chemical (ammonia) by a margin of ~3:1 in the United States annually. For that reason sulfuric acid is often called the "old horse of chemistry."

tions are mixed, a white precipitate of AgCl(s) forms, leaving the $Na^+(aq)$ and $NO_3^-(aq)$ ions in solution. Therefore, AgCl is an insoluble salt and $NaNO_3$ is a soluble salt in water. A proper equation expressing the two separate solutions, combined solutions, and the observation would then be:

$$Ag^+(aq) + NO_3^-(aq) + Na^+(aq) + Cl^-(aq) \rightarrow$$
$$AgCl(s, white) + Na^+(aq) + NO_3^-(aq) \quad (6.9)$$

This is called an **ionic equation.** A **net ionic equation** represents only the ions involved in the observed chemical reaction, in this case, the formation of a precipitate. For this example, the net ionic equation would be:

$$Ag^+(aq) + Cl^-(aq) \rightarrow AgCl(s) \quad (6.10)$$

Since $Na^+(aq)$ and $NO_3^-(aq)$ are not involved in the observed chemical reaction, they are called **spectator ions.**

Soluble acids and bases are also ionic compounds. For example HCl(aq) exists as $H_3O^+(aq)$ and $Cl^-(aq)$ in solution . . . no molecules of HCl are present!

$$HCl(aq) + H_2O(l) \rightarrow H_3O^+(aq) + Cl^-(aq) \quad (6.11)$$

Bottom line . . . if an ionic substance is "soluble," then only ions exist in an aqueous solution . . . no molecular units of the ionic compound are present!

Ionic equations were first introduced in Experiment 3.

Figure 6.2 Common household acids.

Figure 6.3 Common household bases.

(a) (b) (c)

Figure 6.4 The progression of a reaction between solutions of silver nitrate and sodium chloride.

The acidity of most aqueous solutions is frequently the result of *low* concentrations of hydronium ion, H_3O^+. "pH" is a convenient mathematical expression[2] used to express low concentrations of hydronium ion. pH is defined as the negative logarithm of the molar concentration of hydronium ion.

pH

$$pH = -\log[H_3O^+] \quad (6.12)$$

[ ]: Brackets placed around an ion indicate the molar concentration of that ion.
$pH = -\log[1 \times 10^{-7}] = 7.0$.

At 25°C, pure (and neutral) water has a hydronium concentration (and a hydroxide concentration) of 1×10^{-7} mol/L: $[H_3O^+] = 1.0 \times 10^{-7}$; pH = 7.0. Solutions having higher concentrations of hydronium ion, (e.g., 1.0×10^{-2} mol/L) have lower pH's (pH = 2.0). Solutions having lower hydronium concentrations or higher hydroxide concentrations (e.g., $[H_3O^+] = 1 \times 10^{-10}$ mol/L) have higher pH's (pH = 10.0). The pH ranges of some familiar solutions are shown in Figure 6.5.

The pH of solutions is often measured with an acid-base indicator. Litmus being red in an acidic solution and blue in a basic solution is one of the more common laboratory indicators. Different indicators change colors over different pH ranges. Mixtures of acid-base indicators can also be used to approximate the pH of a solution. Various pH test papers (Figure 6.6) and the universal indicator are examples of mixed indicators.

---

[2]Søren P. L. Sørensen, a Danish biochemist and brewmaster is credited for instituting this expression.

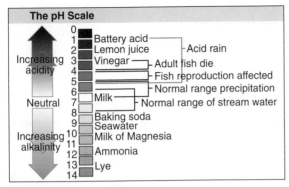

Figure 6.5 pH ranges of various common solutions.
Courtesy of Environment Canada (www.ns.ec.ca)

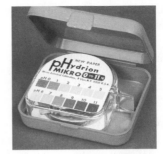

Figure 6.6 The pH of a solution can be estimated with pH paper, a strip of paper that is impregnated with a mixture of acid–base indicators. A gradation of colors on the pH paper approximates the pH of the solution.

EXPERIMENTAL PROCEDURE

The chemical properties of a range of acids, bases, and salts are observed. Chemical (ionic and net ionic) equations are used to account for the observations. The pH of selected acids, bases, and salts are estimated with pH test paper or universal indicator.

Perform the experiment with a partner. At each circled superscript⁽¹⁻¹⁴⁾ in the procedure, *stop,* and record your observation on the Report Sheet. Discuss your observation with your partner and write an equation expressing your observation. **Caution:** *Dilute and concentrated (conc) acids and bases cause severe skin burns and irritation to mucous membranes. Be very careful in handling these chemicals. Clean up all spills immediately with excess water, followed by a covering of baking soda,* $NaHCO_3$*. Refer to the Laboratory Safety section at the beginning of this manual.*

A. Acids and Acidic Solutions

1. **Action of Acids on Metals.** a. Place a small (~1 cm) polished (with steel wool or sandpaper) strip of Mg, Zn, and Cu into separate small test tubes. To each test tube add just enough 6 *M* HCl to submerge the metal and observe for several minutes. Record your observations on the Report Sheet.[1]

   b. Repeat the test of the three metals with 6 *M* H_3PO_4[2] and then again with 6 *M* CH_3COOH.[3]

2. **Effect of Acid Concentration on Reaction Rate.** Set up six small clean test tubes containing about $1\frac{1}{2}$ mL of the acid solutions shown in Figure 6.7. Add a small (~1 cm) polished strip of magnesium to each solution and observe. Explain your observations. Account for any similarities and differences between HCl and CH_3COOH. What effect do differences in the acids *and* the concentrations of the acids have on the reaction rate?[4][5]

Disposal: Discard the acid solutions in the "Waste Acids" container

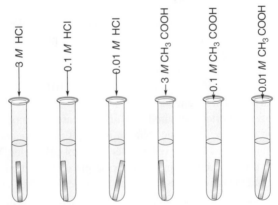

Figure 6.7 A setup for testing the effect of different acids and acid strengths on their reactivity with a metal.

CLEANUP: Rinse all of the used test tubes twice with tap water and twice with deionized water.

1. **Reaction of Aqueous Sodium Hydroxide with Acid.** Place about 1 mL of 1 M NaOH in a small test tube. Test the solution with litmus. Add and count drops of 6 M HCl—after each drop, agitate the mixture and test the solution with litmus—until the litmus changes color. Record your data.[6]

2. **Dissolution of Sodium Hydroxide.** Place a small "BB-sized" sample of NaOH in a clean small test tube. Hold the test tube in your hand containing the NaOH to detect a heat change and add drops of water to the sample. Test the solution with litmus paper and record your observations.[7]

3. **Dissolution of Sodium Carbonate.** Place a small "BB-sized" sample of **anhydrous** Na_2CO_3 in a clean small test tube. Hold the test tube in your hand containing the Na_2CO_3 and add drops of water to the sample. Test the solution with litmus paper and record your observations.[8]

B. Bases and Basic Solutions

Anhydrous salt: a salt having no hydrated water molecules in its solid structure

Disposal: Discard the base solutions in the "Waste Bases" container

---

Boil 10–20 mL of deionized water for about ~5 minutes to expel any dissolved gases, specifically carbon dioxide. Set the water aside to cool.

1. **pH of Water.** Clean a 24-well plate (small test tubes may also be used) with soap and water and *thoroughly* rinse with deionized water. Half-fill wells A1–A3 with boiled (but cooled) deionized water, unboiled deionized water, and tap water, respectively. Add 1–2 drops of universal indicator to each. Account for any differences in the pH.[9]

2. **Estimate the pH of Acids and Bases.** Place 1 mL of each solution listed on the Report Sheet into wells B1–B4 and add 1–2 drops of universal indicator to each. Estimate the pH of each solution. Write an equation showing the origin of the free H_3O^+ or OH^- in each solution.[10]

3. **Estimate the pH of "Common" Solutions.** By the same method as in Part C.2, use wells C1–C4 to determine the pH of vinegar, lemon juice, household ammonia, detergent solution, or substitutes designated by your instructor.[11]

C. pH Measurements

---

1. **Estimate the pH of Each Solution.** Transfer 1 mL of each solution listed on the Report Sheet into wells D1–D5 and add 1–2 drops of universal indicator to each. Estimate the pH of each solution as in Part C. Identify the ion that causes the acidity/basicity of the solution.[12]

D. pH of Salt Solutions

Disposal and Cleanup: Discard the test solutions in Parts C and D into the "Waste Salts" container. Rinse the 24-well plate twice with tap water and twice with deionized water.

---

1. **Neutralization of Acids.** Into three test tubes, successively pipet 2 mL of 0.10 M HCl, 0.10 M H_2SO_4, and 0.10 M CH_3COOH. Add 1–2 drops of phenolphthalein indicator to each solution. Add (and count) drops of 0.5 M NaOH to each acid until a color change occurs (agitate the solution after each drop). Record the drops added. Compare the available hydronium ion of the three acids.[13]

E. Acids and Bases

Disposal and Cleanup: Discard the test solutions into the "Waste Salts" container. Rinse the test tubes twice with tap water and twice with deionized water.

Table 6.1 An Organization of the Reactants for a Series of Metathesis Reactions

Test Tube No. or Well No.	Reactant Solution or Preparation
1	Several crystals of $FeCl_3 \cdot 6H_2O$ in 2 mL of water *or* 2 mL of 0.1 M $FeCl_3$
2	Several crystals of $CoCl_2 \cdot 6H_2O$ in 2 mL of water *or* 2 mL of 0.1 M $CoCl_2$
3	Several crystals of Na_2CO_3 in 2 mL of water *or* 2 mL of 0.1 M Na_2CO_3
4	Several crystals of $CuSO_4 \cdot 5H_2O$ in 2 mL of water *or* 2 mL of 0.1 M $CuSO_4$
5	Several crystals of $Na_3PO_4 \cdot 12H_2O$ in 2 mL of water *or* 2 mL of 0.1 M Na_3PO_4
6	Several crystals of $NiCl_2 \cdot 6H_2O$ in 2 mL of water *or* 2 mL of 0.1 M $NiCl_2$

F. Reactions of Salt Mixtures

This experiment requires either a set of 6 clean, small test tubes or a 24-well plate. Consult with your instructor. If you use test tubes, label them in accordance with the number designations in Table 6.1.

1. **Test Solutions.** Set up the test tubes (Figure 6.8) or the 24-well plate (Figure 6.9) with the reactant solutions or preparations listed in Table 6.1. Volumes of solutions only need to be approximate.[3] Note the color of the cations in solution. Your instructor may substitute, add, or delete chemicals from the table.

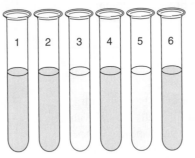

Figure 6.8 The arrangement of 6 small, labeled test tubes for test solutions.

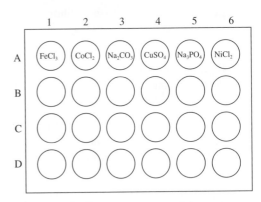

Figure 6.9 The arrangement of the test solutions in a 24-well plate.

2. **Reaction Mixtures of Salt Solutions.** Pairs of reactant solutions in Table 6.1 are combined according Figure 6.8/6.9. Look for evidence of chemical change. Record your observations on the Report Sheet, followed by ionic and net ionic equations.[14] Ask your instructor for assistance if necessary. Use *clean* dropping pipets or Beral pipets to slowly make the following solution transfers:

   a. Transfer one-half of solution 2 to solution 1 in (test tube/well) 1.

   b. Transfer one-half of solution 3 to the remainder of solution 2 in (test tube/well) 2.

   c. Transfer one-half of solution 4 to the remainder of solution 3 in (test tube/well) 3.

   d. Transfer one-half of solution 5 to the remainder of solution 4 in (test tube/well) 4.

   e. Transfer one-half of solution 6 to the remainder of solution 5 in (test tube/well) 5.

Universal indicator provides for a determination of the pH of a soil sample.

> *Disposal:* Dispose of the waste solutions, as identified by your instructor, in the "Waste Salts" container.

CLEANUP: Rinse the test tubes or well plate with tap water twice and with deionized water twice. Discard each rinse in the "Waste Salts" container.

The Next Step

The acidity and basicity (and pH) of endless aqueous systems are informative to most everyone, from the existence of acid indigestion to alkaline soils. Design a systematic study where the measurement of pH may account for the properties of an environmental system, e.g., a watershed, soil samples, citrus fruits, . . .

---

[3]The volume of a small (75-mm) test tube is about 3 mL; the volume of each well is about 3.4 mL.

Experiment 6 *Prelaboratory Assignment*

Acids, Bases, and Salts

Date ________ Lab Sec. ______ Name ______________________________ Desk No. ________

1. In an aqueous solution,
   a. identify the "species" that makes a solution acidic.

   b. identify the "species" that makes a solution basic.

2. Aqueous salt solutions often are *not* neutral with respect to pH. Explain.

3. a. Milk of magnesia is used as a laxative and to treat upset stomachs. What is the formula of milk of magnesia?

   b. Washing soda is often added to detergent formulations to make the wash water more basic. What is the formula of the anhydrous form of washing soda? Does it increase or decrease the pH of the wash water? Explain.

4. Three solutions have the following pH:

   - Solution 1: pH 12.1
   - Solution 2: pH 6.2
   - Solution 3: pH 10.2

   a. Which solution contains the highest H_3O^+ ion concentration? ____________

   b. Which solution is the most acidic? ____________

   c. Which solution is the most basic? ____________

5. When a solution resulting from a stoichiometric reaction mixture of an unknown acid, HA, and sodium hydroxide, NaOH, is tested with litmus paper, the litmus is blue.
   a. Write a balanced equation. What *salt* (write its formula) is present from the reaction between the unknown acid and the sodium hydroxide?

   b. Why is the litmus blue?

   c. Which ion of the salt is responsible for the observation of the litmus test?

6. a. Nickel(II) chloride dissolves in water to form a light green solution. What species (ions) are present in solution, i.e., what is "swimming around" in the aqueous solution? Write the formulas.

   b. Sodium phosphate dissolves in water to form a colorless solution. What species (ions) are in solution? Write the formulas.

   c. When solutions of nickel(II) chloride and sodium phosphate are combined, a green nickel(II) phosphate precipitate forms. What species remain in solution (spectator ions)? What is the color of the solution (assume stoichiometric amounts of nickel(II) chloride and sodium phosphate). Explain.

7. Aqueous solutions of water-soluble $AgNO_3$ and FeI_2 are mixed, forming a dark-yellow precipitate.
   a. What is the formula of the precipitate? See Appendix G.

   b. Write the **ionic** equation representing the reactants, *as they exist in solution **before** the reaction,* and the products, *as they exist in solution **after** the reaction.*

   c. What are the spectator ions in the reaction mixture?

   d. Write the **net ionic** reaction that accounts for the appearance of the precipitate.

Experiment 6 *Report Sheet*

Acids, Bases, and Salts

Date __________ Lab Sec. ______ Name ______________________________________ Desk No. __________

A. Acids and Acidic Solutions

1. **Action of Acids on Metals.** In the following table, state whether or not a reaction is observed. Indicate the relative reaction rate (i.e., fast, moderate, slow) of the metals with the corresponding acid.

	Mg	Zn	Cu
① 6 *M* HCl	______	______	______
② 6 *M* H_3PO_4	______	______	______
③ 6 *M* CH_3COOH	______	______	______

2. **Effect of Acid Concentration on Reaction Rate.** ④ Explain how changes in HCl concentration affect the reaction rate with Mg.

⑤ Explain how changes in CH_3COOH concentration affect the reaction rate with Mg.

Discuss the similarities and differences of HCl and CH_3COOH in the observed reactions with magnesium.

What effect do concentration changes have on the reaction rate?

B. Bases and Basic Solutions

1. **Reaction of Aqueous Sodium Hydroxide with Acid.**[6] On the basis of your observations of the litmus test, describe the progress of the reaction of NaOH with HCl.

2. **Dissolution of Sodium Hydroxide.**[7] Describe your observations.

3. **Dissolution of Sodium Carbonate.**[8] Describe your observations. Write an equation that accounts for your observations.

C. pH Measurements

[9] Water	Approximate pH	Briefly account for the pH if *not* equal to 7
Boiled, deionized		
Unboiled, deionized		
Tap		

[10] Solution	Approximate pH	Balanced equation showing acidity/basicity
0.010 M HCl		
0.010 M CH$_3$COOH		
0.010 M NaOH		
0.010 M NH$_3$		
[11] Vinegar		
Lemon juice		
Household ammonia		
Detergent solution		

D. pH of Salt Solutions

[12] Salt Solution	Approximate pH	Ion Affecting pH	Spectator Ion (Ions)
0.10 M NaCl			
0.10 M Na_2CO_3			
0.10 M Na_3PO_4			
0.10 M NH_4Cl			
0.10 M $AlCl_3$			

E. Acids and Bases

1. [13] Neutralization of Acids

	0.10 M HCl	0.10 M H_2SO_4	0.10 M CH_3COOH
Drops of NaOH for color change			

Discuss the relative acidity (availability of hydronium ion) of the three acids.

F. Reactions of Salt Mixtures [14]

Reactants	Observation(s)	Equations
a. $FeCl_3$ + $CoCl_2$		ionic: net ionic:
b. $CoCl_2$ + Na_2CO_3		ionic: net ionic:
c. Na_2CO_3 + $CuSO_4$		ionic: net ionic:
d. $CuSO_4$ + Na_3PO_4		ionic: net ionic:
e. Na_3PO_4 + $NiCl_2$		ionic: net ionic:

Laboratory Questions

Circle the questions that have been assigned

1. Part A.1. The observation for the reaction of 6 M HCl was obviously different from that of 6 M CH$_3$COOH. What were the contrasting observations? How do the two acids differ? Explain.

2. Part A.2. As the molar concentration of an acid decreases, the reaction rate with an active metal, such as magnesium, is expected to _____. Explain.

3. Part B.1. 6 M H$_2$SO$_4$ is substituted for 6 M HCl. Will more or less drops of 6 M H$_2$SO$_4$ be required for the litmus to change color? Explain.

4. Part B.3. Sodium carbonate dissolved in water produces a basic solution. Water-soluble potassium phosphate, K$_3$PO$_4$, also produces a basic solution. Write an equation that justifies its basicity.

*5. Part B.3. The setting of mortar is a time-consuming process that involves a chemical reaction of quicklime, CaO, with the carbon dioxide and water of the atmosphere, forming CaCO$_3$ and Ca(OH)$_2$ respectively. Write the two balanced equations that represent the setting of mortar.

6. Part C.1. The unboiled, deionized water has a measured pH less than seven. Explain.

7. Part E.1. In Part A.2 there were obvious differences in the chemical reactivity of 0.1 M HCl and 0.1 M CH$_3$COOH. However in Part E.1, there seemed to be no difference. Distinguish between the acidic properties of HCl and CH$_3$COOH.

8. Part F.2. a. Identify the color of salts containing the copper(II) ion.
   b. Identify the color of salts containing the cobalt(II) ion.
   c. Identify the color of salts containing the nickel(II) ion.

9. Part F. Write a net ionic equation for the reaction of hydrochloric acid with sodium hydroxide.

Experiment 7

Empirical Formulas

Crucibles are "fired" at high temperatures to volatilize impurities.

OBJECTIVES

- To determine the empirical formulas of two compounds by combination reactions
- To determine the mole ratio of the decomposition products of a compound

TECHNIQUES

The following techniques are used in the Experimental Procedure

INTRODUCTION

The **empirical formula** of a compound is the simplest whole-number ratio of moles of elements in the compound. The experimental determination of the empirical formula of a compound from its elements requires three steps:

- determine the mass of each element in the sample
- calculate the number of moles of each element in the sample
- express the ratio of the moles of each element as small whole numbers

For example, an analysis of a sample of table salt shows that 2.75 g of sodium and 4.25 g of chlorine are present. The moles of each element are

$$2.75 \text{ g} \times \frac{\text{mol Na}}{22.99 \text{ gNa}} = 0.120 \text{ mol Na}, \qquad 4.25 \text{ g} \times \frac{\text{mol Cl}}{35.45 \text{ g Cl}} = 0.120 \text{ mol Cl}$$

The mole ratio of sodium to chlorine is 0.120 to 0.120. As the empirical formula *must* be expressed in a ratio of small whole numbers, the whole-number ratio is 1 to 1 and the empirical formula of sodium chloride is Na_1Cl_1 or simply NaCl.

The empirical formula also provides a mass ratio of the elements in the compound. The formula NaCl states that 22.99 g (1 mol) of sodium combines with 35.45 g (1 mol) of chlorine to form 58.44 g (1 mol) of sodium chloride.

The mass percentages of sodium and chlorine in sodium chloride are:

$$\frac{22.99}{22.99 + 35.45} \times 100 = 39.34\% \text{ Na} \qquad \frac{35.45}{22.99 + 35.45} \times 100 = 60.66\% \text{ Cl}$$

The mass percent of sodium and chlorine in sodium chloride is always the same (constant and definite), a statement of the **law of definite proportions.**

We can determine the empirical formula of a compound from either a combination reaction or a decomposition reaction. In the **combination reaction,** a known mass of one reactant and the mass of the product are measured. An example of a combination

Combination reaction: two elements combine to form a compound

reaction is the reaction of iron with chlorine: the initial mass of the iron and the final mass of the iron chloride product are determined. From the difference between the masses, the mass of chlorine that reacts and, subsequently, the moles of iron and chlorine that react to form the product are calculated. This mole ratio of iron to chlorine yields the empirical formula of the iron chloride (see Prelaboratory Assignment).

Decomposition reaction: a compound decomposes into two or more elements or simpler compounds

In the **decomposition reaction,** the initial mass of the compound used for the analysis and the final mass of at least one of the products are measured. An example would be the decomposition of a mercury oxide to mercury metal and oxygen gas: the initial mass of the mercury oxide and the final mass of the mercury metal are determined. The difference between the measured masses is the mass of oxygen in the mercury oxide. The moles of mercury and oxygen in the original compound are then calculated to provide a whole-number mole ratio of mercury to oxygen (see Prelaboratory Assignment).

In Part B of this experiment a combination reaction of magnesium and oxygen is used to determine the empirical formula of magnesium oxide. The initial mass of the magnesium and the mass of the product are measured.

In Part C of this experiment a decomposition reaction of a pure compound into calcium oxide, CaO, and carbon dioxide, CO_2, is analyzed. The masses of the compound and the calcium oxide are measured. Further analysis of the data provides the mole ratio of CaO to CO_2 in the compound.

In Part D of this experiment a combination reaction of tin and oxygen is used to determine the empirical formula of a tin oxide. Since tin forms more than one oxide, a difference in laboratory technique and persistence may lead to different determinations of the reported empirical formula. The initial measured mass of tin is not reacted directly with the oxygen of the air, but rather with nitric acid to produce the oxide. Because the mass of the final product consists only of tin and oxygen its empirical formula is calculated.

EXPERIMENTAL PROCEDURE

Procedure Overview: A crucible of constant mass is used to thermally form (Part B) or decompose (Part C) a compound of fixed composition or form a compound of variable composition (Part D). Mass measurements before and after the heating procedure are used to calculate the empirical formula of the compound.

Ask your instructor which part(s) of the experiment you are to complete. If you are to perform more than one part, you will need to organize some of your data on a separate sheet of paper for the Report Sheet.

You are to complete at least two trials for each part of the experiment that you are assigned.

A. Preparation of a Crucible

Fired: heated to a very high temperature to volatilize impurities

1. **Prepare a Clean Crucible.** Obtain a clean crucible and lid. Often-used crucibles tend to form stress fractures or fissures. Check the crucible for flaws; if any are found obtain a second crucible. Support the crucible and lid on a clay triangle and heat with an intense flame for ~5 minutes. Allow them to cool to room temperature.[1] If the crucible remains dirty after heating, *upon the advice of your instructor,* move the apparatus to the fume hood, add 1–2 mL of 6 M HNO_3, and gently evaporate to dryness. (**Caution:** *Avoid skin contact, flush immediately with water.*)

    Measure the mass of the **fired,** *cool* crucible and lid. Use only clean, dry crucible tongs to handle the crucible and lid for the remainder of the experiment; do *not* use your fingers (**Caution:** *hot and cold crucibles look the same—do not touch!*).

B. Combination Reaction of Magnesium and Oxygen

1. **Prepare the Sample.** Polish (with steel wool or sandpaper) 0.15–0.20 g of magnesium ribbon; curl the ribbon to lie in the crucible. Measure and record the mass (± 0.001 g) of the magnesium sample, crucible, and lid.

[1]Cool the crucible and lid to room temperature (and perform all other cooling processes in the Experimental Procedure) in a desiccator if one is available. When cool, remove the crucible and lid from the desiccator with crucible tongs and measure the mass of the crucible and lid.

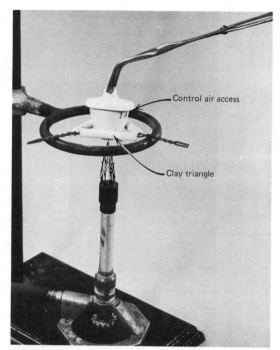

Control air access

Clay triangle

Figure 7.1 Controlling access of air to the magnesium ribbon.

2. **Heat the Sample in Air.** Place the crucible containing the Mg ribbon and lid on the clay triangle. Heat *slowly,* occasionally lifting the lid to allow air to reach the Mg ribbon (Figure 7.1).

   If too much air comes in contact with the Mg ribbon, rapid oxidation of the Mg occurs and it burns brightly. (**Caution:** *You do not want this to happen. If it does, do not watch the burn; it may cause temporary blindness!*) Immediately return the lid to the crucible, allow the apparatus to cool, and return to Part A to repeat the experiment.

3. **Heat for Complete Reaction.** Continue heating the crucible until no visible change is apparent in the magnesium ash at the bottom of the crucible. Remove the lid; continue heating the open crucible and ash for ~30 seconds. Remove the heat and allow the crucible to *cool to room temperature.*[2] *Do not touch!* Measure the mass of the contents in the crucible and with the lid on the same balance that was used earlier and record.

4. **Test for Complete Reaction.** Add a few drops of water to decompose any magnesium nitride[3] that may have formed during combustion. Reheat the sample for ~1 minute, but do not intensify the flame. Allow the apparatus to cool and again measure the mass of the contents. If this second mass differs by more than ±1% from that recorded in Part B.3, repeat Part B.4.

5. **Calculations.** Determine the mole ratio of magnesium to oxygen, and thus the empirical formula, of the pure compound.

Disposal and Cleanup: Wash the cool crucible with a dilute solution of hydrochloric acid and discard in the "Waste Acids" container. Rinse twice with tap water and twice with deionized water.

---

[2]Place the crucible and lid in a desiccator (if available) for cooling.
[3]At higher temperatures magnesium metal reacts with the nitrogen in the air to form magnesium nitride:

$$3 \text{ Mg}(s) + \text{N}_2(g) \rightarrow \text{Mg}_3\text{N}_2(s)$$

C. Decomposition Reaction of a Pure Compound

The pure compound decomposes to calcium oxide and carbon dioxide. Complete Part A.1 for preparing the apparatus.

1. **Prepare the Sample.** Place about 1 g of the pure compound in the clean crucible prepared in Part A.1 and measure the mass (± 0.001 g) of the sample and crucible. The lid need *not* be used for Part C. Record the mass of the compound.

2. **Heat the Sample.** Heat the crucible, gradually intensifying the heat.[4] Maintain the intense flame for 20–25 minutes. Allow the sample to cool (in a desiccator if available). Determine the mass of the contents and crucible. *Do not touch!*

3. **Analyze the Product.** Repeat Part C.2 until $\pm 1\%$ reproducibility of the mass is obtained.

4. **Calculations.** Determine the mole ratio of calcium oxide to carbon dioxide in the pure compound.

> *Disposal and Cleanup:* Wash the cool crucible with a dilute solution of hydrochloric acid and discard in the "Waste Acids" container. Rinse twice with tap water and twice with deionized water.

D. Combination Reaction of Tin and Oxygen

Cool flame. If you can feel the heat of the flame with a hand held beside the crucible, the flame it too hot!

1. **Prepare the Sample.** Place about 0.5 g of granulated tin in the clean crucible prepared in Part A.1, cover with the crucible lid. Record the combined mass (± 0.001 g) of the tin sample, crucible and lid. Only use crucible tongs to handle the crucible and lid.

2. **Transfer the Sample to a Fume Hood.** Add drops of 6 M HNO_3 (**Caution:** *HNO_3 is very corrosive and a severe skin irritant. If it contacts the skin wash immediately with excess water*) to the tin sample until no further reaction is apparent with the tin (**Caution:** *do not inhale the gaseous, toxic vapors!*). Add 4–5 additional drops of the 6 M HNO_3. Keep the crucible and sample in the fume hood until no further gaseous vapors from the reaction are visible.

3. **Heat to Dryness.** Upon approval of your laboratory instructor, return the sample to the laboratory bench.
   a. *Initial dryness.* See Figure T.15e. *Slowly,* and with a "**cool flame,**" heat the sample until the solid first appears dry (*avoid any popping or spattering of the sample throughout the heating process*).
   b. *Final dryness.* Break up the solid with a stirring rod and resume heating, now with a more intense flame, until the solid appears a pale yellow.
   c. *Cool.* Allow the crucible, lid, and sample to cool (in a desiccator if available) to room temperature. Determine the mass of the tin compound, crucible and lid.

4. **Constant Mass.** Repeat Part D.3 until the combined mass of the tin compound, crucible and lid has a $\pm 1\%$ reproducibility.

5. **Calculations.** Determine the mole ratio of tin to oxygen in the compound and the empirical formula of the tin oxide.

> *Disposal and Cleanup:* Wash the cool crucible with a dilute solution of hydrochloric acid and discard in the "Waste Acids" container. Rinse twice with tap water and twice with deionized water.

The Next Step

Many compounds are thermally unstable or pass through various phases upon heating, just as water goes through different phases upon heating. A quantitative technique for determining/measuring the thermal stability of a compound or its phases is called, "differential thermal analysis (DTA)." The Internet has various sites for the theory and applications of DTA . . . What are the details of this technique?

---

[4]The pure compound decomposes near 825°C, about the temperature of a well-adjusted Bunsen flame.

Empirical Formulas

Date _____ Lab Sec. _____ Name _____ Desk No. _____

1. Elemental mercury was first discovered when a mercury oxide was decomposed with heat, forming mercury metal and oxygen gas. When a 0.667-g sample of the mercury oxide is heated, 0.618 g of mercury metal remains. *Note:* do not attempt this experiment in the laboratory because of the release of toxic mercury vapor.
   a. What is the mole ratio of mercury to oxygen in the sample?

   b. What is the empirical formula of the mercury oxide?

2. A 2.60-g sample of iron chemically combines with chlorine gas to form 7.55 g of an iron chloride.
   a. What is the empirical formula of the iron chloride?

   b. What is the percent by mass of chlorine in the sample?

3. a. Experimental Procedure, Part A. List two reasons for using crucible tongs to handle the crucible and lid after their initial firing.

   b. Why is it best to cool the crucible and lid (and sample) in a desiccator rather than on the laboratory bench?

4. Experiment Procedure, Part D.2. State the reason for the use of the fume hood.

5. Experiment Procedure, Part D.3. Characterize a "cool flame."

6. A sample of pure copper is covered with an excess of powdered elemental sulfur. The following data were collected:

Mass of crucible and lid (g)   <u>   19.914   </u>

Mass of copper, crucible, and lid (g)   <u>   21.697   </u>

The mixture was heated to a temperature where a reaction occurred and the excess sulfur was volatilized. Upon cooling the following was recorded.

Mass of compound, crucible, and lid (g)   <u>   23.044   </u>

Complete the following data analysis

Mass of sulfur in compound (g)*   <u>   </u>

Moles of sulfur in compound (mol)*   <u>   </u>

Mass of copper in compound (g)*   <u>   </u>

Moles of copper in compound (mol)*   <u>   </u>

Empirical formula of the copper compound <u>   </u>

*Show calculations

Empirical Formulas

Date _____ Lab Sec. _____ Name _____ Desk No. _____

Indicate whether the following data are for Part B, Part C, or Part D. _____

	Trial 1	*Trial 2*
1. Mass of crucible and lid (*g*)	_____	_____
2. Mass of crucible, lid, and sample	_____	_____
3. Mass of sample (*g*)	_____	_____
4. Instructor's approval	_____	_____
5. Mass of crucible, lid, and product		
1st mass measurement (*g*)	_____	_____
2nd mass measurement (*g*)	_____	_____
3rd mass measurement (*g*)	_____	_____
6. Final mass of crucible, lid, and of product (*g*)	_____	_____
7. Mass of product (*g*)	_____	_____

Organize your calculations showing the empirical formula of the compound for each trial (Parts B and D) or the mole ratio of calcium oxide to carbon dioxide in the compound (Part C).

8. Empirical formula of compound (Parts B and D) _____

9. Percent by mass of the two components: _____% _____% in the final compound (Parts B and D) or initial compound (Part C)

Laboratory Questions

Circle the questions that have been assigned.

1. Part A.1. The crucible is not fired, as the procedure suggests, but had retained some impurities from previous use (or it could be oily smudges from fingers). The mass of the "dirty" crucible is recorded. However the impurities are burned off in the experiment. How does this affect the reported mass of the final product? Explain.

2. Part B.2. The burning of the magnesium becomes uncontrolled (it burns brightly). Oops! How will this procedural error affect the reported mole ratio of magnesium to oxygen in the analysis? Explain.

3. Part B.3. In a hurry to complete the experiment, Josh did not allow all of the magnesium to react. Will his reported magnesium to oxygen ratio be reported too high or too low? Explain.

*4. Part B.4. Elizabeth forgot to add the few drops of water resulting in the presence of some Mg_3N_2 in the final solid product. Will her reported magnesium to oxygen mole ratio be reported too high or too low? Explain. *Hint:* The Mg:N mass ratio is 1:0.38, and the Mg:O mass ratio is 1:0.66.

5. Part C.2. The sample is *not* completely thermally decomposed in the procedure. Will the mole ratio of CaO to CO_2 be too high or too low? Explain.

6. Part C.2. The original sample is *not* pure, but is contaminated with a thermally stable compound. Will the reported mole ratio of CaO to CO_2 be too high, too low, or unaffected? Explain.

7. Part D.2. In an oversight in the experiment the 4–5 additional drops of 6 M HNO_3 are not added! How might this affect the reported number of moles of oxygen in the final product? Explain.

8. Part D.3. Spattering does occur and some of the sample is lost onto the laboratory bench. Will the reported mole ratio of tin to oxygen be too high, too low, or unaffected? Explain.

Experiment 8

Limiting Reactant

Calcium oxalate crystals contribute to the formation of kidney stones.

- To determine the limiting reactant in a mixture of two soluble salts
- To determine the **percent composition** of each substance in a salt mixture

TECHNIQUES

The following techniques are used in the Experimental Procedure

Percent composition: the mass ratio of a component of a mixture or compound to the total mass of the sample times 100

INTRODUCTION

Percent yield:

$$\left(\frac{\text{actual yield}}{\text{theoretical yield}}\right) \times 100$$

Stoichiometrically: by a study of a chemical reaction using a balanced equation

Two factors affect the yield of products in a chemical reaction: (1) the amounts of starting materials (reactants) and (2) the **percent yield** of the reaction. Many experimental conditions, for example, temperature and pressure, can be adjusted to increase the yield of a desired product in a chemical reaction, but because chemicals react according to fixed mole ratios (**stoichiometrically**), only a limited amount of product can form from measured amounts of starting materials. The reactant determining the amount of product generated in a chemical reaction is called the **limiting reactant** in the chemical system.

To better understand the concept of the limiting reactant, let us look at the reaction under investigation in this experiment, the reaction of calcium chloride dihydrate, $CaCl_2 \cdot 2H_2O$, and potassium oxalate monohydrate, $K_2C_2O_4 \cdot H_2O$, in an aqueous solution.

$$CaCl_2 \cdot 2H_2O(aq) + K_2C_2O_4 \cdot H_2O(aq) \rightarrow$$
$$CaC_2O_4 \cdot H_2O(s) + 2\,KCl(aq) + 2\,H_2O(l) \quad (8.1)$$

Calcium oxalate monohydrate, $CaC_2O_4 \cdot H_2O$, is an insoluble compound, but is found naturally in a number of diverse locations. It is found in plants, such as rhubarb leaves, agave, and (in small amounts) spinach, and is the cause of most kidney stones. In small doses, it causes a severe reaction to the lining of the digestive tract. However, the handling of calcium oxalate in the laboratory is safe, so long as it isn't transferred to the mouth.

For the reaction system in this experiment, both the calcium chloride and potassium oxalate are soluble salts but the calcium oxalate is insoluble. The **ionic equation** for the reaction is

$$Ca^{2+}(aq) + 2\,Cl^-(aq) + 2\,K^+(aq) + C_2O_4{}^{2-}(aq) + 3\,H_2O(l) \rightarrow$$
$$CaC_2O_4 \cdot H_2O(s) + 2\,Cl^-(aq) + 2\,K^+(aq) + 2\,H_2O(l) \quad (8.2)$$

Ionic equation: A chemical equation that presents ionic compounds in the form in which they exist in aqueous solution. See Experiment 6.

Spectator ions: *cations or anions that do not participate in any observable or detectable chemical reaction*

Net ionic equation: *an equation that includes only those ions that participate in the observed chemical reaction*

Presenting only the ions that show evidence of a chemical reaction (i.e., the formation of a precipitate) and by removing the **spectator ions** (i.e., no change of ionic form during the reaction), we have the **net ionic equation** for the observed reaction.

$$Ca^{2+}(aq) + C_2O_4^{2-}(aq) + H_2O(l) \rightarrow CaC_2O_4 \cdot H_2O(s) \qquad (8.3)$$

Calcium oxalate monohydrate is thermally stable below ~90°C, but forms the anhydrous salt at temperatures above 110°C.

Therefore, one mole of Ca^{2+} (from one mole of $CaCl_2 \cdot 2H_2O$, molar mass = 147.02 g/mol) reacts with one mole of $C_2O_4^{2-}$ (from one mole of $K_2C_2O_4 \cdot H_2O$, molar mass = 184.24 g/mol) to produce one mole of $CaC_2O_4 \cdot H_2O$ (molar mass = 146.12 g/mol). If the calcium oxalate is heated to temperatures greater than 110°C for drying, then anhydrous CaC_2O_4 (molar mass = 128.10 g/mol) is the product.

In Part A of this experiment the solid salts $CaCl_2 \cdot 2H_2O$ and $K_2C_2O_4 \cdot H_2O$ form a heterogeneous mixture of unknown composition. The mass of the solid mixture is measured and then added to water—insoluble $CaC_2O_4 \cdot H_2O$ forms. The $CaC_2O_4 \cdot H_2O$ precipitate is collected, via gravity filtration and dried, and its mass is measured.

The percent composition of the salt mixture is determined by first testing for the limiting reactant. In Part B, the limiting reactant for the formation of solid calcium oxalate monohydrate is determined from two precipitation tests of the solution: (1) the solution is tested for an excess of calcium ion with an oxalate reagent—observed formation of a precipitate indicates the presence of an excess of calcium ion (and a limited amount of oxalate ion) in the salt mixture; (2) the solution is also tested for an excess of oxalate ion with a calcium reagent—observed formation of a precipitate indicates the presence of an excess of oxalate ion (and a limited amount of calcium ion) in the salt mixture.

Calculations

The calculations for the analysis of the salt mixture require some attention. "How do I proceed to determine the percent composition of a salt mixture of $CaCl_2 \cdot 2H_2O$ and $K_2C_2O_4 \cdot H_2O$ by only measuring the mass of the $CaC_2O_4 \cdot H_2O$ precipitate?"

Example: A 0.538-g sample of the salt mixture is added to water and after drying (to less than 90°C) 0.194 g of $CaC_2O_4 \cdot H_2O$ is measured. Tests reveal that $K_2C_2O_4 \cdot H_2O$ is the limiting reactant. What is the percent composition of the salt mixture? How many grams of the excess $CaCl_2 \cdot 2H_2O$ were in the salt mixture?

Solution: Since $K_2C_2O_4 \cdot H_2O$ is the limiting reactant, then, according to Equation (8.1), the moles of $K_2C_2O_4 \cdot H_2O$ in the salt mixture equals the moles of $CaC_2O_4 \cdot H_2O$ formed. Therefore, the calculated mass of $K_2C_2O_4 \cdot H_2O$ in the original salt mixture is

$$\text{grams } K_2C_2O_4 \cdot H_2O = 0.194 \text{ g } CaC_2O_4 \cdot H_2O \times \frac{1 \text{ mol } CaC_2O_4 \cdot H_2O}{146.12 \text{ g } CaC_2O_4 \cdot H_2O}$$

$$\times \frac{1 \text{ mol } K_2C_2O_4 \cdot H_2O}{1 \text{ mol } CaC_2O_4 \cdot H_2O} \times \frac{184.24 \text{ } K_2C_2O_4 \cdot H_2O}{1 \text{ mol } K_2C_2O_4 \cdot H_2O}$$

$$= 0.245 \text{ g } K_2C_2O_4 \cdot H_2O \text{ in the salt mixture.}$$

The percent by mass of $K_2C_2O_4 \cdot H_2O$ in the original salt mixture is

$$\% \text{ } K_2C_2O_4 \cdot H_2O = \frac{0.245 \text{ g } K_2C_2O_4 \cdot H_2O}{0.538 \text{ g sample}} \times 100 = 45.5\% \text{ } K_2C_2O_4 \cdot H_2O$$

The mass of the $CaCl_2 \cdot 2H_2O$ in the salt mixture is the difference between the mass of the sample and the mass of $K_2C_2O_4 \cdot H_2O$ or (0.538 g − 0.245 g =) 0.293 g. The percent by mass of $CaCl_2 \cdot 2H_2O$ in the original salt mixture is

$$\% \text{ } CaCl_2 \cdot 2H_2O = \frac{0.538 \text{ g} - 0.245 \text{ g}}{0.538 \text{ g sample}} \times 100 = 54.5\% \text{ } CaCl_2 \cdot 2H_2O$$

According to Equation 8.1, the moles of $CaCl_2 \cdot 2H_2O$ that react equals the moles of $K_2C_2O_4 \cdot H_2O$ (the limiting reactant) that react equals the moles of $CaC_2O_4 \cdot H_2O$ that precipitate. Therefore the mass of $CaCl_2 \cdot 2H_2O$ (the *excess* reactant) that reacts is

$$\text{mass } CaCl_2 \cdot 2H_2O = 0.194 \text{ g } CaC_2O_4 \cdot H_2O \times \frac{1 \text{ mol } CaC_2O_4 \cdot H_2O}{146.12 \text{ g } CaC_2O_4 \cdot H_2O}$$

$$\times \frac{1 \text{ mol } CaCl_2 \cdot 2H_2O}{1 \text{ mol } CaC_2O_4 \cdot H_2O} \times \frac{147.02 \text{ g } CaCl_2 \cdot 2H_2O}{1 \text{ mol } CaCl_2 \cdot 2H_2O}$$

$$= 0.195 \text{ g } CaCl_2 \cdot 2H_2O \text{ reacted}$$

The mass of *excess* $CaCl_2 \cdot 2H_2O$ is 0.293 g $-$ 0.195 g $=$ 0.098 g *xs* $CaCl_2 \cdot 2H_2O$

Procedure Overview: In Part A a measured mass of a solid $CaCl_2 \cdot 2H_2O/K_2C_2O_4 \cdot H_2O$ salt mixture of unknown composition is added to water. The precipitate that forms is digested, filtered, and dried, and its mass measured. Observations from tests on the **supernatant** solution in Part B determine which salt in the mixture is the limiting reactant. An analysis of the data provides the determination of the percent composition of the salt mixture.

Two trials are recommended for this experiment. To hasten the analyses, measure the mass of duplicate unknown solid salt mixtures in *clean* 150- or 250-mL beakers and simultaneously follow the procedure for each. Label the beakers accordingly for Trial 1 and Trial 2 to avoid the intermixing of samples and solutions.

Obtain about 2–3 g of an unknown $CaCl_2 \cdot 2H_2O/K_2C_2O_4 \cdot H_2O$ salt mixture.

EXPERIMENTAL PROCEDURE

Supernatant: the clear solution that exists after the precipitate has settled

1. **Prepare the Salt Mixture.** a. *Mass of salt mixture.* Measure the mass (± 0.001 g) of Beaker #1 and record on the Report Sheet for Trial 1. Transfer ~1 g of the salt mixture to the beaker, measure and record the combined mass. Repeat for Trial 2.

   b. *Adjust pH of deionized water.* Fill a 400 mL beaker with deionized water. Test with pH paper . . . if the water is acidic, adjust it to basic with drops of 6 *M* NH_3. If already basic to pH paper, then no addition of NH_3 is necessary.[1]

   c. *Mix deionized water and salt.* Add ~150 mL of the deionized water from Part A.1b to the salt mixture in Beaker #1. Stir the mixture with a stirring rod for 2–3 minutes and then allow the precipitate to settle. Leave the stirring rod in the beaker.

2. **Digest the Precipitate.** a. *Heat.* Cover the beaker with a watch glass and warm the solution on a hot plate (Figure 8.1) to a temperature not to exceed 75°C for 15 minutes. Periodically, stir the solution and in the meantime, proceed to Part A.3.

   b. *Cool.* After 15 minutes, remove the heat and allow the precipitate to settle; the solution does *not* need to cool to room temperature.

   c. *Wash water.* While the precipitate is settling, heat (70–80°C) about 30 mL of deionized water for use as wash water in Part A.5.

A. Precipitation of $CaC_2O_4 \cdot H_2O$ from the Salt Mixture

---

[1] Calcium oxalate does *not* precipitate in an acidic solution because of the formation of $H_2C_2O_4^{-}$, an ion that does not precipitate with Ca^{2+}.

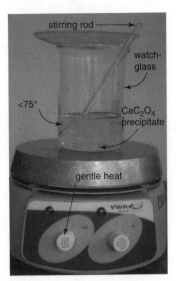

Figure 8.1 Warming and digesting the precipitate.

3. **Set Up a Gravity (or Vacuum[2]) Filtering Apparatus.** Place your initials (in pencil) and Trial #1 on a piece of Whatman No. 42 or Fisher*brand* Q2 filter paper,[3] fold, and tear off its corner. Measure and record its mass (±0.001 g). Seal the filter paper into the filter funnel with a small amount of deionized water. Discard the deionized water from the receiving flask. Have your instructor inspect your apparatus before continuing. Return to Part A.2b.

4. **Withdraw and Save Supernatant.** Once the precipitate has settled and the supernatant has cleared in Part A.2b, use a dropping pipet to withdraw enough supernatant to half-fill two 75-mm test tubes, labeled "1" and "2." Save for Part B.

Rubber policeman: a spatula-like rubber tip attached to a stirring rod

5. **Filter the $CaC_2O_4 \cdot H_2O$ Precipitate.** While the remaining solution of the salt mixture from Part A.4 is still warm, quantitatively transfer the precipitate to the filter (Figure 8.2). Transfer any precipitate on the wall of the beaker to the filter with the aid of a **rubber policeman**; wash any remaining precipitate onto the filter with three or four 5-mL volumes of warm water (from Part A.2c).

6. **Dry and Measure the Amount of $CaC_2O_4 \cdot H_2O$ Precipitate.** Remove the filter paper and precipitate from the filter funnel. Air-dry the precipitate on the filter paper until the next laboratory period or dry in a $<110°C$ constant temperature drying oven for at least 1 hour or overnight. Determine the combined mass (±0.001 g) of the precipitate and filter paper. Record. Repeat for Trial 2.

7. **Formula of the Precipitate.** If the precipitate is air-dried, the precipitate is $CaC_2O_4 \cdot H_2O$; if oven-dried at $\geq110°C$, the precipitate is the anhydrous CaC_2O_4. Enter the mass of the dried precipitate as *either* item A.6 (air-dried) or A.7 (oven-dried) on the Report Sheet.

---

[2]A vacuum filtering apparatus (Technique 11E) can also be used; the filtering procedure will be more rapid, but more precipitate may pass through the filter paper.
[3]Whatman No. 42 and Fisher*brand* Q2 filter papers are both fine-porosity filter papers; a fine-porosity filter paper is used to reduce the amount of precipitate passing through the filter.

Figure 8.2 Gravity filtration is used to filter finely divided precipitates.

From the following two tests (Figure 8.3) you can determine the limiting reactant in the original salt mixture. Some cloudiness may appear in both tests, but one will show a definite formation of precipitate.

B. Determination of the Limiting Reactant

1. **Clarify the Supernatant.** Centrifuge the two collected supernatant samples from Part A.4.

2. **Test for *Excess* $C_2O_4^{2-}$.** Add 2 drops of the test reagent 0.5 M CaCl₂ to the supernatant liquid in test tube 1. If a precipitate forms, the $C_2O_4^{2-}$ is *in excess* and Ca^{2+} is the limiting reactant in the original salt mixture.

3. **Test for *Excess* Ca^{2+}.** Add 2 drops of the test reagent 0.5 M K₂C₂O₄ to the supernatant liquid in test tube 2. If a precipitate forms, the Ca^{2+} is *in excess* and $C_2O_4^{2-}$ is the limiting reactant in the original salt mixture.

An obvious formation of precipitate should appear in only one of the tests. Repeat for Trial 2.

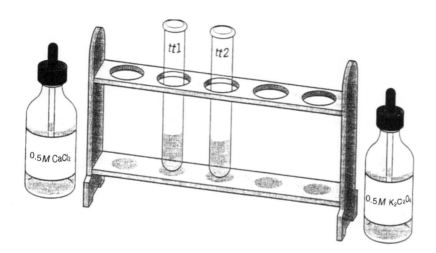

Figure 8.3 Testing for the excess (and the limiting) reactant.

Disposal: Dispose of the calcium oxalate, including the filter paper, in the "Waste Solids" container. Dispose of the waste solutions in the "Waste Liquids" container.

CLEANUP: Rinse each beaker with small portions of warm water and discard in the "Waste Liquids" container. Rinse twice with tap water and twice with deionized water and discard in the sink.

The Next Step

All reactions other than decomposition reactions have limiting reactants! From the combustion of fossil fuels to the many integrated chemical processes of biochemical reactions in living organisms, there is one reactant that limits the process. For examples, what is the limiting reactant in the combustion of gasoline in the cylinder of an engine; what is the limiting reactant in the eutrophication of a body of water; what is the limiting reactant in making bread rise; what is the limiting reactant in the precipitation of a salt, . . . ? Research the limiting reactant concept in upcoming experiments.

Date _____ Lab Sec. _____ Name _____ Desk No. _____

1. The limiting reactant is determined in this experiment.
   a. What are the reactants (and their molar masses) in the experiment?

   b. What is the product (and its molar mass) that is used for determining the limiting reactant?

   c. How is the limiting reactant determined in the experiment?

2. Experimental Procedure, Part A.2. What is the procedure and purpose of "digesting the precipitate?"

3. Two special steps in the Experimental Procedure are incorporated to reduce the loss of the calcium oxalate precipitate. Identify the steps in the procedure and the reason for each step.

4. A 0.972-g sample of a $CaCl_2 \cdot 2H_2O/K_2C_2O_4 \cdot H_2O$ solid salt mixture is dissolved in ~150 mL of deionized water, previously adjusted to a pH that is basic. The precipitate, after having been filtered and *air-dried,* has a mass of 0.375 g. The limiting reactant in the salt mixture was later determined to be $CaCl_2 \cdot 2H_2O$.
   a. What is the percent by mass of $CaCl_2 \cdot 2H_2O$ in the salt mixture?

b. How many grams of the excess reactant, $K_2C_2O_4 \cdot H_2O$, *reacted* in the mixture?

c. How many grams of the $K_2C_2O_4 \cdot H_2O$ in the salt mixture remain *unreacted?*

5. A 1.009-g mixture of the solid salts Na_2SO_4 (molar mass = 142.04 g/mol) and $Pb(NO_3)_2$ (molar mass = 331.20 g/mol) forms an aqueous solution with the precipitation of $PbSO_4$ (molar mass = 303.26 g/mol). The precipitate was filtered and dried, and its mass was determined to be 0.471 g. The limiting reactant was determined to be Na_2SO_4.

a. Write the molecular form of the equation for the reaction.

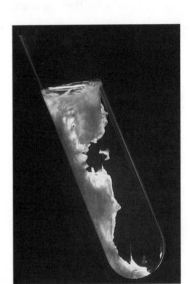

b. Write the net ionic equation for the reaction.

c. How many moles and grams of Na_2SO_4 are in the reaction mixture?

d. How many moles and grams of $Pb(NO_3)_2$ reacted in the reaction mixture?

e. What is the percent by mass of each salt in the mixture?

Limiting Reactant

Date _____ Lab Sec. _____ Name _____ Desk No. _____

A. Precipitation of CaC$_2$O$_4\cdot$H$_2$O from the Salt Mixture

	Trial 1	*Trial 2*
Unknown number _____		
1. Mass of beaker (*g*)	_____	_____
2. Mass of beaker and salt mixture (*g*)	_____	_____
3. Mass of salt mixture (*g*)		
4. Mass of filter paper (*g*)	_____	_____
5. Mass of filter paper and CaC$_2$O$_4\cdot$H$_2$O (*g*)		
6. Mass of air-dried CaC$_2$O$_4\cdot$H$_2$O (*g*)		
or		
7. Mass of oven-dried CaC$_2$O$_4$ (*g*)	_____	_____

B. Determination of Limiting Reactant

1. Limiting reactant in salt mixture (write complete formula)   _____

2. Excess reactant in salt mixture (write complete formula)   _____

Data Analysis

1. Moles of CaC$_2$O$_4\cdot$H$_2$O (or CaC$_2$O$_4$) precipitated (*mol*)   *

2. Moles of limiting reactant in salt mixture (*mol*)
   - formula of limiting hydrate _____

3. Mass of limiting reactant in salt mixture (*g*)
   - formula of limiting hydrate _____

4. Mass of excess reactant in salt mixture (*g*)
   - formula of excess hydrate _____

5. Percent limiting reactant in salt mixture (%)
   - formula of limiting hydrate _____

6. Percent excess reactant in salt mixture (%)
   - formula of excess hydrate _____

7. Mass of excess reactant that reacted (*g*)
   - formula of excess reactant _____

8. Mass of excess reactant, unreacted (*g*)

*Show calculations for trial 1 on the next page.

Show all calculations for Trial 1.

Laboratory Questions

Circle the questions that have been assigned.

1. Part A.2. If the step for digesting the precipitate were omitted, what would be the probable consequence of reporting the "percent limiting reactant" in the salt mixture? Explain.

2. Part A.3. A couple of drops of water were accidentally placed on the properly folded filter paper before its mass was measured. However, in Part A.6, the $CaC_2O_4 \cdot H_2O$ precipitate and the filter paper were dry. How does this sloppy technique affect the reported mass of the limiting reactant in the original salt mixture? Explain.

3. Part A.5. Because of the porosity of the filter paper some of the $CaC_2O_4 \cdot H_2O$ precipitate passes through the filter paper. Will the reported percent of the limiting reactant in the original salt mixture be reported too high or too low? Explain.

4. Part A.5. Excessive quantities of wash water are added to the $CaC_2O_4 \cdot H_2O$ precipitate. How does this affect the mass of $CaC_2O_4 \cdot H_2O$ precipitate reported in Part A.6?

5. Part A.6. The $CaC_2O_4 \cdot H_2O$ precipitate is not completely air-dried when its mass is determined. Will the reported mass of the limiting reactant in the original salt mixture be reported too high or too low? Explain.

6. Part A.6, 7. The drying oven, although thought (and assumed) to be set at 125°C, had an inside temperature of 84°C. How will this error affect the reported percent by mass of the limiting reactant in the salt mixture . . . too high, too low, or unaffected? Explain.

A Volumetric Analysis

A titrimetric analysis requires the careful addition of titrant.

- To prepare and standardize a sodium hydroxide solution
- To determine the molar concentration of a strong acid

The following techniques are used in the Experimental Procedure

A chemical analysis that is performed primarily with the aid of volumetric glassware (e.g., pipets, burets, volumetric flasks) is called a **volumetric analysis.** For a volumetric analysis procedure, a known quantity or a carefully measured amount of one substance reacts with a to-be-determined amount of another substance with the reaction occurring in aqueous solution. The volumes of all solutions are carefully measured with volumetric glassware.

The known amount of the substance for an analysis is generally measured and available in two ways:

1. As a **primary standard:** An accurate mass (and thus, moles) of a solid substance is measured on a balance, dissolved in water, and then reacted with the substance being analyzed.

Primary standard: a substance that has a known high degree of purity, a relatively large molar mass, is nonhygroscopic, and reacts in a predictable way

2. As a **standard solution:** A measured number of moles of substance is present in a measured volume of solution—a solution of known concentration, generally expressed as the molar concentration (or molarity) of the substance. A measured volume of the standard solution then reacts with the substance being analyzed.

Standard solution: a solution having a very well known concentration of a solute

The reaction of the known substance with the substance to be analyzed, occurring in aqueous solution, is generally conducted by a titration procedure.

The titration procedure requires a buret to dispense a liquid, called the **titrant,** into a flask containing the **analyte** (Figure 9.1*a*). The titrant may be a solution of known or unknown concentration. The analyte may be a solution whose volume is measured with a pipet or it may be a dissolved solid with a very accurately measured mass. For the acid–base titration studied in this experiment, the titrant is a standard solution of sodium hydroxide and the analyte is an acid.

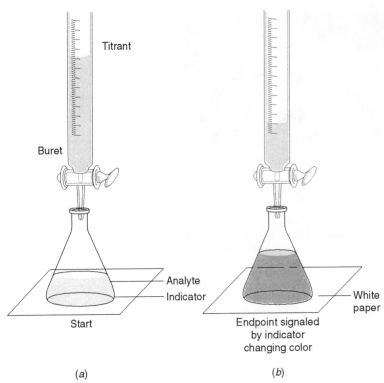

Titrant

Buret

Analyte
Indicator

White
paper

Start

Endpoint signaled
by indicator
changing color

(a)   (b)

Figure 9.1 (a) Titrant in the buret is dispensed into the analyte until (b) the indicator changes color at its endpoint.

Stoichiometric amounts: amounts corresponding to the mole ratio of the balanced equation

Acid–base indicator: a substance having an acidic structure with a different color than its basic structure

pH: the negative logarithm of the molar concentration of H_3O^+, pH = $-log[H_3O^+]$. Refer to Experiment 6.

A reaction is complete when **stoichiometric amounts** of the reacting substances are combined. In a titration this is the **stoichiometric point**.[1] In this experiment the stoichiometric point for the acid–base titration is detected using a phenolphthalein **indicator.** Phenolphthalein is colorless in an acidic solution but pink in a basic solution. The point in the titration at which the phenolphthalein changes color is called the **endpoint** of the indicator (Figure 9.1b). Indicators are selected so that the stoichiometric point in the titration coincides (at approximately the same **pH**) with the endpoint of the indicator.

Standardization of a Sodium Hydroxide Solution

Hygroscopic: able to absorb water vapor readily

potassium hydrogen phthalate

Solid sodium hydroxide is very **hygroscopic;** therefore its mass cannot be measured to prepare a solution with an accurately-known molar concentration (a primary standard solution). To prepare a NaOH solution with a very well known molar concentration, it must be standardized with an acid that *is* a primary standard.

In Part A of this experiment, *dry* potassium hydrogen phthalate, $KHC_8H_4O_4$, is used as the primary acid standard for determining the molar concentration of a sodium hydroxide solution. Potassium hydrogen phthalate is a white, crystalline, acidic solid. It has the properties of a primary standard because of its high purity, relatively high molar mass, and because it is only *very slightly* hygroscopic. The moles of $KHC_8H_4O_4$ used for the analysis is calculated from its measured mass and molar mass (204.44 g/mol):

$$\text{mass (g) } KHC_8H_4O_4 \times \frac{\text{mol } KHC_8H_4O_4}{204.44 \text{ g } KHC_8H_4O_4} = \text{mol } KHC_8H_4O_4 \qquad (9.1)$$

From the balanced equation for the reaction, one mole of $KHC_8H_4O_4$ reacts with one mole of NaOH according to the equation:

$$KHC_8H_4O_4(aq) + NaOH(aq) \rightarrow H_2O(l) + NaKC_8H_4O_4(aq) \qquad (9.2)$$

---

[1]The stoichiometric point is also called the **equivalence point**, indicating the point at which stoichiometrically equivalent quantities of the reacting substances are combined.

In the experimental procedure an accurately measured mass of dry potassium hydrogen phthalate is dissolved in deionized water. A prepared NaOH solution is then dispensed from a buret into the $KHC_8H_4O_4$ solution until the stoichiometric point is reached, signaled by the colorless to pink change of the phenolphthalein indicator. At this point the dispensed volume of NaOH is noted and recorded.

The molar concentration of the NaOH solution is calculated by determining the number of moles of NaOH used in the reaction (Equation 9.2) and the volume of NaOH dispensed from the buret.

$$\text{molar concentration } (M) \text{ of NaOH (mol/L)} = \frac{\text{mol NaOH}}{\text{L of NaOH solution}} \quad (9.3)$$

Once the molar concentration of the sodium hydroxide is calculated, the solution is said to be "standardized" and the sodium hydroxide solution is called a **secondary standard** solution.

Molar Concentration of an Acid Solution

In Part B, an unknown molar concentration of an acid solution is determined. The standardized NaOH solution is used to titrate an accurately measured volume of the acid to the stoichiometric point. By knowing the volume and molar concentration of the NaOH, the number of moles of NaOH used for the analysis is

$$\text{volume (L)} \times \text{molar concentration (mol/L)} = \text{mol NaOH} \quad (9.4)$$

From the stoichiometry of the reaction, the moles of acid neutralized in the reaction can be calculated. If your acid of unknown concentration is a monoprotic acid, HA (as is HCl(*aq*)), then the mole ratio of acid to NaOH will be 1:1 (Equation 9.5). However, if your acid is diprotic, H_2A (as is H_2SO_4), then the mole ratio of acid to NaOH will be 1:2 (Equation 9.6). Your instructor will inform you of the acid type, HA or H_2A.

$$HA(aq) + NaOH(aq) \rightarrow NaA(aq) + 2\,H_2O(l) \quad (9.5)$$
$$H_2A(aq) + 2\,NaOH(aq) \rightarrow Na_2A(aq) + 2\,H_2O(l) \quad (9.6)$$

From the moles of the acid that react and its measured volume, the molar concentration of the acid is calculated:

$$\text{molar concentration of the acid (mol/L)} = \frac{\text{mol acid}}{\text{volume of acid (L)}} \quad (9.7)$$

EXPERIMENTAL PROCEDURE

Procedure Overview: A NaOH solution is prepared with an approximate concentration. A more accurate molar concentration of the NaOH solution (as the titrant) is determined using dry potassium hydrogen phthalate as a primary standard. The NaOH solution, now a secondary standard solution, is then used to determine the "unknown" molar concentration of an acid solution.

A. The Standardization of a Sodium Hydroxide Solution

You are to complete at least three "good" trials ($\pm 1\%$ reproducibility) in standardizing the NaOH solution. Prepare three clean 125-mL or 250-mL Erlenmeyer flasks for the titration.

You will need to use approximately one liter of boiled, deionized water for this experiment. Start preparing that first.

1. **Prepare the Stock NaOH Solution.**[2] One week before the scheduled laboratory period, dissolve about 4 g of NaOH (pellets or flakes) (**Caution:** *NaOH is very corrosive—do not allow skin contact. Wash hands thoroughly with water.*) in 5 mL of deionized water in a 150-mm rubber-stoppered test tube. Thoroughly

---

[2]Check with your laboratory instructor to see if the NaOH solution is prepared for Part A.1 (and/or Part A.3) and to see if the $KHC_8H_4O_4$ is dried for Part A.2.

Tared mass: mass of a sample without regard to its container

mix and allow the solution to stand for the precipitation of sodium carbonate, Na_2CO_3.[3]

2. **Dry the Primary Standard Acid.** Dry 2–3 g of $KHC_8H_4O_4$ at 110°C for several hours in a constant temperature drying oven. Cool the sample in a desiccator.

3. **Prepare the Diluted NaOH Solution.** Decant about 4 mL of the NaOH solution prepared in Part A.1 into a 500-mL polyethylene bottle (Figure 9.2). (**Caution:** *Concentrated NaOH solution is extremely corrosive and will cause severe skin removal!*) Dilute to 500 mL with previously boiled,[4] deionized water cooled to room temperature. Cap the polyethylene bottle to prevent the absorption of CO_2. Swirl the solution and label the bottle.

   Calculate an *approximate* molar concentration of your diluted NaOH solution.

4. **Prepare the Primary Standard Acid.** a. Calculate the mass of $KHC_8H_4O_4$ that will require about 15–20 mL of your diluted NaOH solution to reach the stoichiometric point. Show the calculations on the Report Sheet.

   b. Measure this mass (± 0.001 g) of $KHC_8H_4O_4$ on a **tared** piece of weighing paper (Figure 9.3) and transfer it to a clean, labeled Erlenmeyer flask. Similarly, prepare all three samples while you are occupying the balance. Dissolve the $KHC_8H_4O_4$ in about 50 mL of previously boiled, deionized water and add 2 drops of phenolphthalein.

Figure 9.2 A 500-mL polyethylene bottle for the NaOH solution.

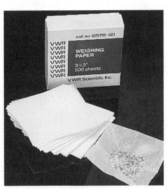

Figure 9.3 Weighing paper for the $KHC_8H_4O_4$ measurements.

5. **Prepare a Clean Buret.** Wash a 50-mL buret and funnel thoroughly with soap and water using a long buret brush. Flush the buret with tap water and rinse several times with deionized water. Rinse the buret with three 5-mL portions of the diluted NaOH solution, making certain that the solution wets the entire inner surface. Drain each rinse through the buret tip. Discard each rinse in the "Waste Bases" container. Have the instructor approve your buret and titration setup before continuing.

---

[3]Carbon dioxide, CO_2, from the atmosphere is an **acidic anhydride** (meaning that when CO_2 dissolves in water, it forms an acidic solution). The acid CO_2 reacts with the base NaOH to form the less soluble salt, Na_2CO_3.

$$CO_2(g) + 2\,NaOH(aq) \rightarrow Na_2CO_3(s) + H_2O(l)$$

[4]Boiling the water removes traces of CO_2 that would react with the sodium hydroxide in solution.

6. **Fill the Buret.** Using a clean funnel, fill the buret with the NaOH solution.[5] After 10–15 seconds, read the volume by viewing the bottom of the meniscus with the aid of a black line drawn on a white card (the buret can be removed from the stand or moved up or down in the buret clamp to make this reading; you need not stand on a lab stool to read the meniscus). Record this initial volume according to the guideline in Technique 16A.2, "using all certain digits (from the labeled calibration marks on the glassware) *plus* one uncertain digit (the last digit which is the best estimate between the calibration marks)." Place a sheet of white paper beneath the Erlenmeyer flask.

7. **Titrate the Primary Standard Acid #1.** Slowly add the NaOH titrant to the first acid sample prepared in Part A.4. Swirl the flask (with the proper hand[6]) after each addition. Initially, add the NaOH solution in 1- to 2-mL increments. As the stoichiometric point nears, the color fade of the indicator occurs more slowly. Occasionally rinse the wall of the flask with (previously boiled, deionized) water from your wash bottle. Continue addition of the NaOH titrant until the endpoint is reached. *The endpoint in the titration should be within one-half drop of a slight pink color* (see opening photo). The color should persist for 30 seconds. After 10–15 seconds, read (Figure 9.4) and record the final volume of NaOH in the buret.

8. **Repeat the Analysis with the Remaining Standard Acid Samples.** Refill the buret and repeat the titration *at least* two more times with varying, but accurately known, masses of $KHC_8H_4O_4$.

Figure 9.4 Read the volume of titrant with a black background.

9. **Do the Calculations.** Calculate the molar concentration of the diluted NaOH solution. The molar concentrations of the NaOH solution from the three analyses should be within ±1%. Place a corresponding label on the 500-mL polyethylene bottle.

Disposal: Dispose of the neutralized solutions in the Erlenmeyer flasks in the "Waste Acids" container.

Three samples of the acid having an unknown concentration are to be analyzed. Ask your instructor for the acid type of your unknown (i.e., HA or H_2A). Prepare three *clean* 125- or 250-mL Erlenmeyer flasks for this determination.

B. Molar Concentration of an Acid Solution

1. **Prepare the Acid Samples of Unknown Concentration.** In an Erlenmeyer flask, pipet 25.00 mL of the acid solution. Add 2 drops of phenolphthalein.

2. **Fill the Buret and Titrate.** Refill the buret with the (now) standardized NaOH solution and, after 10–15 seconds, read and record the initial volume. Refer to Parts A.6 and A.7. Titrate the acid sample to the phenolphthalein endpoint. Read and record the final volume of titrant.

3. **Repeat.** Similarly titrate the other samples of the acid solution.

4. **Calculations.** Calculate the average molar concentration of your acid unknown.

Save. Save your standardized NaOH solution in the *tightly capped* 500-mL polyethylene bottle for Experiments 10, 17, 18, and/or 19. Consult with your instructor.

Disposal: Dispose of the neutralized solutions in the "Waste Acids" container. Consult with your instructor.

---

[5]Be certain all air bubbles are removed from the buret tip.
[6]Check Technique 16C.3 for this procedure.

CLEANUP: Rinse the buret and pipet several times with tap water and discard through the tip into the sink. Rinse twice with deionized water. Similarly clean the Erlenmeyer flasks.

Check and clean the balance area. All solids should be discarded in the "Waste Solid Acids" container.

The Next Step

What are the acid concentrations for various noncarbonated soft drinks? the acid of vinegar (Experiment 10), the acids used for treating swimming pools? the acid of fruit juices? the antacids (Experiment 17), of aspirin (Experiment 19), . . . specifically, what are those acids? Design a procedure for determining the acidity for a select grouping of foods, drinks, or other familiar commercial products.

NOTES AND CALCULATIONS

A Volumetric Analysis

Date _____ Lab Sec. _____ Name _____ Desk No. _____

1. a. Define the analyte in a titration.

   b. Is the indicator generally added to the titrant or the analyte in a titration?

2. a. What is the primary standard used in this experiment (name and formula)? Define a primary standard.

   b. What is the secondary standard used in this experiment (name and formula)? Define a secondary standard.

3. Distinguish between a stoichiometric point and an endpoint in an acid–base titration.

4. a. How do you know that glassware (e.g., a buret or pipet) is clean?

   b. When rinsing a buret after cleaning it with soap and water, should the rinse be dispensed through the buret tip or the top opening of the buret? Explain.

   c. Experimental Procedure, Part A.5. In preparing the buret for titration the final rinse is with the NaOH titrant rather than with deionized water. Explain.

   d. Experimental Procedure, Part A.7. How is a "half-drop" of titrant dispensed from a buret?

5. Experimental Procedure, Part A.1. A 4-g mass of NaOH is dissolved in 5 mL of water.
   a What is the approximate molar concentration of the NaOH?

   b. In Part A.3, a 4-mL aliquot of this solution is diluted to 500 mL of solution. What is the approximate molar concentration of NaOH in the diluted solution? Enter this information on your Report Sheet.

   c. Part A.4. Calculate the mass of $KHC_8H_4O_4$ (molar mass = 204.44 g/mol) that reacts with 15 mL of the NaOH solution in Part A.3.

6. a. A 0.4040-g sample of potassium hydrogen phthalate, $KHC_8H_4O_4$ (molar mass = 204.44 g/mol) is dissolved with 50 mL of deionized water in a 125-mL Erlenmeyer flask. The sample is titrated to the phenolphthalein endpoint with 14.71 mL of a sodium hydroxide solution. What is the molar concentration of the NaOH solution?

   b. A 25.00-mL aliquot of a nitric acid solution of unknown concentration is pipetted into a 125-mL Erlenmeyer flask and 2 drops of phenolphthalein are added. The *above* sodium hydroxide solution (the titrant) is used to titrate the nitric acid solution (the analyte). If 18.92 mL of the titrant is dispensed from a buret in causing a color change of the phenolphthalein, what is the molar concentration of the nitric acid solution?

Experiment 9 *Report Sheet*

A Volumetric Analysis

Date ________ Lab Sec. ______ Name ______________________________ Desk No. ________

Maintain at least three significant figures when recording data and performing calculations.

A. Standardization of a Sodium Hydroxide Solution

Calculate the approximate molar concentration of diluted NaOH solution (Part A.3).

Calculate the approximate mass of $KHC_8H_4O_4$ for the standardization of the NaOH solution (Part A.4).

	Trial 1	*Trial 2*	*Trial 3*
1. Tared mass of $KHC_8H_4O_4$ (g)	________	________	________
2. Molar mass of $KHC_8H_4O_4$		204.44 g/mol	
3. Moles of $KHC_8H_4O_4$ (mol)	________	________	________
Titration apparatus approval		____________	
4. Buret reading of NaOH, *initial* (mL)	________	________	________
5. Buret reading of NaOH, *final* (mL)	________	________	________
6. Volume of NaOH dispensed (mL)	________	________	________
7. Molar concentration of NaOH (mol/L)	________	________	________
8. Average molar concentration of NaOH (mol/L)		____________	
9. Standard deviation of molar concentration		____________	
10. Relative standard deviation of molar concentration (%RSD)		____________	

B. Molar Concentration of an Acid Solution

Acid type: _____ Unknown No. _____

Balanced equation for neutralization of acid with NaOH.

	Sample 1	Sample 2	Sample 3
1. Volume of acid solution (mL)	25.0	25.0	25.0
2. Buret reading of NaOH, *initial* (mL)			
3. Buret reading of NaOH, *final* (mL)			
4. Volume of NaOH dispensed (mL)			
5. Molar concentration of NaOH (mol/L), Part A			
6. Moles of NaOH dispensed (mol)			
7. Molar concentration of acid solution (mol/L)			
8. Average molar concentration of acid solution (mol/L)			
9. Standard deviation of molar concentration			
10. Relative standard deviation of molar concentration (%RSD)			

Laboratory Questions

Circle the questions that have been assigned.

1. Part A.2. Pure potassium hydrogen phthalate is used for the standardization of the sodium hydroxide solution. Suppose that the potassium hydrogen phthalate is *not* completely dry. Will the reported molar concentration of the sodium hydroxide solution be too high, too low, or unaffected because of the moistness of the potassium hydrogen phthalate? Explain.

2. Part A.3. The student "forgot" to prepare any boiled, deionized water for the preparation of the NaOH solution and *then* "forgot" to cap the bottle. Will the concentration of the NaOH solution be greater than, less than, or unaffected by this carelessness? Explain.

3. Part A.4. Phenolphthalein is a weak organic acid, being colorless in an acidic solution and pink in a basic solution. The Experimental Procedure suggests the addition of 2 drops of phenolphthalein for the standardization of the sodium hydroxide solution. Explain why the analysis will be less accurate with the addition of a larger amount, e.g., 20 drops, of phenolphthalein.

4. Part A.7. A drop of the NaOH titrant adheres to the side of the buret (because of a dirty buret) between the initial and final readings for the titration. How does this "clean glass" error affect the reported molar concentration of the NaOH solution? Explain.

5. Part B.2. The wall of the Erlenmeyer flask is occasionally rinsed with water from the wash bottle (see Part A.7) during the analysis of the acid solution. How does this affect the reported molar concentration of the acid solution? Explain.

6. Parts A.7 and B.2. For the standardization of the NaOH solution in Part A.7, the endpoint was consistently reproduced to a dark pink color. However, the endpoint for the titration of the acid solution in Part B.2 was consistently reproduced to a faint pink color. Will the reported molar concentration of the acid solution be too high, too low, or unaffected by the differences in the colors of the endpoints. Explain.

Vinegar is a 4–5% (by mass) solution in acetic acid.

*   To determine the percent by mass of acetic acid in vinegar

The following techniques are used in the Experimental Procedure

INTRODUCTION

Household vinegar is a 4–5% (by mass) acetic acid, CH_3COOH, solution (4% is the minimum federal standard). Generally, caramel flavoring and coloring are also added to make the product aesthetically more appealing. This experiment compares the acetic acid concentrations of at least two vinegars.

acetic acid

A volumetric analysis using the titration technique (see Experiment 9 and Technique 16C) is the method used for the determination of the percent by mass of acetic acid in vinegar. A measured mass of vinegar is titrated to the phenolphthalein endpoint with a measured volume of a standardized sodium hydroxide solution. As the volume and molar concentration of the standardized NaOH solution are known, the moles of NaOH used for the analysis are also known.

The moles of CH_3COOH are calculated from the balanced equation:

$$CH_3COOH(aq) + NaOH(aq) \rightarrow NaCH_3CO_2(aq) + H_2O(l) \qquad (10.1)$$

The mass of CH_3COOH in the vinegar is calculated from the measured moles of CH_3COOH neutralized in the reaction and its molar mass, 60.05 g/mol:

$$\text{mass (g) of } CH_3COOH = \text{mol } CH_3COOH \times \frac{60.05 \text{ g } CH_3COOH}{\text{mol } CH_3COOH} \qquad (10.2)$$

Finally, the percent by mass of CH_3COOH in vinegar is calculated:

$$\% \text{ by mass of } CH_3COOH = \frac{\text{mass (g) of } CH_3COOH}{\text{mass (g) of vinegar}} \times 100 \qquad (10.3)$$

Procedure Overview: Samples of two vinegars are analyzed for the amount of acetic acid in the sample. A titration setup is used for the analysis, using a standardized NaOH solution as the titrant and phenolphthalein as the indicator. Stoichiometry calculations are used to determine the percent of acetic acid in the vinegars.

EXPERIMENTAL
PROCEDURE

Two samples of two vinegars are to be analyzed. At the beginning of the laboratory period, obtain 10 mL of each vinegar in separate 10-mL graduated cylinders.

A standardized NaOH solution was prepared in Experiment 9. If that solution was saved, it is to be used for this experiment. If the solution was not saved, you either must again prepare and standardize the solution (Experiment 9, Part A) or obtain about 150 mL of a standardized NaOH solution prepared by stockroom personnel. Record the *exact* molar concentration of the NaOH solution on the Report Sheet. Your instructor will advise you.

A. Preparation of Vinegar Sample

Clean at least two 125- or 250-mL Erlenmeyer flasks.

1. **Calculate the Volume of Vinegar.** Calculate the volume of vinegar that would be needed for the neutralization of 25 mL of the standardized NaOH solution. Assume the vinegar has a density of 1 g/mL and a percent acetic acid of 5% by mass, and the standardized NaOH solution is 0.1 *M* NaOH. Show the calculation on the Report Sheet (see Prelaboratory Assignment).

2. **Prepare the Vinegar Sample.** Add the (approximate) calculated volume (from Part A.1) of one brand of vinegar to a clean *dry* 125- or 250-mL Erlenmeyer flask with a previously measured mass (±0.01 g) or a flask that has already been tared on the balance. Record the tared mass of the vinegar sample. Add 2 drops of phenolphthalein and rinse the wall of the flask with 20 mL of previously boiled, deionized water.

3. **Prepare the Buret and Titration Setup.** Rinse twice a clean 50-mL buret with ~5 mL of the standardized NaOH solution, making certain no drops cling to the inside wall. Fill the buret with the standardized NaOH solution, eliminate all air bubbles in the buret tip, and, after 10–15 seconds, read (Figure 10.1) and record the initial volume, "using all certain digits (from the labeled calibration marks on the buret) *plus* one uncertain digit (the last digit which is the best estimate between the calibration marks)." Place a sheet of white paper beneath the flask containing the vinegar sample.

B. Analysis of Vinegar Sample

Figure 10.1 Read the volume of titrant with a black background.

1. **Titrate the Vinegar Sample.** Slowly add the NaOH solution from the buret to the acid, swirling the flask (with the proper hand[1]) after each addition. Occasionally, rinse the wall of the flask with (previously boiled, deionized) water from your wash bottle. Continue addition of the NaOH titrant until the endpoint is reached.[2] The endpoint in the titration should be within one-half drop of a slight pink color. Be careful *not* to surpass the endpoint. The color should persist for 30 seconds. After 10–15 seconds, read (Figure 10.1) and record the final volume of NaOH titrant in the buret (see Technique 16A.2).

2. **Repeat with the Same Vinegar.** Refill the buret and repeat the titration *at least* once more with another sample of the same vinegar.

3. **Repeat with Another Vinegar.** Perform at least two analyses of a different vinegar and determine its acetic acid content.

4. **Calculations.** Determine the average percent by mass of acetic acid in the two vinegars.

Disposal: All test solutions and the NaOH solution in the buret can be discarded in the "Waste Bases" container.

CLEANUP: Rinse the buret twice with tap water and twice with deionized water, discarding each rinse through the buret tip into the sink. Similarly, rinse the flasks.

---

[1]Review Technique 16C.3 for this procedure.
[2]The endpoint (and the stoichiometric point) is near when the color fade of the phenolphthalein indicator occurs more slowly with each successive addition of smaller volumes of NaOH solution to the vinegar.

Vinegar Analysis

Date _____ Lab Sec. _____ Name _____ Desk No. _____

1. Assuming the density of a 5% acetic acid solution is 1.0 g/mL, determine the volume of the acetic acid solution necessary to neutralize 25.0 mL of 0.10 M NaOH. Also record this calculation on your Report Sheet.

2. A 31.43-mL volume of 0.108 M NaOH is required to reach the phenolphthalein endpoint in the titration of a 4.441-g sample of vinegar. Calculate the percent acetic acid in the vinegar.

3. a. A chemist often uses a white card with a black mark to aid in reading the meniscus of a clear liquid. Why does this technique make the reading more accurate? Explain.

   b. A chemist should wait 10–15 seconds after dispensing a volume of titrant before a reading is made. Explain why the "wait" is good laboratory technique.

   c. The color change at the endpoint should persist for 30 seconds. Explain why the time lapse is a good titration technique.

4. For the titration of an acid analyte with a NaOH titrant, a phenolphthalein endpoint will change from colorless to pink. If the "pink" resulted from perhaps only a half-drop (or less) of NaOH titrant and the Erlenmeyer flask is set aside on the laboratory bench, the pink color may disappear after an hour or so. Explain.

5. Explain why it is quantitatively *not* acceptable to titrate each of the vinegar samples with the NaOH titrant to the same *dark pink* endpoint.

6. Lemon juice has a pH of about 2.5. *Assuming* that the acidity of lemon juice is due solely to citric acid, that citric acid is a monoprotic acid, and that the density of lemon juice is 1.0 g/mL, then the citric acid concentration calculates to 0.5% by mass. Estimate the volume of 0.0100 *M* NaOH required to neutralize a 3.71-g sample of lemon juice. The molar mass of citric acid is 190.12 g/mol.

7. Oxalic acid, $H_2C_2O_4 \cdot 2H_2O$ (molar mass = 126.07 g/mol) is often used as a primary standard for the standardization of a NaOH solution. If 0.147 g of oxalic acid dihydrate is neutralized by 23.64 mL of a NaOH solution, what is the molar concentration of the NaOH solution? Oxalic acid is a diprotic acid. (Hint: what is the balanced equation?)

Vinegar Analysis

Date _____ Lab Sec. _____ Name _____ Desk No. _____

A. Preparation of Vinegar Sample

Calculate the approximate volume of the vinegar sample needed for the analyses.

Brand of Vinegar or Unknown No.   _____   _____

	Trial 1	Trial 2	Trial 1	Trial 2
1. Mass of flask (*g*)	_____	_____	_____	_____
2. Mass of flask + vinegar (*g*)	_____	_____	_____	_____
3. Mass of vinegar (*g*)				

B. Analysis of Vinegar Sample

	Trial 1	Trial 2	Trial 1	Trial 2
1. Buret reading of NaOH, *initial* (*mL*)	_____	_____	_____	_____
2. Buret reading of NaOH, *final* (*mL*)	_____	_____	_____	_____
3. Volume of NaOH used (*mL*)				
4. Molar concentration of NaOH (*mol/L*)	_____		_____	
5. Moles of NaOH added (*mol*)		*		
6. Moles of CH_3COOH in vinegar (*mol*)				
7. Mass of CH_3COOH in vinegar (*g*)				
8. Percent by mass of CH_3COOH in vinegar (*%*)				
9. Average percent by mass of CH_3COOH in vinegar (*%*)				

*Calculations for Trial 1 of the first vinegar sample on next page.

Calculations for Trial 1.

Discuss briefly a comparison of the two vinegars.

Laboratory Questions

Circle the questions that have been assigned.

1. Part A.2. *Previously boiled*, deionized water is unavailable. In a hurry to pursue the analysis, deionized water (not boiled) is added. How does this attempt to expedite the analysis affect the reported percent acetic acid in the vinegar? Explain.

2. Part A.2 a. In determining the percent acetic acid in vinegar, the mass of each vinegar sample is measured rather than the volume. Explain.
   b. If the vinegar were measured volumetrically (e.g., a pipet), what additional piece of data would be needed to complete the calculations for the experiment?

3. Part A.2. If *too much* phenolphthalein is added to the vinegar in the flask, will the reported percent acetic acid in vinegar be reported too high, too low, or unchanged? Explain.

4. Part A.3. The buret is filled with the NaOH titrant and the initial volume reading is immediately recorded without waiting the recommended 10–15 seconds. However in Part B.1, the 10–15 second time lapse does occur before the reading is made. How does this technique error affect the reported percent acetic acid in the vinegar? Explain.

5. Part B.1. The endpoint of the titration is overshot! How will this error affect the reported percent acetic acid in the vinegar? Explain.

6. Part B.1. The wall of the flask is periodically rinsed with the previously boiled, deionized water from the wash bottle. How does this titrimetric technique affect the reported percent acetic acid in the vinegar? Explain.

7. Part B.1. A drop of NaOH titrant, dispensed from the buret, adheres to the wall of the Erlenmeyer flask but is not washed into the vinegar with the wash bottle. Does this error in technique result in the reported percent of acetic acid being too high, too low, or unaffected? Explain.

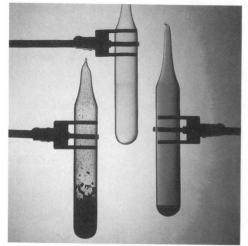

The halogens are, from left to right, solid iodine crystals, chlorine gas, and liquid bromine.

Periodic Table and Periodic Law

- To become more familiar with the periodic table
- To observe and to generalize the trends of various atomic properties within groups and periods of elements
- To observe from experiment the trends of the chemical properties within groups and periods of elements

OBJECTIVES

The following techniques are used in the Experimental Procedure

TECHNIQUES

INTRODUCTION

Similarities between the chemical and physical properties of elements were well known early in the nineteenth century. Several reports of grouping elements with like properties provided the background from which the modern periodic table finally evolved.

However, it was in 1869, nearly simultaneously, when *two* masterful organizations of all known elements were revealed, one by Dmitri Mendeleev from Russia and the other from Lothar Meyer from Germany. From their independent research, their arrangement of the elements established the modern periodic table. Mendeleev showed that with the elements arranged in order of increasing atomic mass, their *chemical* properties recur periodically. When Meyer arranged the elements in order of increasing atomic mass, he found that their *physical* properties recur periodically. The two tables, however, were virtually identical. Because he drafted his table earlier in 1869 and because his table included "blanks" for yet-to-be-discovered elements to fit, Mendeleev is considered the "father of the modern periodic table."

In 1913, H. G. J. Moseley's study of the X-ray spectra of the elements refined the periodic table to its current status: **when the elements are arranged in order of increasing atomic *number*, certain chemical and physical properties repeat periodically.** See the inside back cover for a modern version of the periodic table. Photos of most all of the elements can be seen on *www.periodictabletable.com*.

The periodic table continues to expand with the synthesis of new elements (the transactinides[1]), primarily at the Lawrence Livermore National Laboratory (California), the Joint Institute for Nuclear Research (Dubna, Russia), and the Institute for Heavy-Ion

Dmitri Mendeleev (1834–1907).

Atomic number: the number of protons in the nucleus

[1]en.wikipedia.org/wiki/Transactinide_element

Glenn T. Seaborg (1912–1999)

Research (Darmstadt, Germany). The most recent confirmation is the synthesis of element 118.[2] Glenn T. Seaborg was the most prominent of the U.S. chemists involved in the synthesis of the *transuranium* elements. He was the recipient of the 1951 Nobel Prize in chemistry and was honored with the naming of element 106, Seaborgium.

In the periodic table each horizontal row of elements is a **period** and each column is a **group** (or **family**). All elements within a group have similar chemical and physical properties. Common terms associated with various sections of the periodic table are

- Representative elements: Group "A" elements
- Transition elements: Groups 3–12
- Inner transition elements: the lanthanide (at. no. 58-71) and actinide (at. no. 90-103) series
- Metallic elements: elements to the left of the "stairstep line" that runs diagonally from B to At
- Nonmetallic elements: elements to the right of the "stairstep line"
- Metalloids: elements that lie adjacent to the "stairstep line," excluding Al
- Post-transition metals (or Poor metals): metals to the right of the transition metals
- Alkali metals: Group 1A elements
- Alkaline earth metals: Group 2A elements
- Chalcogens: Group 6A elements
- Halogens: Group 7A elements
- Noble gases: Group 8A elements
- Rare earth metals: the lanthanide series of elements
- Coinage metals: Cu, Ag, Au
- Noble metals: Ru, Os, Rh, Ir, Pd, Pt, Ag, Au, Hg

The periodicity of a physical property for a series of elements can be shown by plotting the experimental value of the property versus increasing atomic number. Physical properties studied in this experiment are

- Ionization energy (Figure 11.1): the energy required to remove an electron from a gaseous atom
- Atomic radius (Figure 11.2): the radius of an atom of the element
- Electron affinity (Figure 11.3): the *energy released* when a neutral gaseous atom accepts an electron
- Density (Figure 11.4): the mass of a substance per unit volume

For a very complete look at the properties and periodic trends of the elements, go to *www.webelements.com*.

Other physical properties that show trends in groups and periods of elements are listed for chlorine in the table.

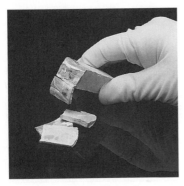

Sodium metal is a shiny, but very reactive metal, typical of Group 1A elements.

Trends in the *chemical* properties of the bold-face elements are studied in this experiment.

Properties of Chlorine

Atomic number	17
Molar mass	35.453 g/mol
Density at 293 K	3.214 g/L
Molar volume	22.7 cm³/mol
Melting point	172.22 K
Boiling point	239.2 K
Heat of fusion	3.203 kJ/mol
Heat of vaporization	10.20 kJ/mol
First ionization energy	1251.1 kJ/mol
Second ionization energy	2297.3 kJ/mol
Third ionization energy	3821.8 kJ/mol
Electronegativity	3.16
Electron affinity	349 kJ/mol
Specific heat	0.48 J/g·K
Heat of atomization	121 kJ/mol atoms
Atomic radius	100 pm
Ionic radius (−1 ion)	167 pm
Thermal conductivity	0.01 J/m·s·K

2	Li	Be	B	C	N	O	**F**
3	**Na**	**Mg**	**Al**	Si	P	**S**	**Cl**
4	K	**Ca**	Ga	Ge	As	Se	**Br**
5	Rb	**Sr**	In	Sn	Sb	Te	**I**

In this experiment, the relative acidic and/or basic strength of the hydroxides or oxides in the third period of the periodic table, the relative chemical reactivity of the halogens, and the relative solubility of the hydroxides and sulfates of magnesium, calcium, and strontium are observed through a series of qualitative tests. Observe closely the results of each test before generalizing your information.[3]

---

[2]*www.cen-online.org*, link to October 23, 2006 issue of C&EN.
[3]For trends in chemical properties, go to *http://en.wikipedia.org/wiki/category:chemistry*

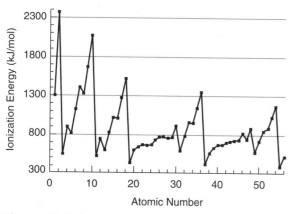

Figure 11.1 Ionization energies (kJ/mol) plotted against atomic number.

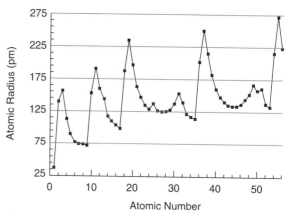

Figure 11.2 Atomic radii (pm) plotted against atomic number.

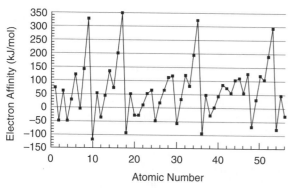

Figure 11.3 Electron affinities (kJ/mol) plotted against atomic number.

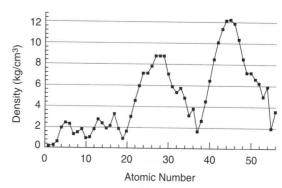

Figure 11.4 Density (kg/m³) plotted against atomic number.

Procedure Overview: General trends in the physical properties of the elements are observed and studied in Figures 11.1–11.4. Experimental observations of the physical and chemical properties of a number of representative elements are made. Special attention is paid to the chemical properties of the halogens.

Ask your instructor about a working relationship, individuals or partners, for Parts A and B. For Parts C, D, and E perform the experiment with a partner. At each circled superscript⁽¹⁻¹⁸⁾ in the procedure, *stop,* and record your observation on the Report Sheet. Discuss your observations with your lab partner and your instructor.

EXPERIMENTAL PROCEDURE

Figures 11.1 through 11.4 plot the experimental data of four physical properties of the elements as a function of their atomic number. While actual values cannot be readily obtained from the graphical data, the periodic trends are easily seen. All values requested for the analyses of the graphical data need only be given as your "best possible" estimates from the figure.

The periodic trends for the elements are analyzed through a series of questions on the Report Sheet.

A. Periodic Trends in Physical Properties (Dry Lab)

(Optional) Figure 11.5 is a graphical plot of the ionization energies for the thirteen electrons of aluminum. An interpretation of the experimental data is requested on the Report Sheet.

B. Ionization Energies for Aluminum (Dry Lab)

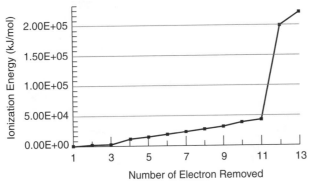

Figure 11.5 The 13 ionization energies for aluminum.

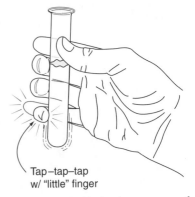

Tap–tap–tap w/ "little" finger

Figure 11.6 Shake the contents of the test tube with the "little" finger.

C. The Appearance of Some Representative Elements

1. **Samples of Elements.** Samples of the third period elements sodium, magnesium, aluminum, silicon, and sulfur are on the reagent table. Note that the Na metal is stored under a nonaqueous liquid to prevent rapid air oxidation. Polish the Mg and Al metal strips with steel wool for better viewing. Record your observations on the Report Sheet.

   Since some chlorine, bromine, and iodine vapors may escape the test tubes in Parts C.2–4 and D.1–3, *you may want to conduct the experiments in the fume hood.* Consult with your laboratory instructor.

2. **Chlorine.** In a clean, 150-mm test tube, place 2 mL of a 5% sodium hypochlorite, NaClO, solution (commercial laundry bleach) and 10 drops of cyclohexane. Agitate the mixture (Figure 11.6). Which layer is the cyclohexane layer[4]?

   Add 10 drops of 6 *M* HCl. (**Caution:** 6 *M* HCl *is very corrosive. Wash immediately from the skin.*) Swirl/agitate the mixture (with a stirring rod) so that the HCl mixes with the NaClO solution. Note the color of the chlorine in the cyclohexane layer. Record your observation.[①] Do not discard—save for Part D.1.

3. **Bromine.** In a second, clean test tube, place 2 mL of 3 *M* KBr solution and 3 drops of cyclohexane. Add 5–10 drops of 8 *M* HNO_3. (**Caution:** HNO_3 *attacks skin tissue; flush the affected area immediately with water*.) Agitate/swirl the mixture so that the 8 *M* HNO_3 mixes with the KBr solution. Note the color of the bromine in the cyclohexane layer. Record,[②] but do not discard—save for Part D.2.

4. **Iodine.** Repeat Part C.3 in a third test tube, substituting 3 *M* KI for 3 *M* KBr. Record. Compare the appearance of the three halogens dissolved in the cyclohexane.[③] Save for Part D.3.

D. The Chemical Properties of the Halogens

Pinch: a solid mass about the size of a grain of rice

For Parts D.1–3, six clean, small (~75-mm), test tubes (or 24-well plate) are required. Summarize your observations at the conclusion of Part D.3. Number each test tube (Figure 11.7).

1. **Chlorine and its Reactions with Bromide and Iodide Ions.** Clean two small test tubes; add a **pinch** (on the end of a spatula) of solid KBr to the *first* test tube and a pinch of KI to the *second*. Use a dropping pipet to withdraw the chlorine/cyclohexane layer from Part C.2 and add an (approximately) equal portion to the two test tubes. Swirl/agitate the solution, observe, and record. Write appropriate net ionic equations.[④]

[4]Mineral oil, or any colorless cooking oil, may be substituted for cyclohexane.

2. **Bromine and its Reactions with Chloride and Iodide Ions.** Add a pinch of solid NaCl to a *third,* small clean test tube and a pinch of KI to the *fourth* test tube. Use a dropping pipet to withdraw the bromine/cyclohexane layer from Part C.3 and add an (approximately) equal portion to the two test tubes. Swirl/agitate the solution, observe and record. Write appropriate net ionic equations.[5]

3. **Iodine and its Reactions with Chloride and Bromide Ions.** Add a pinch of solid NaCl to a *fifth,* small clean test tube and a pinch of KBr to the *sixth* test tube. Use a dropping pipet to withdraw the iodine/cyclohexane layer from Part C.3 and add an (approximately) equal portion to the two test tubes. Swirl/agitate the solution, observe and record. Write appropriate net ionic equations.[6]

What can you conclude about the relative chemical reactivity of the halogens?

Figure 11.7 Six labeled test tubes to test the relative reactivity of the halogens.

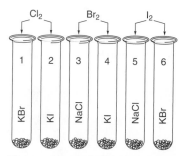

> *Disposal:* Dispose of the waste water/halogen mixtures in the "Waste Halogens" container.

E. The Chemical Properties of the Halides

Appendix G

Twelve clean, small (~75-mm) test tubes (or a 24-well plate) are required for the chemical reactions observed in Part E. Number each test tube (Figure 11.8).

1. **The Reactions of the Halides with Various Metal Ions.** Label 12 clean, small test tubes and transfer the following to each:

   • Test tubes 1, 2, and 3: a pinch of NaF and 10 drops of water
   • Test tubes 4, 5, and 6: a pinch of NaCl and 10 drops of water
   • Test tubes 7, 8, and 9: a pinch of KBr and 10 drops of water
   • Test tubes 10, 11, and 12: a pinch of KI and 10 drops of water

   a. Slowly add 10 drops of 2 M Ca(NO$_3$)$_2$ to test tubes 1, 4, 7, and 10. Observe closely and over a period of time. Vary the color of the background of the test tubes for observation.[7]

   b. Slowly add 10 drops of 0.1 M AgNO$_3$ to test tubes 2, 5, 8, and 11. Observe. After about 1 minute, add 10 drops of 3 M NH$_3$.[8]

   c. Add 1 drop of 6 M HNO$_3$ (**Caution!**) and slowly add 10 drops of 0.1 M Fe(NO$_3$)$_3$ to test tubes 3, 6, 9, and 12. Observe closely and over a period of time.[9]

   d. Summarize your observations of the chemical activity for the halides with the Ca^{2+}, Ag$^+$, and Fe^{3+} ions.

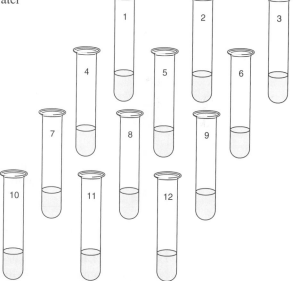

Figure 11.8 Twelve labeled test tubes to test the reactivity of the halides.

> *Disposal:* Dispose of the waste water/halogen mixtures in the "Waste Halogens" container.

CLEANUP: Rinse the test tubes with copious amounts of tap water and twice with deionized water. Discard the rinses in the sink.

F. Chemical Reactivity of Some Representative Elements

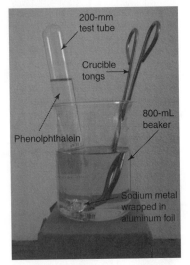

Figure 11.9
Collection of hydrogen gas from the reaction of sodium and water.

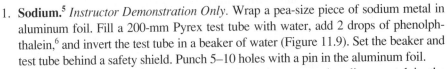

1. **Sodium.**[5] *Instructor Demonstration Only.* Wrap a pea-size piece of sodium metal in aluminum foil. Fill a 200-mm Pyrex test tube with water, add 2 drops of phenolphthalein,[6] and invert the test tube in a beaker of water (Figure 11.9). Set the beaker and test tube behind a safety shield. Punch 5–10 holes with a pin in the aluminum foil.

    With a pair of tongs or tweezers, place the wrapped sodium metal in the mouth of the test tube, keeping it under water. What is the evolved gas? Test the gas by holding the mouth of the inverted test tube over a Bunsen flame.⑩ A loud "pop" indicates the presence of hydrogen gas. Account for the appearance of the color change in the solution.⑪

2. **Magnesium and Aluminum.** a. *Reaction with acid.* Polish strips of Mg and Al metal; cut 5-mm pieces and place them into separate small test tubes. Add 10 drops of 3 M HCl to each test tube. (**Caution:** *Do not allow the* HCl *to touch the skin. Wash the affected area immediately.*) Which metal reacts more rapidly?⑫ What is the gas that is evolved?⑬

    b. *Reaction with base.* Add (and count) drops of 6 M NaOH to each test tube until a precipitate appears. Continue to add NaOH to the test tube containing the aluminum ion until a change in appearance occurs. Add the same number of drops to the test tube containing the magnesium ion. Record your observations.[7] ⑭

    Add drops of 6 M HCl until both solutions are again colorless. Observe closely as each drop is added. Record and explain.

3. **Solubilities of Alkaline-Earth Cations.** a. *Solubility of the hydroxides.* Place 10 drops of 0.1 M MgCl$_2$, 0.1 M CaCl$_2$, and 0.1 M Sr(NO$_3$)$_2$ in three, separate, clean test tubes. Count and add drops of 0.050 M NaOH until a cloudiness appears in each test tube. Predict the trend in the solubility of the hydroxides of the Group 2A cations.⑮

    b. *Solubility of the sulfates.* Place 10 drops of 0.1 M MgCl$_2$, 0.1 M CaCl$_2$, and 0.1 M Sr(NO$_3$)$_2$ in three, separate, clean test tubes. Count and add drops of 0.10 M Na$_2$SO$_4$ until a cloudiness appears in each test tube. Predict the trend in the solubility of the sulfates of the Group 2A cations.⑯

4. **Sulfurous Acid and Sulfuric Acid.** Because of the possible evolution of a foul-smelling gas, you may want to conduct this part of the experiment in the fume hood. Consult with your instructor.

    a. Place a "double" pinch of solid sodium sulfite, Na$_2$SO$_3$, into a clean, small or medium-sized test tube. Add 5–10 drops of 6 M HCl. Test the evolved gas with wet, blue litmus paper. Write a balanced equation for the reaction.⑰

    b. Repeat the test, substituting solid sodium sulfate, Na$_2$SO$_4$, for the Na$_2$SO$_3$. Account for any differences or similarities in your observations.⑱

> *Disposal:* Discard the solutions as directed by your instructor.

CLEANUP: Rinse the test tubes twice with tap water and with deionized water. Discard the rinses in the sink.

The Next Step

The periodic trends for many chemical and physical properties of the elements can be found on the Internet. There are many web sites, and the data are plotted for many properties using various coordinates. A first reference is *www.acs.org*. Seek a data plot, for example, of atomic radii vs. ionization energy or for any properties, as listed in the table for chlorine in the **Introduction.**

---

[5]Calcium metal can be substituted for sodium metal with the same results.
[6]Phenolphthalein is an acid-base indicator; it is colorless in an acidic solution but pink in a basic solution.
[7]Magnesium ion precipitates as magnesium hydroxide, Mg(OH)$_2$; aluminum ion also precipitates as the hydroxide, Al(OH)$_3$, but redissolves in an excess of OH$^-$ to produce Al(OH)$_4{}^-$, the aluminate ion.

Experiment 11 *Prelaboratory Assignment*

Periodic Table and Periodic Law

Date __________ Lab Sec. ______ Name ____ Desk No. __________

1. On the blank periodic table, *clearly* identify six of the following element sections, using your own color code.
   a. Representative elements
   b. Transition elements
   c. Inner transition elements
   d. Chalcogens
   e. Coinage metals
   f. Metalloids
   g. Post-transition metals
   h. Alkali metals
   i. Alkaline earth metals
   j. Halogens
   k. Noble gases
   l. Noble metals
   m. Lanthanide series
   n. Actinide series

Periodic Table

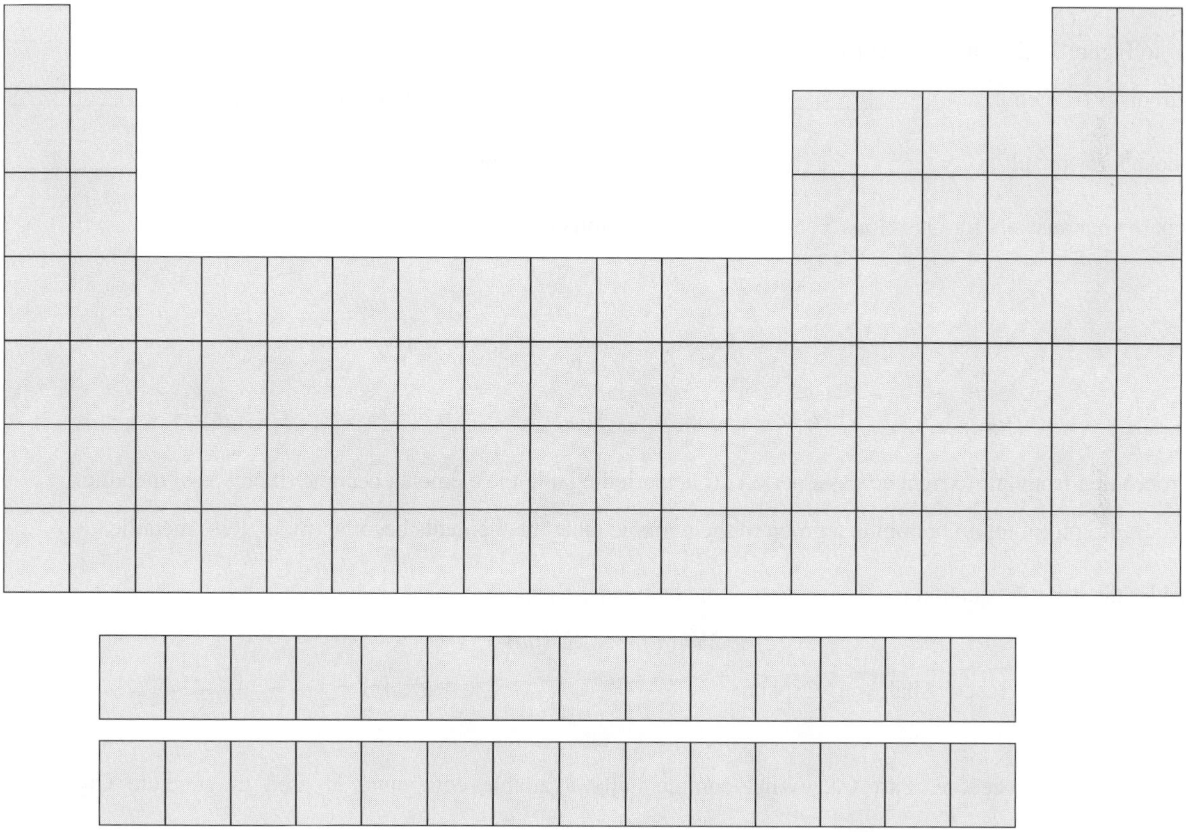

2. Sketch in the "stairstep" line that separates the metals from the nonmetals on the periodic table.

3. Identify the atomic numbers of the elements that would be called the "transactinide" elements.

4. Classify each of the following elements according to the categories of elements identified in Question 1:

a. Lithium _____   f. Calcium _____

b. Plutonium _____   g. Element 118 _____

c. Argon _____   h. Copper _____

d. Zirconium _____   i. Iodine _____

e. Indium _____   j. Cerium _____

5. Refer to Figure 11.1. Which of the following has the highest ionization energy?

a. carbon or oxygen _____   c. magnesium or aluminum _____

b. phosphorus or sulfur _____   d. magnesium or calcium _____

6. Refer to Figure 11.2. Which of the following has the largest atomic radius?

a. carbon or oxygen _____   d. magnesium or aluminum _____

b. phosphorus or sulfur _____   e. magnesium or calcium _____

Compare your answers for Questions 5 and 6. What correlation can be made?

7. a. Proceeding from left to right across a period of the periodic table the elements become (more, less) metallic.

b. Proceeding from top to bottom in a group of the periodic table the elements become (more, less) metallic.

8. Consider the generic equation for the reaction of the halogens, X_2 and Y_2:

$$X_2(g) + 2Y^-(aq) \rightarrow 2X^-(aq) + Y_2(g)$$

Is X_2 or Y_2 the more reactive halogen? Explain.

9. a. Experimental Procedure, Part C.2. What commercially available compound is used to generate Cl_2 in the experiment?

b. Experimental Procedure, Part C. The observation for the presence of the elemental form of the halogens is in a solvent other than water. Identify the solvent.

c. Experimental Procedure, Part F.3. Identify the tests used to observe the periodic trends in the chemical properties of the alkaline–earth metal ions.

Periodic Table and Periodic Law

Date _____ Lab Sec. _____ Name _____ Desk No. _____

A. Periodic Trends in Physical Properties (Dry Lab)

Consult with your laboratory instructor as to the procedure and schedule for submitting your responses to the following questions about periodic trends.

1. Figure 11.1: graphical data for the ionization energies of the elements show sawtooth trends across the periods of the elements.
   a. Locate the noble gas group of elements. What appears to be the periodic trend in ionization energies "down" the noble gas group (i.e., with increasing atomic number)?
   b. Scanning the graphical data for elements adjacent to and then further removed from the halogens, what *general statement* can summarize the trend in the ionization energies when moving down a group of elements?
   c. Which element has the highest ionization energy?
   d. What *general statement* can summarize the trend in the ionization energies when moving "across" a period of elements?

2. Figure 11.2: graphical data for the atomic radii of the elements show generally decreasing trends across a period of elements. The noble gases are an anomaly.
   a. Which *group* of elements has the largest atomic radii?
   b. Moving down a group of elements (increasing atomic number), what is the general trend for atomic radii?
   c. Which element has the largest atomic radius?
   d. What *general statement* can summarize the correlation of ionization energies to atomic radii for the elements?

3. Figure 11.3: graphical data for the electron affinities of the elements show a number of irregularities, but a general increasing trend in values exists across a period of elements.
   a. Which *group* of elements has the highest electron affinities?
   b. Is the trend in electron affinities repetitive for Periods 2 and 3? Cite examples.
   c. Which element has the highest electron affinity?
   d. Is there a correlation of electron affinities to atomic radii for the elements? If so, what is it? Cite examples.

4. Figure 11.4 shows repeated trends in density for the periods of elements.
   a. What is the general trend in densities for Periods 2 and 3?
   b. What is the trend in the densities moving down a group of elements?
   c. Which section of the periodic table (see Introduction) has elements with greater densities?
   d. Which element has the greatest density?

B. Ionization Energies for Aluminum (Dry Lab)

Consult with your laboratory instructor as to the procedure and schedule for submitting your responses to the following question.

1. (Optional) See your laboratory instructor. Figure 11.5 graphs the experimental values for the 13 ionization energies of aluminum.
   a. On the basis of the electron configuration of aluminum, explain why there is a noticeable increase in the fourth ionization energy relative to the third for aluminum.
   b. On the basis of the electron configuration of aluminum, explain why there is an even more significant increase in the twelfth ionization energy for aluminum.
   c. Where might you see similar increases in ionization energies for magnesium? For silicon?

C. The Appearance of Some Representative Elements

Element	Physical State (*g, l, s*)	Physical Appearance and Other Observations	Color
Na	_____	_____	_____
Mg	_____	_____	_____
Al	_____	_____	_____
Si	_____	_____	_____
S_8	_____	_____	_____
①Cl_2	_____	_____	_____
②Br_2	_____	_____	_____
③I_2	_____	_____	_____

D. The Chemical Properties of the Halogens

Observations in the cyclohexane layer

	$Cl_2$④	$Br_2$⑤	$I_2$⑥	Net Ionic Equation(s)
KCl	XX*	_____	_____	_____
KBr	_____	XX	_____	_____
KI	_____	_____	XX	_____

*No reaction mixture.

What can you conclude about the relative reactivity of Cl_2, Br_2, and I_2?

E. The Chemical Properties of the Halides

1. **The Reactions of the Halides with Various Metal Ions**
   Describe the appearance of each mixture in tubes 1–12.

	⑦$Ca(NO_3)_2$(*aq*)	⑧$AgNO_3$(*aq*)	⑨$Fe(NO_3)_3$(*aq*)
(Test tube no.) (1)		(2)	(3)
NaF(*aq*)	_____	_____	_____
(Test tube no.) (4)		(5)	(6)
NaCl(*aq*)	_____	_____	_____
(Test tube no.) (7)		(8)	(9)
NaBr(*aq*)	_____	_____	_____
(Test tube no.) (10)		(11)	(12)
NaI(*aq*)	_____	_____	_____

From the observed data, answer the following questions.

    a. Which of the metal fluorides are insoluble? _____

    b. Which of the metal chlorides are insoluble? _____

    c. Which of the metal bromides are insoluble? _____

    d. Which of the metal iodides are insoluble? _____

F. Chemical Reactivity of Some Representative Elements

1. Sodium (or Calcium)

    a. [10] Name the gas evolved in the reaction of Na with water: _____

    b. [11] What is produced from the reaction of Na with water as indicated by the action of phenolphthalein?

    c. Write the chemical equation for the reaction of Na with water:

2. Magnesium and Aluminum

    a. [12] Which metal reacts more rapidly with HCl? _____

       [13] What is the gas that is evolved when Mg and Al react with HCl? _____

    b. Reaction of metal ion with NaOH

[14] Observations	6 M NaOH	Excess 6 M NaOH
Mg^{2+} (drops to precipitate)	_____	_____
Al^{3+} (drops to precipitate)	_____	_____

    c. Explain the differences in chemical behavior of the magnesium and aluminum hydroxides. Use chemical equations in your discussion.

3. Solubilities of Alkaline-Earth Cations

Observations	[15] 0.050 M NaOH	[16] 0.10 M Na$_2$SO$_4$
Mg^{2+} (drops to precipitate)	_____	_____
Ca^{2+} (drops to precipitate)	_____	_____
Sr^{2+} (drops to precipitate)	_____	_____

List the hydroxide salts in order of increasing solubility: _____ < _____ < _____

List the sulfate salts in order of increasing solubility: _____ < _____ < _____

What can you conclude about the general trend in solubilities of the Group 2A metal hydroxides and sulfates?

4. Sulfurous Acid and Sulfuric Acid

a. ⑰What does the litmus test indicate about the chemical stability of sulfurous acid?

b. What is the gas that is generated? _____

c. Write an equation for the decomposition of sulfurous acid.

d. ⑱What does the litmus test indicate about the chemical stability of sulfuric acid? _____

e. What chemical property differentiates sulfurous acid from sulfuric acid?

_____

Laboratory Questions

Circle the questions that have been assigned.

1. Part A. a. Which group of elements has the highest ionization energies?
   b. Which group of elements has the lowest ionization energies?

2. Part A. a. Which group of elements has the largest atomic radii?
   b. Which group of elements has the smallest atomic radii?

3. Part A. a. Which group of elements has the highest electron affinity?
   b. Which group of elements has the lowest electron affinity?

4. Part D. Chlorine is used extensively as a disinfectant and bleaching agent. Without regard to adverse effects or costs, would bromine be a more or less effective disinfectant and bleaching agent? Explain.

5. Part D. Is fluorine gas predicted to be more or less reactive than chlorine gas? Explain.

6. Part D. Not much is known of the chemistry of astatine. Would you expect astatine to more or less reactive than iodine? Explain.

7. Part F.2. Predict the reactivity of silicon metal relative to that of magnesium and aluminum. Explain.

8. Part F.3. a. Is barium hydroxide predicted to be more or less soluble that $Ca(OH)_2$? Explain.
   b. Is barium sulfate predicted to be more or less soluble that $CaSO_4$? Explain.

Atomic and Molecular Structure

Metallic cations heated to high temperatures produce characteristic colors that appear in the starbursts.

- To view and calibrate visible line spectra
- To identify an element from its visible line spectrum
- To identify a compound from its infrared spectrum
- To predict the three-dimensional structure of molecules and molecular ions

INTRODUCTION

Visible light, as we know it, is responsible for the colors of nature—the blue skies, the green trees, the red roses, the orange-red rocks, and the brown deer. Our eyes are able to sense and distinguish the subtleties and the intensities of those colors through a complex naturally designed detection system, our eyes. The "beauty" of nature is a result of the interaction of sunlight with matter. Since all matter consists of atoms and molecules, then it is obvious that sunlight, in some way, interacts with them to produce nature's colors.

Internally within molecules of compounds there are electrons, vibrating bonds, and rotating atoms, all of which can absorb energy. Since every compound is different, every molecule of a given compound possesses its own "unique" set of electronic, vibrational, and rotational **energy states,** which are said to be **quantized.** When incident electromagnetic (EM) radiation falls on a *molecule,* the radiation absorbed (the absorbed light) is an energy equal to the difference between two energy states, placing the molecule in an "excited state." The remainder of the EM radiation passes through the molecule unaffected.

Atoms of elements interact with EM radiation in much the same way, except there are no bonds to vibrate or atoms to rotate. Only electronic energy states are available for energy absorption.

EM radiation is energy as well as light, and light has wavelengths and frequencies. The relationship between the energy, *E,* and its **wavelength,** λ, and **frequency,** ν, is expressed by the equation:

$$E = \frac{hc}{\lambda} = h\nu \qquad (D3.1)$$

where h is Planck's constant, 6.63×10^{-34} J•s/photon; c equals the velocity of the EM radiation, 3.00×10^8 m/s; λ is the wavelength of the EM radiation in meters; and ν (pronounced "new") is its frequency in reciprocal seconds (s^{-1}).

EM radiation includes not only the wavelengths of the visible region (400 to 700 nm), but also those that are shorter (e.g., the ultraviolet and X-ray regions) and longer (e.g., the infrared, microwave, and radio wave regions). See Figure D3.1. From Equation D3.1, shorter wavelength EM radiation has higher energy.

When an atom or molecule absorbs EM radiation from the visible region of the spectrum, it is usually an electron that is excited from a lower to a higher energy state.

Energy state: the amount of energy confined within an atom or molecule, which can be changed by the absorption or emission of discrete (quantized) amounts of energy

Quantized: only a definitive amount (of energy)

Wavelength: the distance between two crests of a wave

Frequency: the number of crests that pass a given point per second

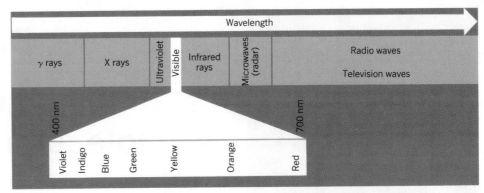

Figure D3.1 Visible light (400–700 nm) is a small part of the electromagnetic spectrum.

White light: EM radiation containing all wavelengths of visible light

Spectrophotometer: an instrument used to detect and monitor the interaction of electromagnetic radiation with matter. The instrument has an EM radiation source, a grating to sort wavelengths, a sample cell, and an EM radiation detector.

When **white light** passes through a sample, our eyes (and the EM detector of a **spectrophotometer**) detect the wavelengths of visible light *not* absorbed, i.e., the transmitted light. Therefore, the colors we see are *complementary* to the ones absorbed. If, for example, the atom or molecule absorbs energy exclusively from the violet region of the visible spectrum, the transmitted light (and the substance) appears yellow (violet's complementary color) (Figure D3.2) In reality, atoms and molecules of a substance absorb a range of wavelengths, some wavelengths more so than others, resulting in a mix of transmitted colors leading to various "shades" of colors.

Table D3.1 lists the colors corresponding to wavelength regions of light (and their complements) in the visible region of the EM spectrum.

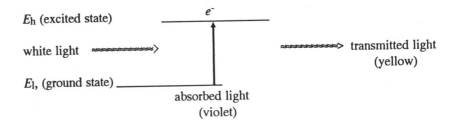

Figure D3.2 White light that is *not* absorbed is the transmitted light that we detect with our eyes.

Atomic Structure

Photon: a particlelike quantity of electromagnetic radiation, often associated with electron transitions

When an atom of an element absorbs EM radiation, it is the electrons that absorb energy to reach excited states. When the electrons return to the lowest energy states **(the ground state of the atom)** by various pathways, the same amount of energy absorbed is now emitted as **photons.**[1] The photons have unique energies and wavelengths that represent the difference in the energy states of the atom.

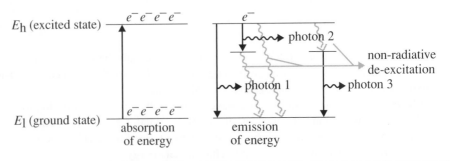

---

[1]Many electrons de-exite to the ground state by only non-radiative pathways, others through a combination of photons *and* nonradiative pathways.

Table D3.1 Color and Wavelengths in the Visible Region of the Electromagnetic Spectrum

Color Absorbed	Wavelength (nm)	Color Transmitted
red	750–610	green-blue
orange	610–595	blue-green
yellow	595–580	violet
green	580–500	red-violet
blue	500–435	orange-yellow
violet	435–380	yellow

Since electrons can have a large number of excited states, a large collection of excited state electrons returning to the ground state produces an array of photons. When these emitted photons pass through a prism, an emission **line spectrum** is produced; each line in the spectrum corresponds to photons of fixed energy and wavelength. The line spectrum for hydrogen is shown in Figure D3.3.

Each element exhibits its own characteristic line spectrum because of the unique electronic energy states in its atoms. For example, the 11 electrons in sodium have a different set of electron energy states than do the 80 electrons in mercury. Therefore when an electron in an excited state of a sodium atom moves to a lower energy state, the emitted photon has a different energy and wavelength from one that is emitted when an electron de-excites (i.e., moves to a lower energy state) in a mercury atom.

The different wavelengths of the emitted photons produce different, yet characteristic, colors of light. Light emitted from an excited sodium atom is characteristically yellow-orange but mercury emits a blue light. Flame tests (Experiment 38) and exploding, aerial fireworks attest to the uniqueness of the electronic energy states of the atoms for different elements.

Much of the modern theory of atomic structure, which we call quantum theory or quantum mechanics, is based on the emission spectra of the elements.

Molecular Structure

The structures of molecules are much more complex than of atoms. Molecules may contain thousands of atoms, each in a unique (energy and three-dimensional) environment. Since each molecule is unique, so must be its energy levels for EM absorption.

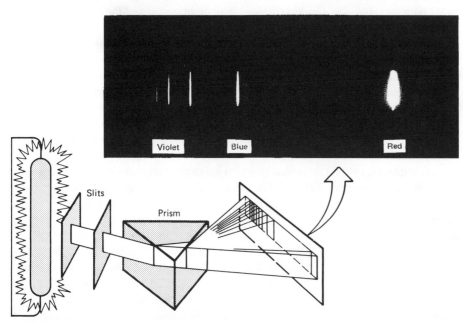

Violet Blue Red

Slits

Prism

Figure D3.3 Formation of the line spectrum for hydrogen.

Table D3.2 Infrared Absorption Bands for Specific Atoms-in-Bond Arrangements in Molecules

Atoms-in-Bonds	Wavenumbers	Wavelengths
O—H	3700 to 3500 cm^{-1}	2.7 to 2.9 μm
C—H	3000 to 2800 cm^{-1}	3.3 to 3.6 μm
C=O	1800 to 1600 cm^{-1}	5.6 to 6.2 μm
C—O	1200 to 1050 cm^{-1}	8.3 to 9.5 μm
C—C	1670 to 1640 cm^{-1}	6.0 to 6.1 μm

Using EM radiation to elucidate the three-dimensional structure of a molecule can be painstakingly tedious, especially for the structures of the "living" (biochemical) molecules, such as DNA and hemoglobin. Oftentimes, EM radiation is only one of many tools used to determine the molecular structure of a compound.

Infrared EM radiation is used as a "probe" for the identification of specific atom arrangements in molecules. For example, the O—H bond, as in water and alcohols, absorbs infrared radiation for a principal vibrational energy transition between 3700 and 3500 **cm^{-1}** or wavelengths of 2.7 to 2.9 μm. Other characteristic infrared absorption bands for specific atom arrangements in molecules are listed in Table D3.2.

A first view of an infrared spectrum of a molecule seems confusing, as more absorption bands appear than what might be anticipated from looking at the structure of the molecule and Table D3.2. Other absorptions occur as a result of multi-atom interactions (stretches, bends, etc.) or orientations, but the primary bands appear as expected. Therefore, in an analysis of an infrared spectrum a search of the primary bands in Table D3.2 is first and foremost.

For other molecules and molecular ions, other regions of EM radiation are most effective. For example, the FeNCS^{2+} ion has a major absorption of radiation at 447 nm, and this property is used to measure its concentration in an aqueous solution in Experiments 34 and 35—the higher the concentration, the more EM radiation that is absorbed.

*cm^{-1}: Infrared spectroscopists often indicate absorption bands in units of reciprocal centimeters, called **wavenumbers**, rather than as wavelengths.*

Lewis Theory

Valence electrons: electrons in the highest energy state (outermost shell) of an atom in the ground state

Isoelectronic: two atoms are isoelectronic if they have the same number of electrons.

Gilbert Newton Lewis (1875–1946).

In 1916, G. N. Lewis developed a theory that focused on the significance of **valence electrons** in chemical reactions and in bonding. He proposed the "octet rule" in which atoms form bonds by losing, gaining, or sharing valence electrons until each atom of the molecule has the same number of valence electrons (eight) as the nearest noble gas in the periodic table. The resulting arrangement of atoms formed the Lewis structure of the compound.

The Lewis structure for water shows that by sharing the one valence electron on each of the hydrogen atoms with the six valence electrons on the oxygen atom, all three atoms obtain the same number of valence electrons as the nearest noble gas. Thus, in water the hydrogen atoms are **isoelectronic** with helium atoms and the oxygen atom is isoelectronic with the neon atom.

The Lewis structures for ions are written similarly, except that electrons are removed from (for cations) or added to (for anions) the structure to account for the ion's charge.

An extension of the Lewis structure also exists for molecules or molecular ions in which the central atom is of Period 3 or greater. While the "peripheral" atoms retain the noble gas configuration, the central atom often "extends" its valence electrons to accommodate additional electron pairs. As a consequence the central atom accommodates more than eight valence electrons, an extension of the octet rule. For example, the six valence electrons of sulfur bond to four fluorine atoms in forming SF_4. To do so, four of the six valence electrons on sulfur share with the four fluorine atoms and two remain "nonbonding"—now, ten valence electrons exist for the bonded sulfur atom.

Although a Lewis structure accounts for the bonding based on the valence electrons on each atom, it does not predict the three-dimensional structure for a molecule. The development of the **v**alence **s**hell **e**lectron **p**air **r**epulsion (VSEPR) theory provides insight into the three-dimensional structure of the molecule.

VSEPR theory proposes that the structure of a molecule is determined by the repulsive interaction of electron pairs in the valence shell of its central atom. The three-dimensional orientation is such that the distance between the electron pairs is maximized such that the electron pair–electron pair interactions are minimized. A construction of the Lewis structure of a molecule provides the first link in predicting the molecular structure.

Methane, CH_4, has four bonding electron pairs in the valence shell of its carbon atom (the central atom in the molecule). Repulsive interactions between these four electron pairs are minimized when the electron pairs are positioned at the vertices of a tetrahedron with H—C—H bond angles of 109.5°. On the basis of the VSEPR theory, one can generalize that *all* molecules (or molecular ions) having four electron pairs in the valence shell of its central atom have a tetrahedral arrangement of these electron pairs with *approximate* bond angles of 109.5°. The nitrogen atom in ammonia, NH_3, and the oxygen atom in water, H_2O, also have four electron pairs in their valence shell!

The preferred arrangement of the bonding and nonbonding electron pairs around the central atom gives rise to the corresponding structure of a molecule. The three-dimensional structures for numerous molecules and molecular ions can be grouped into a few basic structures. Based upon a correct Lewis structure, a VSEPR formula summarizes the number and type (bonding and nonbonding) of electron pairs in the compound or ion. The VSEPR formula uses the following notations:

A refers to the central atom

X_m refers to m number of *bonding* pairs of electrons on A

E_n refers to n number of *nonbonding* pairs of electrons on A

If a molecule has the formula AX_mE_n, it means there are $m + n$ electron pairs in the valence shell of A, the central atom of the molecule; m are bonding and n are nonbonding electron pairs. For example, CH_4, SiF_4, $GeCl_4$, PH_4^+, and PO_4^{3-} all have a VSEPR formula of AX_4. Thus, they all have the same three-dimensional structure, that of a tetrahedral structure.

It should be noted here that valence electrons on the central atom contributing to a multiple bond do *not* affect the geometry of a molecule. For example in SO_2, the VSEPR formula is AX_2E, and the geometric shape of the molecule is "V-shaped." See Table D3.3. Further applications are presented in more advanced chemistry courses.[2]

Valence Shell Electron Pair Repulsion (VSEPR) Theory of the Structures of Molecules and Molecular Ions

Table D3.3 VSEPR and Geometric Shapes of Molecules and Molecular Ions

Valence Shell Electron Pairs	Bonding Electron Pairs	Nonbonding Electron Pairs	VSEPR Formula	Three-Dimensional Structure	Bond Angle	Geometric Shape	Examples
2	2	0	AX_2	:—A—:	180°	Linear	$HgCl_2$, $BeCl_2$
3	3	0	AX_3		120°	Planar triangular	BF_3, $In(CH_3)_3$
	2	1	AX_2E		<120°	V-shaped	$SnCl_2$, $PbBr_2$
4	4	0	AX_4		109.5°	Tetrahedral	CH_4, $SnCl_4$
	3	1	AX_3E		<109.5°	Trigonal pyramidal	NH_3, PCl_3, H_3O^+
	2	2	AX_2E_2		<109.5°	Bent	H_2O, OF_2, SCl_2
5	5	0	AX_5		90°/120°	Trigonal bipyramidal	PCl_5, $NbCl_5$
	4	1	AX_4E		>90°	Irregular tetrahedral	SF_4, $TeCl_4$
	3	2	AX_3E_2		<90°	T-shaped	ICl_3
	2	3	AX_2E_3		180°	Linear	ICl_2^-, XeF_2
6	6	0	AX_6		90°	Octahedral	SF_6
	5	1	AX_5E		>90°	Square pyramidal	BrF_5
	4	2	AX_4E_2		90°	Square planar	ICl_4^-, XeF_4

[2]For more information on VSEPR theory and structure, go to *http://winter.group.shef.ac.uk/vsepr*.

Table D3.3 presents a summary of the VSEPR theory for predicting the geometric shape and approximate bond angles of a molecule or molecular ion based on the five basic VSEPR three-dimensional structures of molecules and molecular ions.

Let us refer back to SF_4, the molecule with the "extended" valence shell on the central atom. The sulfur atom has four bonding electron pairs ($m = 4$) and one non-bonding pair ($n = 1$). This gives a VSEPR formula of AX_4E_1, predicting a geometric shape of "irregular tetrahedral" (sometimes also called "seesaw") with bond angles greater than 90°. Molecular models will enable you to envision these properties of SF_4.

DRY LAB PROCEDURE

In Part A, the visible spectra of a number of elements are studied (see color plate on back cover of this manual). The wavelengths of the spectra are to be calibrated relative to the mercury spectrum at the bottom of color plate. The most intense lines of the mercury spectrum are listed in Table D3.4.

In Part B, a spectrum from the color plate will be assigned, and, with reference to Table D3.5, the element producing the line spectrum will be identified.

In Part C, a molecule will be assigned and, with reference to Table D3.2, a spectra will be matched to the molecule.

In Part D, a number of simple molecules and molecular ions will be assigned, and their three-dimensional structure and approximate bond angles will be determined. The Lewis structure and the VSEPR adaptation of the Lewis structure are used for analysis.

Discuss with your instructor which parts of the **Dry Lab Procedure** are to be completed and by when. If there is no advanced preparation and you are to complete all parts, you may be rushed for time.

A. The Mercury Spectrum

1. **The Color Plate.** Notice the various experimental emission line spectra on the color plate (back cover). A continuous spectrum appears at the top, the line spectra for various elements appear in the middle, and the Hg spectrum appears at the bottom.

2. **Calibrate the Spectra of the Color Plate.** Use a ruler to mark off a linear wavelength scale across the bottom of the color plate such that the experimental wavelengths of the mercury spectrum correlate with those in Table D3.4.

   Extend the linear wavelength scale perpendicularly and upward across spectra on the color plate, thus creating a wavelength grid for calibrating all of the emission spectra. A wax marker or "permanent" felt tip pen may be required for marking the wavelength grid.

   Have your instructor approve your calibration of the spectra on the color plate. See the Report Sheet.

Table D3.4 Wavelengths of the Visible Lines in the Mercury Spectrum

Violet	404.7 nm
Violet	407.8 nm
Blue	435.8 nm
Yellow	546.1 nm
Orange	577.0 nm
Orange	579.1 nm

B. The Spectra of Elements

1. **Hydrogen Spectrum.** Use Figure D3.3 to identify which of the emission spectra on the color plate on the back cover is that of hydrogen. Justify your selection.

2. **Unknown Spectra.** Your instructor will assign to you one or two emission spectra from the color plate. Analyze each spectrum by locating the most intense wavelengths in the assigned emission spectrum. Compare the wavelengths of the most intense lines with the data in Table D3.5. Identify the element having the assigned spectrum.

C. Infrared Spectra of Compounds

1. **Match of Molecule with Infrared Spectrum.** Your instructor will assign one or more compounds on page 162 for which you are to determine its infrared spectrum. The absorption bands characteristic of atoms-in-bonds are listed in Table D3.2 to assist in the "match."

Table D3.5 Wavelengths and Relative Intensities of the Emission Spectra of Several Elements

Element	Wavelength (nm)	Relative Intensity	Element	Wavelength (nm)	Relative Intensity	Element	Wavelength (nm)	Relative Intensity
Argon	451.1	100	Helium	388.9	500	Rubidium	420.2	1000
	560.7	35		396.5	20		421.6	500
	591.2	50		402.6	50		536.3	40
	603.2	70		412.1	12		543.2	75
	604.3	35		438.8	10		572.4	60
	641.6	70		447.1	200		607.1	75
	667.8	100		468.6	30		620.6	75
	675.2	150		471.3	30		630.0	120
	696.5	10000		492.2	20	Sodium	466.5	120
	703.0	150		501.5	100		466.9	200
	706.7	10000		587.5	500		497.9	200
	706.9	100		587.6	100		498.3	400
Barium	435.0	80		667.8	100		568.2	280
	553.5	1000	Neon	585.2	500		568.8	560
	580.0	100		587.2	100		589.0	80000
	582.6	150		588.2	100		589.6	40000
	601.9	100		594.5	100		616.1	240
	606.3	200		596.5	100	Thallium	377.6	12000
	611.1	300		597.4	100		436.0	2
	648.3	150		597.6	120		535.0	18000
	649.9	300		603.0	100		655.0	16
	652.7	150		607.4	100		671.4	6
	659.5	3000		614.3	100	Zinc	468.0	300
	665.4	150		616.4	120		472.2	400
Cadmium	467.8	200		618.2	250		481.1	400
	479.9	300		621.7	150		507.0	15
	508.6	1000		626.6	150		518.2	200
	610.0	300		633.4	100		577.7	10
	643.8	2000		638.3	120		623.8	8
Cesium	455.5	1000		640.2	200		636.2	1000
	459.3	460		650.7	150		647.9	10
	546.6	60		660.0	150		692.8	15
	566.4	210	Potassium	404.4	18			
	584.5	300		404.7	17			
	601.0	640		536.0	14			
	621.3	1000		578.2	16			
	635.5	320		580.1	17			
	658.7	490		580.2	15			
	672.3	3300		583.2	17			
				691.1	19			

D. Structure of Molecules and Molecular Ions

1. **Five Basic Structures.** Using an appropriate set of molecular models, construct the five basic three-dimensional structures shown in Table D3.3. Because of the possible limited availability of molecular models, some sharing with other chemists may be necessary. Consult with your laboratory instructor.

2. **Determine Three-Dimensional Structures.** On the Report Sheet are selections of suggested molecules and molecular ions for which their three-dimensional structures (geometric shapes) and approximate bond angles are to be determined. Ask your instructor which are to be completed. Set up the table as suggested on the Report Sheet for your assigned molecules/molecular ions. Refer to your "built" five basic VSEPR structures as you analyze each molecule/molecular ion.

The Next Step

What you complete in this experiment are techniques/tools for atomic and molecular structural analysis. Chemists, biologists, and biochemists are very interested in the identification and structures of compounds . . . electromagnetic radiation is just one of those avenues for determination.

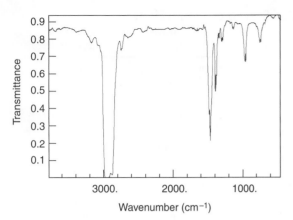

Figure D3.5A

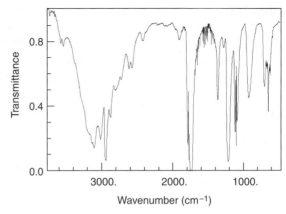

Figure D3.5B

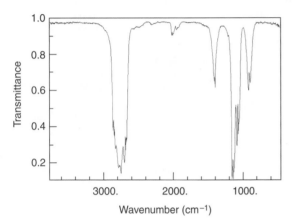

Figure D3.5C

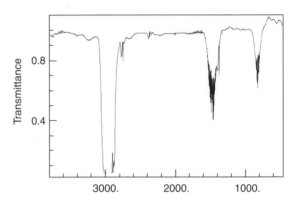

Figure D3.5D

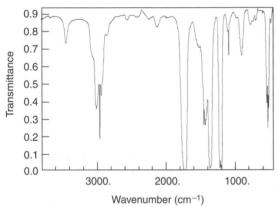

Figure D3.5E

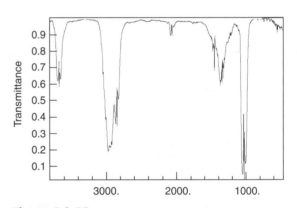

Figure D3.5F

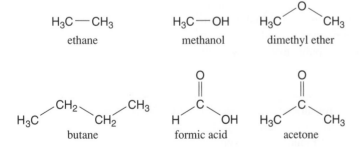

H_3C—CH_3
ethane

H_3C—OH
methanol

H_3C—O—CH_3
dimethyl ether

H_3C—CH_2—CH_2—CH_3
butane

formic acid

acetone

Atomic and Molecular Structure

Date _____ Lab Sec. _____ Name _____ Desk No. _____

A. The Mercury Spectrum

Instructor's approval of the calibration of the color plate (back cover) _____

B. The Spectra of Elements

1. Spectrum number _____ is the emission line spectrum for hydrogen on the color plate.

   What are the wavelengths and colors of the emission lines of its visible spectrum?

2. Identification of Spectra
   a. Unknown spectrum number _____

      Intense lines in the spectrum: _____ nm; _____ nm; _____ nm; _____ nm; _____ nm; _____ nm; _____ nm

      Element producing the spectrum _____

   b. Unknown spectrum number _____

      Intense lines in the spectrum: _____ nm; _____ nm; _____ nm; _____ nm; _____ nm; _____ nm; _____ nm

      Element producing the spectrum _____

C. Match of Molecule with Infrared Spectrum

Complete the following table.

Molecule Assigned	Atom-in-Bond Arrangements	Absorption Bands (cm^{-1})	Infrared Spectrum Figure No.
_____	_____	_____	_____
_____	_____	_____	_____
_____	_____	_____	_____

D. Structure of Molecules and Molecular Ions

On a separate sheet of paper, set up the following table (with eight columns) for each of the molecules/molecular ions that are assigned to you/your group. The central atom of the molecule/molecular ion is italicized.

Molecule or Molecular Ion	Lewis Structure	Valence Shell Electron Pairs	Bonding Electron Pairs	Nonbonding Electron Pairs	VSEPR Formula	Approx. Bond Angle	Geometric Shape
1. CH_4	H:C:H with H above and below	4	4	0	AX_4	109.5°	tetrahedral
2. SF_4							
3. H_2O							

1. Complete the table (as outlined above) for the following molecules/molecular ions, all of which obey the Lewis octet rule. Complete those that are assigned by your laboratory instructor.

   a. H_3O^+ d. CH_3^- g. PO_4^{3-} j. SiF_4

   b. NH_3 e. SnH_4 h. PF_3 k. H_2S

   c. NH_4^+ f. BF_4^- i. AsH_3 l. NH_2^-

2. Complete the table (as outlined above) for the following molecules/molecular ions, *none* of which obey the Lewis octet rule. Complete those that are assigned by your laboratory instructor.

   a. GaI_3 d. XeF_2 g. $XeOF_4$ j. SnF_6^{2-}

   b. PCl_2F_3 e. XeF_4 h. SbF_6^- k. IF_4^-

   c. BrF_3 f. $XeOF_2$ i. SF_6 l. IF_4^+

3. Complete the table (as outlined above) for the following molecules/molecular ions. No adherence to the Lewis octet rule is indicated. Complete those that are assigned by your laboratory instructor.

   a. AsF_3 d. SnF_2 g. PF_5 j. KrF_2

   b. ClO_2^- e. SnF_4 h. SO_4^{2-} k. TeF_6

   c. CF_3Cl f. PF_4^+ i. CN_2^{2-} l. AsF_5

4. Complete the table (as outlined above) for the following molecules/molecular ions. For molecules or molecular ions with two or more atoms considered as central atoms, consider each atom separately in the analysis according to Table D3.3. Complete those that are assigned by your laboratory instructor.

   a. $OPCl_3$ d. Cl_3CCF_3 g. $COCl_2$ j. O_3

   b. H_2CCH_2 e. Cl_2O h. $OCCCO$ k. NO_3^-

   c. $CH_3NH_3^+$ f. $ClCN$ i. CO_2 l. $TeF_2(CH_3)_4$

Dry Lab Questions

Circle the questions that have been assigned.

1. What experimental evidence leads scientists to believe that only "quantized" electronic energy states exist in atoms?

2. a. What is the wavelength range of the visible spectrum for electromagnetic radiation?

   b. What is the color of the "short" wavelength end of the visible spectrum?

   c. If a substance absorbed the wavelengths from the "short" wavelength end of the visible spectrum, what would be its color?

3. Explain why, "roses are red and violets are blue."

4. a. Is the energy absorption associated with "bands" in an infrared spectrum of higher or lower energy than the "lines" appearing in a visible line spectrum? Explain.

   b. Identify the type of energy transition occurring in a molecule that causes a band to appear in an infrared spectrum?

   c. Identify the type of energy transition occurring in an atom that causes a line to appear in a visible line spectrum?

5. Since $FeNCS^{2+}$ has an absorption maximum at 447 nm, what is the color of the $FeNCS^{2+}$ ion in solution?

6. a. Write the Lewis structure for XeF_4.

   b. Write the VSEPR formula for XeF_4.

   c. Sketch (or describe) the three dimensional structure (or geometric shape) of XeF_4.

   d. What are the approximate F–Xe–F bond angles in XeF_4?

7. Glycine, the simplest of the amino acids, has the formula, $CH_2(NH_2)COOH$, and the Lewis structure at right.
   a. Write the VSEPR formula for the nitrogen atom as the central atom in glycine.

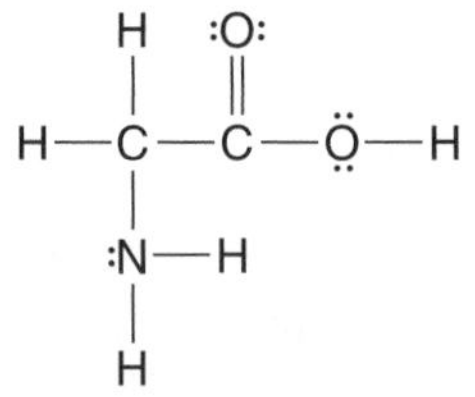

   b. Based upon VSEPR theory, what is the approximate C—N—H bond angle in glycine? Explain.

   c. What is the approximate O—C—O bond angle in glycine? Explain.

   d. Identify at least two absorption bands (and corresponding cm^{-1}) that would likely appear in the infrared spectrum of glycine.

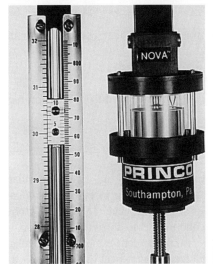

Experiment 12

Molar Mass of a Volatile Liquid

The mercury barometer accurately measures atmospheric pressure in mmHg (or torr).

OBJECTIVES

- To measure the physical properties of pressure, volume, and temperature for a gaseous substance
- To determine the molar mass (molecular weight) of a volatile liquid

TECHNIQUES

The following techniques are used in the Experimental Procedure

INTRODUCTION

Chemists in academia, research, and industry synthesize new compounds daily. To identify a new compound a chemist must determine its properties; physical properties such as melting point, color, density, and elemental composition are all routinely measured. The molar mass of the compound, also one of the most fundamental properties, is often an early determination.

A number of analytical methods can be used to measure the molar mass of a compound; the choice of the analysis depends on the properties of the compound. For example, the molar masses of "large" molecules, such as proteins, natural drugs, and enzymes found in biochemical systems, are often determined with an osmometer, an instrument that measures changes in osmotic pressure of the solvent in which the molecule is soluble. For smaller molecules a measurement of the melting point change of a solvent (Experiment 14) in which the molecule is soluble can be used. Recent developments in **mass spectrometry** have expanded its use to include not only molar mass measurements but also the structures of high molar mass compounds in the biochemical fields.

For **volatile** liquids, molecular substances with low boiling points and relatively low molar masses, the Dumas method (John Dumas, 1800–1884) of analysis can provide a fairly accurate determination of molar mass. In this analytical procedure the liquid is vaporized into a fixed-volume vessel at a measured temperature and barometric pressure. From the data and the use of the ideal gas law equation (assuming ideal gas behavior), the number of moles of vaporized liquid, n_{vapor}, is calculated:

Mass spectrometry: an instrumental method for identifying a gaseous ion according to its mass and charge

Volatile: readily vaporizable

$$n_{vapor} = \frac{PV}{RT} = \frac{P(\text{atm}) \times V(\text{L})}{(0.08206 \text{ L·atm/mol·K}) \times T(\text{K})} \qquad (12.1)$$

In this equation, R is the universal gas constant, P is the barometric pressure in atmospheres, V is the volume in liters of the vessel into which the liquid is vaporized, and T is the temperature in kelvins of the vapor.

R = 0.08206 L·atm/mol·K

The mass of the vapor, m_{vapor}, is determined from the mass difference between the "empty" vessel and the vapor-filled vessel.

$$m_{vapor} = m_{flask + vapor} - m_{flask} \tag{12.2}$$

The molar mass of the compound, $M_{compound}$, is then calculated from the acquired data.

$$M_{compound} = \frac{m_{vapor}}{n_{vapor}} \tag{12.3}$$

Gases and liquids with relatively large intermolecular forces and large molecular volumes do *not* behave according to the ideal gas law equation; in fact, some compounds that we normally consider as liquids, such as H_2O, deviate significantly from ideal gas behavior in the vapor state. Under these conditions, van der Waals' equation, a modification of the ideal gas law equation, can be used to correct for the intermolecular forces and molecular volumes in determining the moles of gas present in the system:

$$\left(P + \frac{n^2 a}{V^2}\right)(V - nb) = nRT \tag{12.4}$$

In this equation, P, V, T, R, and n have the same meanings as in Equation 12.1. a is an experimental value that is representative of the intermolecular forces of the vapor, and b is an experimental value that is representative of the volume (or size) of the molecules.

If a more accurate determination of the moles of vapor, n_{vapor}, in the flask is required, van der Waals' equation can be used instead of the ideal gas law equation. Values of a and b for a number of low-boiling-point liquids are listed in Table 12.1. Others may be found in your textbook or on the Internet.

Table 12.1 Van der Waals' Constants for Some Low-Boiling-Point Compounds

Name	a $\left(\dfrac{L^2 \cdot atm}{mol^2}\right)$	b (L/mol)	Boiling Point (°C)
methanol	9.523	0.06702	65.0
ethanol	12.02	0.08407	78.5
acetone	13.91	0.0994	56.5
propanol	14.92	0.1019	82.4
hexane	24.39	0.1735	69.0
cyclohexane	22.81	0.1424	80.7
pentane	19.01	0.1460	36.0
water	5.46	0.0305	100.0

EXPERIMENTAL PROCEDURE

Procedure Overview: A boiling water bath of measured temperature is used to vaporize an unknown liquid into a flask. The volume of the flask is measured by filling the flask with water. As the flask is open to the atmosphere, you will record a barometric pressure.

You are to complete three trials in determining the molar mass of your low-boiling-point liquid. Initially obtain 15 to 20 mL of liquid from your instructor. The same apparatus is used for each trial.

Since a boiling water bath is needed in Part A.3, you may choose to begin its preparation.

A. Preparing the Sample

1. **Prepare the Flask for the Sample.** Clean a 125-mL Erlenmeyer flask and dry it either in a drying oven or by allowing it to air-dry. Do *not* wipe it dry or heat it over a direct flame. Cover the dry flask with a small piece of aluminum foil (Figure 12.1) and secure it with a rubber band. Determine the mass (± 0.001 g) of the *dry* flask, aluminum foil, and rubber band.

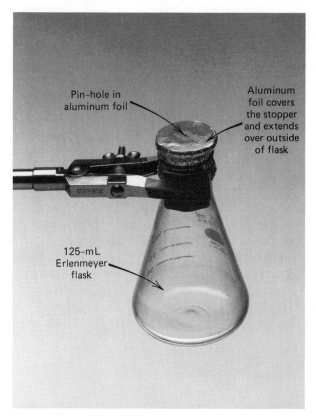

Figure 12.1 Preparation of a flask for the placement of the volatile liquid.

Figure 12.2 Apparatus for determining the molar mass of a volatile liquid.

2. **Place the Sample in the Flask.** Transfer *about* 5 mL of the unknown liquid into the flask; again cover the flask with the aluminum foil and secure the foil with a rubber band. You do *not* need to conduct a mass measurement. With a pin, pierce the aluminum foil several times.

3. **Prepare a Boiling Water Bath.** Half-fill a 400-mL beaker with water. Add one or two **boiling chips** to the water. The heat source may be a hot plate or a Bunsen flame—consult with your instructor. Secure a thermometer (digital or glass) to measure the temperature of the water bath.

Boiling chip: a piece of porous ceramic that releases air when heated (the bubbles formed prevent water from becoming superheated)

1. **Place the Flask/Sample in the Bath.** Lower the flask/sample into the bath and secure it with a utility clamp. Be certain that neither the flask nor the clamp touches the beaker wall. Adjust the water level *high* on the neck of the flask (Figure 12.2).[1]

2. **Heat the Sample to the Temperature of Boiling Water.** Gently heat the water until it reaches a *gentle* boil. (**Caution:** *Most unknowns are flammable; use a hot plate* or *moderate flame for heating.*) When the liquid in the flask and/or the vapors escaping from the holes in the aluminum foil are no longer visible, continue heating for another 5 minutes. Read and record the temperature of the boiling water.

B. Vaporize the Sample

[1]You may choose to wrap the upper portion of the flask and beaker with aluminum foil; this will maintain the upper portion of the flask *not* in the boiling water bath at nearly the same temperature as the boiling water.

3. **Measure the Mass of the Flask/Sample.** Remove the flask and allow it to cool to room temperature. Sometimes the remaining vapor in the flask condenses; that's O.K. *Dry the outside of the flask* and determine the mass (±0.001 g, use the same balance!) of the flask, aluminum foil, rubber band, and the vapor.

4. **Do It Again and Again.** Repeat the experiment for Trials 2 and 3. You only need to transfer another 5 mL of liquid to the flask (i.e., begin with Part A.2) and repeat Parts B.1–B.3.

Disposal: Dispose of the leftover unknown liquid in the "Waste Organics" container.

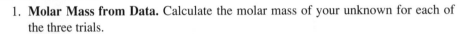

C. Determine the Volume and Pressure of the Vapor

1. **Measure the Volume of the Flask.** Fill the empty 125-mL Erlenmeyer flask to the brim with water. Measure the volume (±0.1 mL) of the flask by transferring the water to a 50- or 100-mL graduated cylinder. Record the total volume.

2. **Record the Pressure of the Vapor in the Flask.** Find the barometer in the laboratory. Read and record the atmospheric pressure in atmospheres, "using all certain digits (from the labeled calibration marks on the barometer) *plus* one uncertain digit (the last digit which is the best estimate between the calibration marks)."

D. Calculations

Appendix B

1. **Molar Mass from Data.** Calculate the molar mass of your unknown for each of the three trials.

2. **Determine the Standard Deviation and the Relative Standard Deviation (%RSD).** Refer to Appendix B and calculate the standard deviation and the %RSD for the molar mass of your unknown from your three trials.

3. **Obtain Group Data.** Obtain the values of molar mass for the same unknown from other chemists. Calculate the standard deviation and the %RSD for the molar mass of the unknown.

The Next Step

There are a number of techniques used to determine the molar mass of a volatile liquid, the most common (if the instrument is available) is mass spectrometry. Describe how your sample's molar mass would be determined using mass spectrometry. Search the Internet for other procedures that can be used to measure the molar mass of volatile substances.

Molar Mass of a Volatile Liquid

Date _____ Lab Sec. _____ Name _____ Desk No. _____

1. A 125-mL Erlenmeyer flask has a measured volume of 152 mL. A 0.199-g sample of an unknown vapor occupies the flask at 98.7°C and a pressure of 754 torr. Assume ideal gas behavior.
   a. How many moles of vapor are present?

   b. What is the molar mass of the vapor?

2. a. If the atmospheric pressure of the flask is assumed to be 760 torr in Question 1, what is the reported molar mass of the vapor?

   b. What is the percent error caused by the error in the recording of the pressure of the vapor?

   $$\% \text{ error} = \frac{M_{\text{difference}}}{M_{\text{actual}}} \times 100$$

3. The ideal gas law equation (Equation 12.1) is an equation used for analyzing "ideal gases." According to the kinetic molecular theory that defines an ideal gas, no ideal gases exist in nature, only "real" gases. Van der Waals' equation is an attempt to make corrections to real gases that do not exhibit ideal behavior. Describe the type of gaseous molecules that are most susceptible to nonideal behavior.

4. a. How is the pressure of the vaporized liquid determined in the experiment?

   b. How is the volume of the vaporized liquid determined in the experiment?

   c. How is the temperature of the vaporized liquid determined in the experiment?

   d. How is the mass of the vaporized liquid determined in the experiment?

5. The molar mass of a compound is measured to be 33.4, 35.2, 34.1, and 33.9 g/mol in four trials.

   a. What is the average molar mass of the compound?

   b. Calculate the standard deviation and the relative standard deviation (as %RSD) (see Appendix B) for the determination of the molar mass.

Molar Mass of a Volatile Liquid

Date _____ Lab Sec. _____ Name _____ Desk No. _____

A. Preparing the Sample

Unknown Number _____   *Trial 1*   *Trial 2*   *Trial 3*

   1. Mass of dry flask, foil, and rubber band (*g*)   _____   _____   _____

B. Vaporize the Sample

   1. Temperature of boiling water (*°C, K*)   _____   _____   _____

   2. Mass of dry flask, foil, rubber band, and vapor (*g*)   _____   _____   _____

C. Determine the Volume and Pressure of the Vapor

   1. Volume of 125-mL flask (*L*)

      _____ + _____ + _____ = total volume

   2. Atmospheric pressure (*torr, atm*)

D. Calculations

   1. Moles of vapor, n_{vapor} (*mol*)

   2. Mass of vapor, m_{vapor} (*g*)

   3. Molar mass of compound (*g/mol*)   *

   4. Average molar mass (*g/mol*)

   5. Standard deviation of molar mass

   6. Relative standard deviation of molecular mass (*%RSD*)

     *Calculation of Trial 1. Show work here.

Class Data/Group	1	2	3	4	5	6
Molar mass						
Sample Unknown No. _____						

Calculate the standard deviation and the relative standard deviation (as %RSD) of the molar mass of the unknown for the class. See Appendix B.

(Optional) Ask your instructor for the name of your unknown liquid. Using van der Waals' equation and the values of a and b for your compound, repeat the calculation for the moles of vapor, n_{vapor} (show for Trial 1 below), to determine a more accurate molar mass of the compound.

E. Calculations (van der Waals' equation)

Unknown Number _____ a = _____, b = _____	Trial 1*	Trial 2	Trial 3
1. Moles of vapor, n_{vapor} (mol)			
2. Mass of vapor, m_{vapor} (g)			
3. Molar mass of compound (g/mol)			
4. Average molar mass (g/mol)			

*Calculation of n_{vapor} from van der Waals' equation for Trial 1. Show work here.

Laboratory Questions

Circle the questions that have been assigned.

1. Part A.1. The mass of the flask is measured when the outside of the flask is wet. However in Part B.3 the outside of the flask is dried before its mass is measured.
   a. Will the mass of vapor in the flask be reported too high, too low, or unaffected? Explain.
   b. Will the molar mass of vapor in the flask be reported too high, too low, or unaffected? Explain.

2. Part A.1. From the time the mass of the flask is first measured in Part A.1 until the time it is finally measured in Part B.3, it is handled a number of times with oily fingers. How does this lack of proper technique affect the reported mass of the vapor in the flask? Explain.

3. Part B.2. The flask is not only completely filled with vapor, but some liquid also remains in the bottom of the flask when it is removed from the hot water bath in Part B.3. How will this oversight affect the reported molar mass of the liquid . . . too high, too low, or unaffected? Explain.

4. Part B.2. The flask is completely filled with vapor *only* when it is removed from the hot water bath in Part B.3. However when the flask cools some of the vapor condenses in the flask. As a result of this observation, will the reported molar mass of the liquid be too high, too low, or unaffected? Explain.

5. Part B.2. Suppose the thermometer is miscalibrated to read 0.3°C higher than actual. What effect does this error in calibration have on the reported molar mass of the compound? Explain.

6. Part C.2. The pressure reading from the barometer is recorded higher than it actually is. How does this affect the reported molar mass of the liquid . . . too high, too low, or unaffected? Explain.

Calcium Carbonate Analysis; Molar Volume of Carbon Dioxide

The reaction of hydrochloric acid on calcium carbonate produces carbon dioxide gas.

OBJECTIVES

- To determine the percent calcium carbonate in a **heterogeneous mixture**
- To determine the molar volume of carbon dioxide gas at 273 K and 760 torr

Heterogeneous mixture: a nonuniform mix of two or more substances, oftentimes in different phases

TECHNIQUES

The following techniques are used in the Experimental Procedure

INTRODUCTION

Calcium carbonate is perhaps the most prevalent simple inorganic compound in the Earth's crust. More commonly known as limestone, calcium carbonate is found in many forms and formulations. Chalk, marble (a dense form of calcium carbonate), shells of shellfish, stalactites, stalagmites (Figure 13.1), caliche, and the minerals responsible for hard water have their origin from fossilized remains of marine life.

Calcium carbonate readily reacts in an acidic medium to produce carbon dioxide gas:

$$CaCO_3(s) + 2\,H_3O^+(aq) \rightarrow Ca^{2+}(aq) + 3\,H_2O(l) + CO_2(g) \qquad (13.1)$$

In this experiment, the percent by mass of calcium carbonate in a heterogeneous mixture is determined. The calcium carbonate of the sample is treated with an excess of hydrochloric acid, and the carbon dioxide gas is collected over water. The moles of carbon dioxide generated in the reaction are measured and, from the stoichiometry of Equation 13.1, the moles and mass of $CaCO_3$ in the sample are calculated. Because carbon dioxide is relatively soluble in water, the water over which the CO_2 is collected is pretreated to saturate it with carbon dioxide.

In addition to analyzing the unknown mixture for percent calcium carbonate and impurities, the **molar volume** of carbon dioxide is also determined. At **standard temperature and pressure** (STP), one mole of an *ideal* gas occupies 22.4 L; that is, its molar volume is 22.4 L at STP. Because carbon dioxide is not an ideal gas, we may expect its molar volume to vary slightly from this number.

To make these two determinations in the experiment, two important measurements are made: (1) the CO_2 gas evolved from the reaction is collected and its volume is measured and (2) the mass difference of the $CaCO_3$ mixture, before and after reaction, is also measured.

Figure 13.1 Stalagmites and stalactites are primarily calcium carbonate.

Molar volume: volume occupied by one mole (6.023×10^{23} atoms or molecules) of gas at defined temperature and pressure conditions

Standard temperature and pressure: 273 K (0°C) and 1 atmosphere (760 torr)

Molar Volume of Carbon Dioxide

The mass loss of the $CaCO_3$ mixture is due to the mass of CO_2 gas evolved in the reaction; the mass of CO_2 is converted to moles of CO_2 evolved. The volume of CO_2 gas, collected over water under the temperature and pressure conditions of the experiment, is calculated at STP conditions (see Equation 13.4). Knowing the number of moles and the volume at STP, the molar volume of CO_2 gas is calculated.

$$\frac{V_{CO_2}(STP)}{n_{CO_2}} = \text{molar volume of } CO_2 \tag{13.2}$$

Volume of Collected Carbon Dioxide at STP

*Dalton's law of partial pressures: the total pressure, P_T, exerted by a mixture of gases is the sum of the individual pressures (called **partial pressures**) exerted by each of the constituent gases*

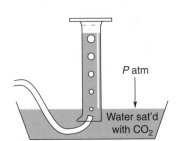

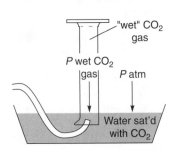

Figure 13.2 The collection of CO_2 gas over water (top) and the pressure measurement of "wet" CO_2 gas (bottom).

The CO_2 gas evolved in the reaction is collected by displacing an equal volume of water (Figure 13.2 (top)). Because the CO_2 gas is bubbled through the water, it is considered "wet," meaning that the volume occupied by the CO_2 gas is also saturated with water vapor at the temperature of the water over which it is collected. Therefore, the total pressure, P_T, in this volume is due to the combined pressures of the CO_2 gas, p_{CO_2}, and the water vapor, p_{H_2O}. The pressure of the "dry" CO_2 is calculated using **Dalton's law of partial pressures.**

For carbon dioxide gas,

$$p_{CO_2} = P_T - p_{H_2O} \tag{13.3}$$

The pressure of the water vapor, p_{H_2O}, at the gas-collecting temperature (obtained from Appendix E) is subtracted from the total pressure of the gases, P_T, in the gas-collecting vessel. Experimentally, the total pressure of the gaseous mixture ($CO_2 + H_2O$) is adjusted to atmospheric pressure ($P_T = P_{atm}$) by adjusting the water levels inside and outside the gas-collecting vessel to be equal (Figure 13.2 (bottom)). Atmospheric pressure is read from the laboratory barometer.

Once the experimental values for the volume, pressure, and temperature of the CO_2 gas are determined, the volume of CO_2 gas at STP conditions is calculated using a combination of Boyle's law ($P \propto 1/V$) and Charles' law ($V \propto T$).

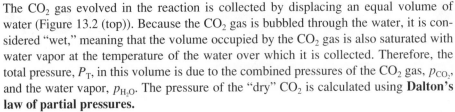

$$V_{CO_2}(\text{at STP}) = V_{CO_2 \, expt} \times \left(\frac{p_{CO_2 \, expt}(torr)}{760 \, torr}\right) \times \left(\frac{273 \, K}{T_{CO_2 \, expt}(K)}\right) \tag{13.4}$$

Boyle's law correction   Charles' law correction

This value is used in Equation 13.2.

In addition to the fact that the collected CO_2 gas is "wet," CO_2 also has an appreciable solubility in water. When the CO_2 gas is bubbled through pure water, some of the CO_2 dissolves in the water and is not measured as gas evolved in the reaction. To minimize any loss of evolved CO_2 as a result of its water solubility, the water is saturated with CO_2 *before* the experiment is conducted. This saturation is completed with the addition of either sodium bicarbonate to slightly acidified water or an antacid tablet that evolves CO_2, such as Alka-Seltzer®.

Percent Calcium Carbonate in Mixture

By use of Equation 13.1, once the moles of CO_2 gas evolved in the reaction is known, the moles and mass of $CaCO_3$ in the mixture can be calculated. The percent $CaCO_3$ in the mixture is calculated by dividing this mass of $CaCO_3$ by that of the original mixture and multiplying by 100.

$$\frac{\text{mass of } CaCO_3}{\text{mass of mixture}} \times 100 = \% \, CaCO_3 \tag{13.5}$$

EXPERIMENTAL PROCEDURE

Procedure Overview: A gas generator is constructed to collect the CO_2 gas evolved from a reaction. The masses of the sample in the gas generator before and after the reaction are measured; the volume of CO_2 gas evolved in the reaction is collected and measured.

Two trials are required in this experiment. To hasten the analyses, prepare two samples for Part A and perform the experiment with a partner.

Obtain an unknown calcium carbonate sample from your instructor. Record the sample number.

A. Sample Preparation and Setup of Apparatus

1. **Water Saturated with CO_2.** Fill a 1-L beaker with tap water and saturate the water with CO_2, using one Alka-Seltzer® tablet.[1] Save for Part A.3.

2. **Sample Preparation.** a. *Mass of sample.* Calculate the mass of $CaCO_3$ that would produce ~40 mL of CO_2 at STP. *Show the calculation on the Report Sheet.* On weighing paper, measure this calculated mass (±0.001 g) of the sample mixture, one that contains $CaCO_3$ and a noncarbonate impurity. Carefully transfer the sample to a 75-mm test tube.

   b. *Set up the CO_2 generator.* Place 10 mL of 3 *M* HCl in a 200-mm test tube; carefully slide the 75-mm test tube into the 200-mm test tube without splashing any of the acid into the sample. **Important:** The HCl(*aq*) is in the 200-mm test tube and separately, but inserted, is the $CaCO_3(s)$ sample in the 75-mm test tube . . . do not mix the two substances until Part B.1!

   c. *Mass of CO_2 generator.* Measure the combined mass (±0.001 g) of this CO_2 generator (Figure 13.3).

   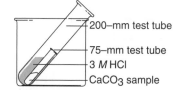

   Figure 13.3 Apparatus for measuring the combined mass of the CO_2 generator.

3. **Setup of CO_2 Collection Apparatus.** a. *Saturated Water with CO_2.* Fill the pan (Figure 13.4) about two-thirds full with the water that is saturated with CO_2 (Part A.1). Do not proceed in the experiment until the CO_2 from the Alka-Seltzer® is no longer evolved.

   b. *Fill $CO_2(g)$-collecting graduated cylinder.* Use the CO_2-saturated water in the pan to fill the 50-mL graduated cylinder that will collect the CO_2 from the sample. Fill the graduated cylinder by laying it horizontal in the water, and then, without removing the mouth of the cylinder from the water, set it upright.

   c. *Connect the gas inlet tube.* Place the gas inlet tube (connect to rubber or Tygon tubing) that connects to the CO_2 generator into the mouth of the 50-mL $CO_2(g)$-collecting graduated cylinder. Support the graduated cylinder with a ring stand and clamp. Read and record the water level in the graduated cylinder (if no air entered the cylinder, it should read zero). Note that the graduation marks on the cylinder are "upside down"—be careful in reading and recording the volume.

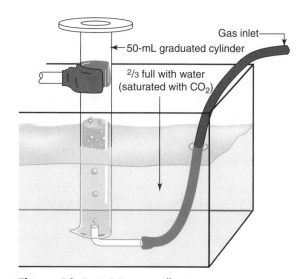

Figure 13.4 A CO_2 gas-collection apparatus.

[1] An acidic solution of sodium bicarbonate can also be used to saturate the water with CO_2.

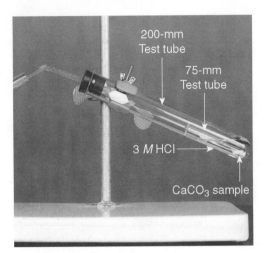

Figure 13.5 A CO_2 gas generator.

200-mm
Test tube

75-mm
Test tube

3 *M* HCl

$CaCO_3$ sample

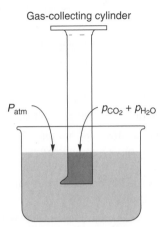

Gas-collecting cylinder

P_{atm}

$p_{CO_2} + p_{H_2O}$

Figure 13.6 Equalizing the pressure in the test tube to atmospheric pressure.

4. **Setup of CO_2 Generator.** Prepare a one-hole rubber stopper (check to be certain there are no "cracks" in the rubber stopper) fitted with a short piece of glass tubing and *firmly* insert it into the 200-mm test tube to avoid any leaking of $CO_2(g)$ from the reaction. Clamp the CO_2 generator (200-mm test tube) to the ring stand at a 45° angle from the horizontal (Figure 13.5). Connect the gas delivery tube from the CO_2 collection apparatus (Figure 13.4) to the CO_2 generator (Figure 13.5).

5. **Obtain Instructor's Approval.** Obtain your instructor's approval before continuing.

B. Collection of the Carbon Dioxide Gas

1. **Generate and Collect the CO_2 Gas.** *Initiate the reaction. Gently* agitate the generator (Figure 13.5) to allow some of the HCl solution to contact the sample mixture. As the evolution rate of CO_2 gas decreases, agitate again and again until CO_2 gas is no longer evolved.

C. Determination of the Volume, Temperature, and Pressure of the Carbon Dioxide Gas

1. **Determine the Volume of $CO_2(g)$ Evolved.** When no further generation of $CO_2(g)$ is evident in the gas-collection apparatus (Figure 13.4), and while still in the water-filled pan, adjust the CO_2-gas collecting graduated cylinder so that the water levels inside and outside of the graduated cylinder are equal (Figure 13.6). Read and record the final volume of gas collected in the graduated cylinder.

2. **Determine the Temperature of $CO_2(g)$.** Read and record the temperature of the water in the pan.

3. **Determine the Pressure of the $CO_2(g)$.** When the water levels inside and outside of the graduated cylinder are equal, the pressure of the "wet" $CO_2(g)$ equals atmospheric pressure.

Appendix E

Read and record the barometric pressure in the laboratory. Obtain the vapor pressure of water at the gas-collecting temperature in Appendix E to calculate the pressure of the "dry" $CO_2(g)$ evolved in the reaction.

D. Mass of Carbon Dioxide Evolved

1. **Determine a Mass Difference.** Determine the mass (± 0.001 g) of the 200-mm CO_2 generator and its remaining contents. Compare this mass with that in Part A.2c. After subtracting for the mass of the generator, calculate the mass loss of the sample.

The Next Step

The analysis for the percent aluminum in an aluminum container can be completed using the same technique, a reaction of aluminum metal with hydrochloric acid. Research and design an experimental procedure for its analysis.

Experiment 13 Prelaboratory Assignment

Calcium Carbonate Analysis; Molar Volume of Carbon Dioxide

Date _____ Lab Sec. _____ Name _____ Desk No. _____

1. In some solid calcium carbonate samples, calcium bicarbonate, $Ca(HCO_3)_2$, is also present. Write a balanced equation for its reaction with hydrochloric acid.

2. A mixture of gases collected over water at 14°C has a total pressure of 0.981 atm and occupies 55 mL. How many grams of water escaped into the vapor state? Hint: What is the vapor pressure of water at 14°C?

3. Experimental Procedure, Part A.2. Complete the calculation required to appear on the Report Sheet (here and on Report Sheet).

4. a. Experimental Procedure, Part A.3b. Explain how a water-filled graduated cylinder is inverted in a pan of water.

   b. Experimental Procedure, Part A.4. What is the proper procedure for inserting a piece of glass tubing into a rubber stopper?

5. Experimental Procedure, Part C.1. How is the volume of the $CO_2(g)$ collected (over water) measured in this experiment?

6. A 37.7-mL volume of an unknown gas is collected *over water* at 19°C and a total pressure of 770 torr. The mass of the gas is 68.6 mg.

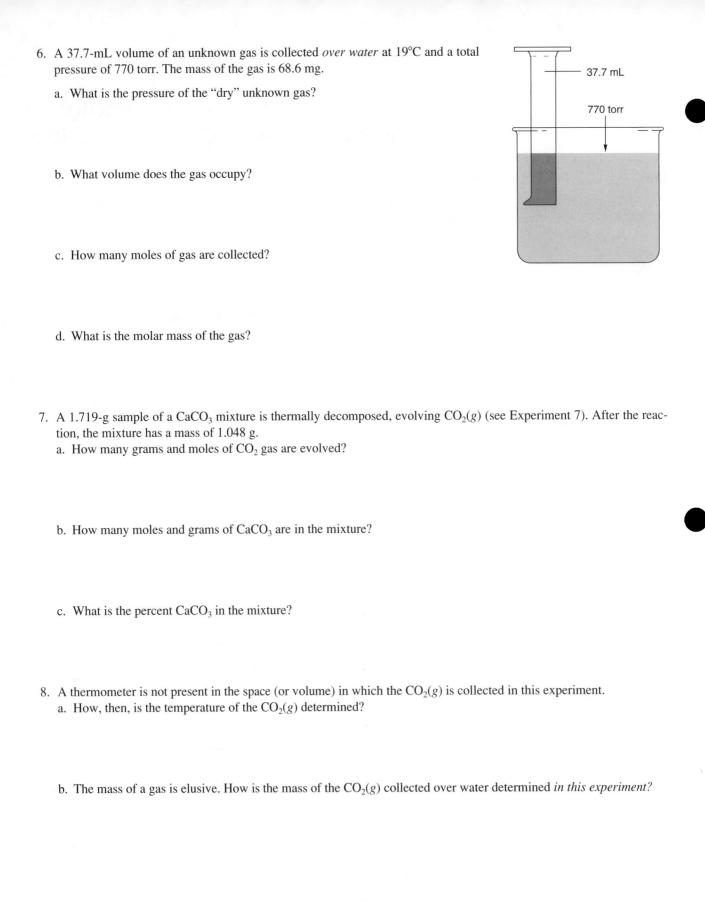

a. What is the pressure of the "dry" unknown gas?

b. What volume does the gas occupy?

c. How many moles of gas are collected?

d. What is the molar mass of the gas?

7. A 1.719-g sample of a $CaCO_3$ mixture is thermally decomposed, evolving $CO_2(g)$ (see Experiment 7). After the reaction, the mixture has a mass of 1.048 g.
   a. How many grams and moles of CO_2 gas are evolved?

   b. How many moles and grams of $CaCO_3$ are in the mixture?

   c. What is the percent $CaCO_3$ in the mixture?

8. A thermometer is not present in the space (or volume) in which the $CO_2(g)$ is collected in this experiment.
   a. How, then, is the temperature of the $CO_2(g)$ determined?

   b. The mass of a gas is elusive. How is the mass of the $CO_2(g)$ collected over water determined *in this experiment?*

Experiment 13 *Report Sheet*

Calcium Carbonate Analysis; Molar Volume of Carbon Dioxide

Date __________ Lab Sec. ______ Name ____ Desk No. __________

A. Sample Preparation and Setup Apparatus

Calculation of mass of $CaCO_3$ sample for analysis

Unknown Sample No. __________	*Trial 1*	*Trial 2*
1. Mass of sample (*g*)	__________	__________
2. Mass of generator + sample before reaction (*g*)	__________	__________
3. Instructor's approval of apparatus	__________	

C. Determination of Volume, Temperature, and Pressure of the Carbon Dioxide Gas

	Trial 1	*Trial 2*
1. *Initial* volume of water in "CO_2-collecting" graduated cylinder (*mL*)	__________	__________
2. *Final* volume of water in "CO_2-collecting" graduated cylinder (*mL*)	__________	__________
3. Volume of $CO_2(g)$ collected (*L*)	__________	__________
4. Temperature of water (°C)	__________	__________
5. Barometric pressure (*torr*)	__________	__________
6. Vapor pressure of H_2O at ___°C (*torr*)	__________	__________
7. Pressure of "dry" $CO_2(g)$ (*torr*)	__________	__________

D. Mass of Carbon Dioxide Evolved

	Trial 1	*Trial 2*
1. Mass of generator + sample after reaction (*g*)	__________	__________
2. Mass loss of generator = mass CO_2 evolved (*g*)	__________	__________

E. Molar Volume of CO$_2$ Gas

	Trial 1	*Trial 2*
1. Pressure of "dry" CO$_2$(g) (*atm*)		
2. Volume of CO$_2$(g) at STP (*L*)		
3. Moles of CO$_2$(g) generated (*mol*)		
4. Molar volume of CO$_2$(g) at STP (*L/mol*)		
5. Average molar volume of CO$_2$(g) at STP (*L/mol*)		

F. Percent CaCO$_3$ in Mixture

	Trial 1	*Trial 2*
1. Moles of CaCO$_3$ in sample from mol CO$_2$ generated (*mol*)		
2. Mass of CaCO$_3$ in sample (*g*)		
3. Mass of original sample (*g*)		
4. Percent of CaCO$_3$ in sample (*%*)		
5. Average percent of CaCO$_3$ in sample (*%*)		

Laboratory Questions

Circle the questions that have been assigned.

1. Part A.1. The water for the pan (Part A.3) is not saturated with CO$_2$. Will the reported percent CaCO$_3$ in the original sample be too high, too low, or unaffected? Explain.

2. Part A.3. A few drops of HCl(*aq*) spilled over into the CaCO$_3$ sample prior to firmly seating the stopper and prior to collecting any CO$_2$(g). What error in the reported percent CaCO$_3$ will result from this poor laboratory technique? Explain.

3. Part A.4. A small crack is present in the rubber stopper. How does this affect the calculated molar volume of the CO$_2$ . . . too high, too low, or unaffected? Explain.

4. Part C.1. The water level in the CO$_2$-gas collecting cylinder is *higher* than the water level outside the cylinder.
   a. Is the "wet" CO$_2$ gas pressure greater or less than atmospheric pressure? Explain.
   b. An adjustment is made to equilibrate the water levels. Will the volume of the "wet" CO$_2$ gas increase or decrease? Explain.
   c. The student chemist chooses *not* to equilibrate the inside and outside water levels. Will the reported number of moles of CO$_2$ generated in the reaction be too high, too low, or unaffected by this carelessness? Explain.

5. Part C.1. An air bubble accidentally enters the CO$_2$-collection graduated cylinder after the completion of the reaction. How does this error affect the reported moles of CO$_2$(g) collected? Explain.

6. Part D. If a large amount of calcium sulfate (does not form a gas with the addition of HCl(*aq*)) is present in the original sample, how does it affect the reported
   a. molar volume of the CO$_2$(g)? Explain.
   b. percent CaCO$_3$ in the sample? Explain.

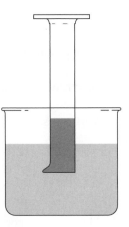

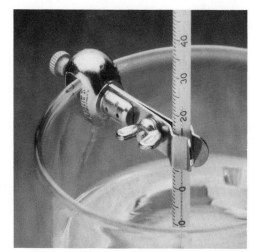

A thermometer is secured with a thermometer clamp to guard against breakage.

Molar Mass of a Solid

- To observe and measure the effect of a solute on the freezing point of a solvent
- To determine the molar mass (molecular weight) of a nonvolatile, nonelectrolyte solute

The following techniques are used in the Experimental Procedure

A pure liquid, such as water or ethanol, has characteristic physical properties: the melting point, boiling point, density, vapor pressure, viscosity, surface tension, and additional data are listed in handbooks of chemistry. The addition of a soluble solute to the liquid forms a homogeneous mixture, called a **solution.** The solvent of the solution assumes physical properties that are no longer definite, but dependent on the amount of solute added. The vapor pressure of the solvent decreases, the freezing point of the solvent decreases, the boiling point of the solvent increases, and the osmotic pressure of the solvent increases. The degree of the change depends on the *number* of solute particles that have dissolved, *not* on the chemical identity of the solute. These four physical properties that depend on the number of solute particles dissolved in a solvent are called **colligative properties.**

For example, one mole of glucose or urea (neither of which dissociates in water) lowers the freezing point of one kilogram of water by 1.86°C; whereas one mole of sodium chloride lowers the freezing point of one kilogram of water by nearly twice that amount ($\sim$3.72°C) because, when dissolved in water, it dissociates into Na^+ and Cl^- providing *twice* as many moles of solute particles per mole of solute as do glucose or urea.

When freezing ice cream at home, a salt/ice/water mixture provides a lower temperature bath than an ice/water mixture alone. Antifreeze (ethylene glycol, Figure 14.1) added to the cooling system of an automobile reduces the probability of freeze-up in the winter and boiling over in the summer because the antifreeze/water solution has a lower freezing point and a higher boiling point than pure water.

These changes in the properties of pure water that result from the presence of a **nonvolatile solute** are portrayed by the phase diagram in Figure 14.2, a plot of vapor pressure versus temperature. The solid lines refer to the equilibrium conditions between the respective phases for pure water; the dashed lines represent the same conditions for an aqueous solution.

Figure 14.1 Ethylene glycol is a major component of most antifreeze solutions.

Colligative properties: properties of a solvent that result from the presence of the number of solute particles in the solution and not their chemical composition

Nonvolatile solute: a solute that does not have a measurable vapor pressure

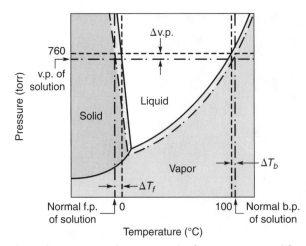

Figure 14.2 Phase diagram (not shown to scale) for water (—) and for an aqueous solution (- - -).

The **vapor pressure** of water is 760 torr at its boiling point of 100°C. When a nonvolatile solute dissolves in water to form a solution, solute molecules occupy a part of the surface area. This inhibits movement of some water molecules into the vapor state, causing a **vapor pressure lowering** of the water (Δv.p. in Figure 14.2), lower than 760 torr. With the vapor pressure less than 760 torr, the solution (more specifically, the water in the solution) no longer boils at 100°C. For the solution to boil, the vapor pressure must be increased to 760 torr; boiling can resume only if the temperature is increased above 100°C. This **boiling point elevation** (ΔT_b in Figure 14.2) of the water is due to the presence of the solute.

A solute added to water also affects its freezing point. The normal freezing point of water is 0°C, but in the presence of a solute, the temperature must be lowered below 0°C before freezing occurs (the energy of the water molecules must be lowered to increase the magnitude of the intermolecular forces, so that the water molecules "stick" together to form a solid); this is called a **freezing point depression** of the water (ΔT_f in Figure 14.2).

The changes in the freezing point, ΔT_f, and the boiling point, ΔT_b, are directly proportional to the molality, m, of the solute in solution. The proportionality is a constant, characteristic of the actual solvent. For water the freezing point constant, k_f, is 1.86°C·kg/mol and the boiling point constant, k_b, is 0.512°C·kg/mol.

$$\Delta T_f = \left| T_{f, \text{solvent}} - T_{f, \text{solution}} \right| = k_f m \tag{14.1}$$

$$\Delta T_b = \left| T_{b, \text{solvent}} - T_{b, \text{solution}} \right| = k_b m \tag{14.2}$$

In Equations 14.1 and 14.2, T_f represents the freezing point and T_b represents the boiling point of the respective system. $\left| T_{f, \text{solvent}} - T_{f, \text{solution}} \right|$ represents the absolute temperature difference in the freezing point change. **Molality** is defined as

$$\text{molality, } m = \frac{\text{mol solute}}{\text{kg solvent}} = \frac{(\text{mass/molar mass})}{\text{kg solvent}} \tag{14.3}$$

k_f and k_b values for various solvents are listed in Table 14.1.

In this experiment the freezing points of a selected pure solvent and of a solute/solvent mixture are measured. The freezing point lowering (difference), the k_f data from Table 14.1 for the solvent, and Equations 14.1 and 14.3 are used to calculate the moles of solute dissolved in solution and, from its measured mass, the molar mass of the solute.

The freezing points of the solvent and the solution are obtained from a **cooling curve**—a plot of temperature versus time. An ideal plot of the data appears in Figure 14.3. The cooling curve for a pure solvent reaches a plateau at its freezing point;

Table 14.1 Molal Freezing Point and Boiling Point Constants for Several Solvents

Substance	Freezing Point (°C)	$k_f\left(\dfrac{°C\cdot kg}{mol}\right)$	Boiling Point (°C)	$k_b\left(\dfrac{°C\cdot kg}{mol}\right)$
H_2O	0.0	1.86	100.0	0.512
cyclohexane	—	20.0	80.7	2.69
naphthalene	80.2	6.9	—	—
camphor	179	39.7	—	—
acetic acid	17	3.90	118.2	2.93
t-butanol	25.5	9.1	—	—

extrapolation of the plateau to the temperature axis determines its freezing point. The cooling curve for the solution does *not* reach a plateau, but continues to decrease slowly as the solvent freezes out of solution. Its freezing point is determined at the intersection of two straight lines drawn through the data points above and below the freezing point (Figure 14.3).

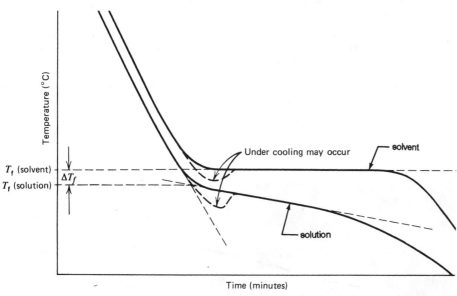

Figure 14.3 Cooling curves for a solvent and solution.

Procedure Overview: Cyclohexane is the solvent selected for this experiment although other solvents are just as effective for the determination of the molar mass of a solute. Another solvent[1] may be used at the discretion of the laboratory instructor. Consult with your instructor. The freezing points of cyclohexane and a cyclohexane *solution* are determined from plots of temperature versus time. The mass of the solute is measured before it is dissolved in a known mass of cyclohexane.

Obtain about 15 mL of cyclohexane. You'll use the cyclohexane throughout the experiment. Your laboratory instructor will issue you about 1 g of unknown solute. Record the unknown number of the solute on the Report Sheet.

The cooling curve to be plotted in Part A.4 can be established by using a thermal probe that is connected directly to either a calculator or computer with the appropriate software. If this thermal sensing/recording apparatus is available in the laboratory, consult with your instructor for its use and adaptation to the experiment. The probe merely replaces the glass or digital thermometer in Figure 14.4.

EXPERIMENTAL PROCEDURE

---

[1] t-Butanol is a suitable substitute for cyclohexane in this experiment.

A. Freezing Point of Cyclohexane (Solvent)

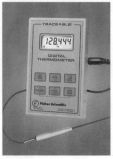

A modern digital thermometer

Appendix C

1. **Prepare the Ice/Water Bath.** Assemble the apparatus shown in Figure 14.4. A 400-mL beaker is placed inside a 600-mL beaker, the latter being an outside, insulating beaker. You may want to place a paper towel between the beakers to further insulate the ice/water bath. Place about 300 mL of an ice/water slurry into the 400-mL beaker.[2]

   Obtain a digital or glass thermometer and thermometer clamp, mount it to the ring stand, and position the thermometer in the test tube. (**Caution:** *If the thermometer is a glass thermometer, handle the thermometer carefully. If the thermometer is accidentally broken, notify your instructor immediately.*)

2. **Prepare the Cyclohexane.** Determine the mass (± 0.01 g)[3] of a *clean, dry* 200-mm test tube in a 250-mL beaker (Figure 14.5). Add approximately 12 mL of cyclohexane (**Caution:** *Cyclohexane is flammable—keep away from flames; cyclohexane is a mucous irritant—do not inhale*) to the test tube. Place the test tube containing the cyclohexane into the ice/water bath (Figure 14.4). Secure the test tube with a utility clamp. Insert the thermometer probe and a wire stirrer into the test tube. *Secure the thermometer* so that the thermometer bulb or thermal sensor is completely submerged into the cyclohexane.

3. **Record Data for the Freezing Point of Cyclohexane.** While stirring with the wire stirrer, record the temperature at timed intervals (15 or 30 seconds) on the Report Sheet. The temperature remains virtually constant at the freezing point until the solidification is complete. Continue collecting data until the temperature begins to drop again.

4. **Plot the Data.** On linear graph paper or by using appropriate software, plot the temperature (°C, vertical axis) versus time (sec, horizontal axis) to obtain the cooling curve for cyclohexane. Have your instructor approve your graph.

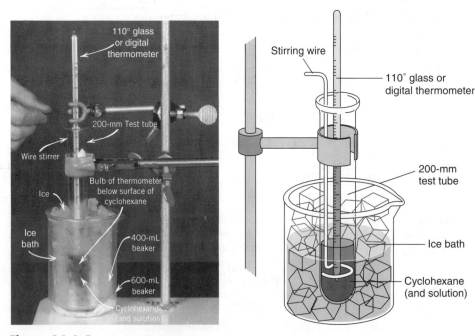

Figure 14.4 Freezing point apparatus.

---

[2]Rock salt may be added to further lower the temperature of the ice/water bath.
[3]Use a balance with ± 0.001 g sensitivity, if available.

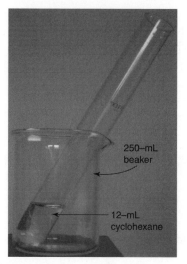

Figure 14.5 Determining the mass of beaker and test tube before and after adding cyclohexane.

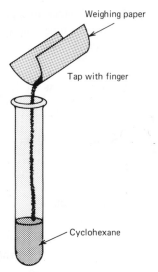

Figure 14.6 Transfer of the unknown solid solute to the test tube containing cyclohexane.

Three freezing point trials for the cyclohexane solution are to be completed. Successive amounts of unknown sample are added to the cyclohexane in Parts B.4 and B.5.

B. Freezing Point of Cyclohexane plus Unknown Solute

1. **Measure the Mass of Solvent and Solid Solute.** Dry the outside of the test tube containing the cyclohexane and measure its mass in the same 250-mL beaker. On weighing paper tare the mass of 0.1–0.3 g of unknown solid solute (ask your instructor for the approximate mass to use) and record. *Quantitatively* transfer the solute to the cyclohexane in the 200-mm test tube (Figure 14.6).[4]

2. **Record Data for the Freezing Point of Solution.** Determine the freezing point of this solution in the same way as that of the solvent (Part A.3). Record the time/temperature data on the Report Sheet. When the solution nears the freezing point of the pure cyclohexane, record the temperature at more frequent time intervals (~15 seconds). A "break" in the curve occurs as the freezing begins, although it may not be as sharp as that for the pure cyclohexane.

3. **Plot the Data on the Same Graph.** Plot the temperature versus time data on the *same* graph (and same coordinates) as those for the pure cyclohexane (Part A.4). Draw straight lines through the data points above and below the freezing point (see Figure 14.3); the intersection of the two straight lines is the freezing point of the solution.

Appendix C

4. **Repeat with Additional Solute.** Remove the test tube/solution from the ice/water bath. Add an additional 0.1–0.3 g of unknown solid solute, using the *same procedure* as in Part B.1. Repeat the freezing point determination and, again, plot the temperature versus time data on the same graph (Parts B.2 and B.3). The total mass of solute in solution is the sum from the first and second trials.

5. **Again, Repeat with Additional Solute.** Repeat Part B.4 with an additional 0.1–0.2 g of unknown solid solute, using the *same procedure* as in Part B.1. Repeat the freezing point determination and again plot the temperature versus time data on the same graph (Parts B.2–4). You now should have four plots on the same graph.

---

[4]In the transfer be certain that *none* of the solid solute adheres to the test tube wall. If some does, roll the test tube until the solute dissolves.

6. **Obtain Instructor's Approval.** Have your instructor approve the three temperature versus time graphs (Parts B.3–5) that have been added to your first temperature versus time graph (Part A.4) for the pure cyclohexane.

Disposal: Dispose of the waste cyclohexane and cyclohexane solution in the "Waste Organic Liquids" container.

CLEANUP: Safely store/return the thermometer. Rinse the test tube once with acetone; discard the rinse in the "Waste Organic Liquids" container.

C. Calculations

1. From the plotted data, determine ΔT_f for Trial 1, Trial 2, and Trial 3. Refer to the plotted cooling curves (see Figure 14.3).

2. From k_f, the mass (in kg) of the cyclohexane, and the measured ΔT_f, calculate the moles of solute for each trial.

3. Determine the molar mass of the solute for each trial (remember the mass of the solute for each trial is different).

4. What is the average molar mass of your unknown solute?

5. Calculate the standard deviation and the relative standard deviation (%RSD) for the molar mass of the solute.

The Next Step

Salts dissociate in water. (1) Design an experiment to determine the percent dissociation for a selection of salts in water—consider various concentrations of the salt solutions. Explain your data. (2) Determine the total concentration of dissolved solids in a water sample using this technique and compare your results to the data in Experiment 3.

Molar Mass of a Solid

Date _____ Lab Sec. _____ Name _____ Desk No. _____

1. This experiment is more about understanding the colligative properties of a solution rather than the determination of the molar mass of a solid.
   a. Define colligative properties.

   b. Which of the following solutes has the greatest affect on the colligative properties for a given mass of pure water? Explain.
      i)   0.01 mol of $CaCl_2$ (an electrolyte)
      ii) 0.01 mol of KNO_3 (an electrolyte)
      iii) 0.01 mol of $CO(NH_2)_2$ (a nonelectrolyte)

2. A 0.517-g sample of a nonvolatile solid solute dissolves in 15.0 g of *t*-butanol. The freezing point of the solution is 22.7°C.
   a. What is the molality of the solute in the solution. See Table 14.1.

   b. Calculate the molar mass of the solute.

   c. The same mass of solute is dissolved in 15.0 g of cyclohexane instead of *t*-butanol. What is the expected freezing point *change* of this solution? See Table 14.1.

3. Explain why ice cubes formed from water of a glacier freeze at a higher temperature than ice cubes formed from water of an underground aquifer.

4. Two students prepare two cyclohexane solutions having the same mass of solute. Student 1 uses 13.0 g of cyclohexane solvent and Student 2 uses 10 g of cyclohexane solvent. Which student will observe the larger freezing point change? Show calculations.

5. Two solutions are prepared using the *same* solute:
   Solution A: 0.14 g of the solute dissolves in 15.4 g of *t*-butanol
   Solution B: 0.17 g of the solute dissolves in 12.7 g of cyclohexane
   Which solution has the greatest freezing point change? Show calculations and explain.

6. Experimental Procedure. a. How many (total) data plots are to be completed for this experiment? Account for each.

   b. What information is to be extracted from each data plot?

7. See Figure 14.3. The temperature of a solvent remains fixed over time during the phase change from liquid to solid. However, the temperature of a solution continually decreases over time during the phase change from liquid to solid. Explain.

Molar Mass of a Solid

Date _____ Lab Sec. _____ Name _____ Desk No. _____

A. Freezing Point of Cyclohexane (Solvent)

1. Mass of beaker, test tube (*g*)   _____

2. Freezing point, from cooling curve (°*C*)   _____

3. Instructor's approval of graph   _____

B. Freezing Point of Cyclohexane plus Unknown Solute

	Trial 1 (*Parts B.1, B.3*)	*Trial 2* (*Part B.4*)	*Trial 3* (*Part B.5*)
Unknown Solute No. __7__			
1. Mass of beaker, test tube, cyclohexane (*g*)	153.012g	153.012g	153.012g
2. Mass of cyclohexane (*g*)	6.944g	6.944g	6.944g
3. Tared mass of added solute, *total* (*g*)	.10 g	.41g	.58
4. Freezing point, from cooling curve (°*C*)	6°C	4°C	3°C
5. Instructor's approval of graph	✓		

Calculations

1. k_f for cyclohexane (pure solvent)	20.0 °C • kg/mol		
2. Freezing point *change*, ΔT_f (°*C*)			
3. Mass of cyclohexane in solution (*kg*)			
4. Moles of solute, *total* (*mol*)			
5. Mass of solute in solution, *total* (*g*)			
6. Molar mass of solute (*g/mol*)		*	
7. Average molar mass of solute			
8. Standard deviation of molar mass			
9. Relative standard deviation of molar mass (%RSD)			

*Show calculation(s) for trial 2 on the next page.

*Calculations for Trial 2.

180	6.5	180	4	180	3.5
200	6	200	4	200	3.5
220	6	220	4	220	3
240	6	240		240	3
260	6	260	4	260	3
280	6	280		280	3
300	6	300		300	3
320	6	320		320	3
340		340		340	3
360	6	360	4	360	3
380	5	380		380	3
400	4	400		400	2.5

(left column: 200 7.7 / 220 7 / 240 7 / 260 7 / 280 7 / 300 7 / 320 7 / 340 6.5 / 360 6 / 380 6 / 400)

A. Cyclohexane		B. Cyclohexane + Unknown Solute					
Time	Temp	Trial 1		Trial 2		Trial 3	
		Time	Temp	Time	Temp	Time	Temp
0	18,26	0	21	0	11	0	9
20	12,14	20	18	20	8	20	8
40	10,5	40	13	40	7	40	7
60	9,11	60	11	60	6	60	5
80	8,10	80	9	80	5	80	4
100	7.8,9	100	8	100	5	100	4
120	7,8	120	7.5	120	5	120	4
140	7,8	140	7	140	4.5	140	3.5
160	7,7	160	6.5	160	4	160	3.5
180	7,7						

Continue recording data on your own paper and submit it with the Report Sheet.

Laboratory Questions

Circle the questions that have been assigned.

1. Part A.3. Some of the cyclohexane solvent vaporized during the temperature vs. time measurements in Part A.3. How will this loss of solvent affect its freezing point determination? Explain.

2. Part A.3. The digital thermometer is miscalibrated by +0.15°C over its entire range. If the same thermometer is used in Part B.2, will the reported moles of solute in the solution be too high, too low, or unaffected? Explain.

3. Part A.4. The experimental freezing point for cyclohexane is found to be ~1°C less than its value listed in the literature. Account for this discrepancy.

4. Part B.1. Some of the solid solute adheres to the side of the test tube during the freezing point determination of the solution in Part B.2. As a result of the oversight, will the reported molar mass of the solute be too high, too low, or unaffected? Explain.

5. Part B.2. Some of the cyclohexane solvent vaporized during the temperature vs. time measurement. How will this loss of solvent affect the freezing point determination of the solution? Explain.

6. Part B.2. The solute dissociates slightly in the solvent. How will the slight dissociation affect the reported molar mass of the solute . . . too high, too low, or unaffected? Explain.

7. Part C.1. Interpretation of the data plots consistently shows that the freezing points of three solutions are too low. As a result of this "misreading of the data," will the reported molar mass of the solute be too high, too low, or unaffected? Explain.

Experiment 15

Synthesis of Potassium Alum

A crystal of potassium alum, $KAl(SO_4)_2 \cdot 12H_2O$.

OBJECTIVES

- To prepare an alum from an aluminum can
- To test the purity of the alum using a melting point measurement

TECHNIQUES

The following techniques are used in the Experimental Procedure

INTRODUCTION

An alum is a hydrated double sulfate salt with the general formula

$$M^+M'^{3+}(SO_4)_2 \cdot 12H_2O$$

Univalent: an ion that has a charge of one

M^+ is a **univalent** cation, commonly Na^+, K^+, Tl^+, NH_4^+, or Ag^+; M'^{3+} is a trivalent cation, commonly Al^{3+}, Fe^{3+}, Cr^{3+}, Ti^{3+}, or Co^{3+}. A common household alum is ammonium aluminum sulfate **dodeca**hydrate (Figure 15.1).

Dodeca: Greek prefix meaning "12"

Some common alums and their uses are listed in Table 15.1.

Preparation of Potassium Alum

Potassium alum, commonly just called **alum,** is widely used in the chemical industry for home and commercial uses. It is extensively used in the pulp and paper industry for **sizing** paper, also for sizing fabrics in the textile industry. Alum is also used in municipal water treatment plants for purifying drinking water.

Sizing: to effect the porosity of paper or fabrics

In this experiment potassium aluminum sulfate dodecahydrate (potassium alum), $KAl(SO_4)_2 \cdot 12H_2O$, is prepared from an aluminum can and potassium hydroxide. Aluminum metal rapidly reacts with hot aqueous and excess KOH producing a soluble potassium aluminate salt solution and hydrogen gas.

$$2\,Al(s) + 2\,K^+(aq) + 2\,OH^-(aq) + 6\,H_2O(l) \rightarrow$$
$$2\,K^+(aq) + 2\,Al(OH)_4^-(aq) + 3\,H_2(g) \quad (15.1)$$

When treated with sulfuric acid, the aluminate ion, $Al(OH)_4^-$, precipitates as aluminum hydroxide, but redissolves with the application of heat.

$$2\,K^+(aq) + 2\,Al(OH)_4^-(aq) + 2\,H^+(aq) + SO_4^{2-}(aq) \rightarrow$$
$$2\,Al(OH)_3(s) + 2\,K^+(aq) + SO_4^{2-}(aq) + 2\,H_2O(l) \quad (15.2)$$

$$2\,Al(OH)_3(s) + 6\,H^+(aq) + 3\,SO_4^{2-}(aq) \xrightarrow{\Delta}$$
$$2\,Al^{3+}(aq) + 3\,SO_4^{2-}(aq) + 6\,H_2O(l) \quad (15.3)$$

Figure 15.1 Ammonium aluminum sulfate dodecahydrate.

Table 15.1 Common Alums

Alum	Formula	Uses
Sodium aluminum sulfate dodecahydrate (sodium alum)	$NaAl(SO_4)_2 \cdot 12H_2O$	Baking powders: hydrolysis of Al^{3+} releases H^+ in water to react with the HCO_3^- in baking soda—this produces CO_2, causing the dough to rise
Potassium aluminum sulfate dodecahydrate (alum or potassium alum)	$KAl(SO_4)_2 \cdot 12H_2O$	Water purification, sewage treatment, fire extinguishers, and "sizing" paper
Ammonium aluminum sulfate dodecahydrate (ammonium alum)	$NH_4Al(SO_4)_2 \cdot 12H_2O$	Pickling cucumbers
Potassium chromium(III) sulfate dodecahydrate (chrome alum)	$KCr(SO_4)_2 \cdot 12H_2O$	Tanning leather and waterproofing fabrics
Ammonium ferric sulfate dodecahydrate (ferric alum)	$NH_4Fe(SO_4)_2 \cdot 12H_2O$	Mordant in dying and printing textiles

Potassium aluminum sulfate dodecahydrate forms octahedral-shaped crystals when the nearly saturated solution cools (see opening photo).

$$K^+(aq) + Al^{3+}(aq) + 2\ SO_4^{2-}(aq) + 12\ H_2O(l) \rightarrow KAl(SO_4)_2 \cdot 12H_2O(s) \quad (15.4)$$

EXPERIMENTAL PROCEDURE

Procedure Overview: A known mass of starting material is used to synthesize the potassium alum. The synthesis requires the careful transfer of solutions and some evaporation and cooling techniques.

Prepare an ice bath by half-filling a 600-mL beaker with ice.

A. Potassium Alum Synthesis

Cool flame: a nonluminous Bunsen flame with a reduced flow of natural gas

1. **Prepare the Aluminum Sample.** Cut an approximate 2-inch square of scrap aluminum from a beverage can and clean both sides (to remove the plastic coating on the inside, a paint covering on the outside) with steel wool or sand paper. Rinse the aluminum with deionized water. Cut the "clean" aluminum into small pieces.[1] Tare a 100-mL beaker and measure about 0.5 g (± 0.01 g) of aluminum pieces.

2. **Dissolve the Aluminum Pieces.** Move the beaker to a well-ventilated area, such as a fume hood. Add 10–12 mL of 4 *M* KOH to the aluminum pieces (**Caution:** *Wear safety glasses; do not splatter the solution, KOH is caustic.*) and swirl the reaction mixture. Warm the beaker *gently* with a **cool flame** or hot plate to initiate the reaction. As the reaction proceeds hydrogen gas is being evolved as is evidenced by the "fizzing" at the edges of the aluminum pieces.

   The dissolution of the aluminum pieces may take up to 20 minutes—it is important to maintain the solution at a level that is one-half to three-fourths of its original volume by adding small portions of deionized water during the dissolution process.[2]

3. **Gravity Filter the Reaction Mixture.** When no further reaction is evident, return the reaction mixture to the laboratory desk. Gravity filter the warm reaction mixture through a cotton plug or filter paper into a 100-mL beaker to remove the insoluble impurities (see Figures T.11d and T.11e). If solid particles appear in the filtrate, repeat the filtration. Rinse the filter with 2–3 mL of deionized water.

---

[1] The smaller the aluminum pieces, the more rapid is the reaction. Aluminum foil may be used in place of the scrap aluminum pieces.
[2] Some impurities, such as the label or the plastic lining of the can, may remain undissolved.

4. **Allow the Formation of Aluminum Hydroxide.** Allow the clear solution (the filtrate) to cool in the 100-mL beaker. While stirring and from a 25-mL graduated cylinder, *slowly* add, in 5-mL increments (because the reaction is exothermic!), approximately 15 mL of 6 M H_2SO_4 (**Caution:** *Avoid skin contact!*).

5. **Dissolve the Aluminum Hydroxide.** When the solution shows evidence of the white, gelatinous $Al(OH)_3$ precipitate in the acidified filtrate, stop adding the 6 M H_2SO_4. Gently heat the mixture until the $Al(OH)_3$ dissolves.

6. **Crystallize the Alum.** Remove the solution from the heat. Cool the solution in an ice bath. Alum crystals should form within 20 minutes. If crystals do *not* form, use a hot plate (Figure 15.2) to gently reduce the volume by one-half (*do not boil!*) and return to the ice bath. For larger crystals and a higher yield, allow the crystallization process to continue until the next laboratory period.

7. **Isolate and Wash the Alum Crystals.** Vacuum filter the alum crystals from the solution. Wash the crystals on the filter paper with two (cooled-to-ice temperature) 5-mL portions of a 50% (by volume) ethanol–water solution.[3] Maintain the vacuum suction until the crystals appear dry. Determine the mass (±0.01 g) of the crystals. Have your laboratory instructor approve the synthesis of your alum.

8. **Percent Yield.** Calculate the percent yield of your alum crystals.

Disposal: Discard the filtrate in the "Waste Salts" container.

CLEANUP: Rinse all glassware twice with tap water and twice with deionized water. All rinses can be discarded as advised by your instructor, followed by a generous amount of tap water.

The melting point of the alum sample can be determined with either a commercial melting point apparatus (Figure 15.5) or with the apparatus shown in Figure 15.6. Consult with your instructor.

B. Melting Point of the Alum

1. **Prepare the Alum in the Melting Point Tube.** Place finely ground, dry alum to a depth of about 0.5 cm in the bottom of a melting point capillary tube. To do this, place some alum on a piece of dry filter paper and "tap–tap" the open end of the capillary tube into the alum until the alum is at a depth of about 0.5 cm (Figure 15.3). Invert the capillary tube and compact the alum at the bottom of the tube—either drop the tube onto the lab bench through a 25-cm piece of glass tubing (Figure 15.4) or vibrate the capillary tube with a triangular file (Figure 15.4).

2. **Determine the Melting Point of the Alum.** Use either the apparatus in Figure 15.5 or 15.6.

   a. *Melting point apparatus, Figure 15.5.* Place the capillary tube containing the sample into the melting point apparatus.

   b. *Melting point apparatus, Figure 15.6.* Mount the capillary tube containing the sample *beside* the thermometer bulb (Figure 15.6 insert) with a rubber band or tubing. Transfer the sample/thermometer into the water bath

   c. *Heat the sample.* Slowly heat the sample at about 3°C per minute while carefully watching the alum sample. When the solid melts, note the temperature. Allow the sample to cool to just below this *approximate* melting point; at a 1°C per minute heating rate, heat again until it melts. Repeat the cooling/heating cycle until reproducibility is obtained—this is the melting point of your alum. Record this on the Report Sheet.

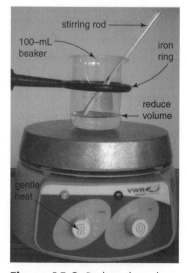

Figure 15.2 Reduce the volume of the solution on a hot plate.

---

[3]The alum crystals are marginally soluble in a 50% (by volume) ethanol–water solution.

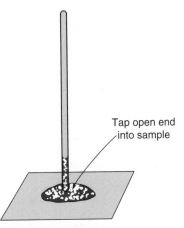

Figure 15.3 Invert the capillary melting point tube into the sample and "tap."

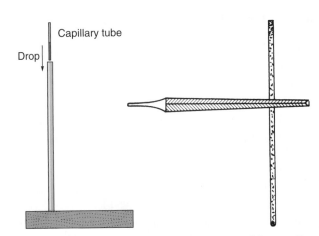

Figure 15.4 Compact the sample to the bottom of the capillary melting point tube by (a) dropping the capillary tube into a long piece of glass tubing or (b) vibrating the sample with a triangular file.

Disposal: Dispose of the melting point tube in the "Glass Only" container.

The Next Step

Other alums (Table 15.1) can be similarly synthesized. (1) Design a procedure for synthesizing other alums. (2) Research the role of alums in soil chemistry, in the dyeing industry, the leather industry, in water purification, or food industry. (3) "Growing" alum crystals can be a very rewarding scientific accomplishment, especially the "big" crystals! How is it done?

Figure 15.5
Electrothermal melting point apparatus.

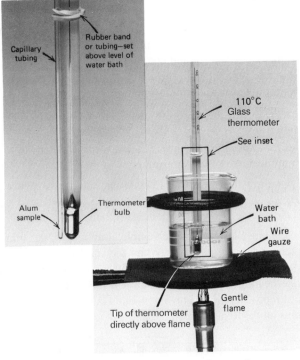

Figure 15.6 Melting point apparatus for an alum.

Synthesis of an Alum

Date _____ Lab Sec. _____ Name _____ Desk No. _____

1. An alum is a "double salt" consisting of a monovalent cation, a trivalent cation, and two sulfate ions with twelve waters of hydration (waters of crystallization) as part of the crystalline structure.
   a. Are the twelve waters of hydration used to calculate the theoretical yield of the alum? Explain.

   b. The 12 waters of hydration are "hydrated (strongly attracted)" to the metal ions in the crystalline alum structure. Are the water molecules more strongly hydrated to the monovalent cation or the trivalent cation? Explain.

   c. What might you expect to happen to the alum if it were heated to a high temperature? Explain.

2. Potassium alum, synthesized in this experiment, has the formula $KAl(SO_4)_2 \cdot 12\ H_2O$, but written as a "double salt" its formula is $K_2SO_4 \cdot Al_2(SO_4)_3 \cdot 24\ H_2O$. Refer to Table 15.1 and write the formula of
   a. chrome alum as a double salt

   b. ferric alum as a double salt

3. a. Experimental Procedure, Part A.3. What is the technique for securing a piece of filter paper into a funnel for gravity filtration?

b. Experimental Procedure, Part A.7. What is the technique for securing a piece of filter paper into a Büchner funnel for vacuum filtration?

c. Experimental Procedure, Part A.7. The alum crystals are washed of contaminants with an alcohol/water mixture rather than with deionized water. Explain.

4. For the synthesis of potassium alum, advantage is taken by the fact that aluminum hydroxide is amphoteric, meaning it can react as an acid (with a base) or as a base (with an acid).
Write equations showing the amphoteric properties of $Al(OH)_3$.

5. An aluminum can is cut into small pieces. A 0.74-g sample of the aluminum "chips" is used to prepare potassium alum according to the procedure described in this experiment. Calculate the theoretical yield (in grams) of potassium alum that could be obtained in the reaction. The molar mass of potassium alum is 474.39 g/mol.

6. A mass of 12.96 g of $(NH_4)_2SO_4$ (molar mass = 132.06 g/mol) is dissolved in water. After the solution is heated, 25.11 g of $Al_2(SO_4)_3 \cdot 18H_2O$ (molar mass = 666.36 g/mol) is added. Calculate the theoretical yield of the resulting alum (refer to Table 15.1 for the formula of the alum). *Hint:* This is a limiting reactant problem.

7. Experimental Procedure, Part B.2. To measure the melting point of the alum, the temperature sample is *slowly* increased. Why does this procedure ensure a more accurate melting point measurement?

Synthesis of an Alum

Date _____ Lab Sec. _____ Name _____ Desk No. _____

A. Potassium Alum Synthesis

1. Mass of aluminum (*g*)　　　_____

2. Mass of alum synthesized (*g*)　_____

3. Instructor's approval of alum　_____

4. Theoretical yield (*g*)

5. Percent yield (%)

B. Melting Point of the Alum

1. Melting point (°C) _____ _____ _____

2. Average melting point (°C)　_____

The melting point of potassium alum is 92.5°C. Comment on the purity of your sample based upon your experimental melting point.

Calculations:

Laboratory Questions

Circle the questions that have been assigned.

1. Part A.1. The aluminum sample is not cut into small pieces, but rather left as one large piece.
    a. How will this oversight affect the progress of completing the experimental procedure? Explain.
    b. Will the percent yield of the alum be too high, too low, or unaffected by the oversight? Explain.

2. Part A.3. Aluminum pieces inadvertently collect on the filter.
    a. If left on the filter, will the percent yield of the alum be reported too high or too low? Explain.
    b. If the aluminum pieces are detected on the filter, what steps would be used to remedy the observation? Explain.

3. Part A.4. Too much sulfuric acid is added. What observation would be expected in Part A.6? Explain.

4. Part A.8. Explain why a percent yield of greater than 100% might be reported for the synthesis of the alum.

5. Part B.2. Explain why the melting point of your prepared alum must either be equal to or be less than the actual melting point of the alum. Consult with your laboratory instructor.

6. Experimentally, how can the moles of the waters of hydration in an alum sample be determined? See Experiment 5.

7. A greater yield and larger alum crystals may be obtained by allowing the alum solution to cool in a refrigerator overnight or for a few days. Explain.

LeChâtelier's Principle; Buffers*

The chromate ion (left) is yellow and the dichromate ion (right) is orange. An equilibrium between the two ions is affected by changes in pH.

OBJECTIVES

- To study the effects of concentration and temperature changes on the position of equilibrium in a chemical system
- To study the effect of strong acid and strong base addition on the pH of buffered and unbuffered systems
- To observe the common-ion effect on a dynamic equilibrium

The following techniques are used in the Experimental Procedure

TECHNIQUES

INTRODUCTION

Most chemical reactions do not produce a 100% yield of product, not because of experimental technique or design, but rather because of the chemical characteristics of the reaction. The reactants initially produce the expected products, but after a period of time the concentrations of the reactants and products *stop* changing.

This apparent cessation of the reaction before a 100% yield is obtained implies that the chemical system has reached a state where the reactants combine to form the products at a rate equal to that of the products re-forming the reactants. This condition is a state of **dynamic equilibrium,** characteristic of all reversible reactions.

For the reaction

$$2\ NO_2(g) \rightleftharpoons N_2O_4(g) + 58\ kJ \tag{16.1}$$

chemical equilibrium is established when the rate at which two NO_2 molecules react equals the rate at which one N_2O_4 molecule dissociates (Figure 16.1).

If the concentration of one of the species in the equilibrium system changes, or if the temperature changes, the equilibrium tends to *shift* in a way that compensates for the change. For example, assuming the system represented by Equation 16.1 is in a state of dynamic equilibrium, if more NO_2 is added, the probability of its reaction with other NO_2 molecules increases. As a result, more N_2O_4 forms and the reaction shifts to the *right,* until equilibrium is re-established.

A general statement governing all systems in a state of dynamic equilibrium follows:

> *If an external stress (change in concentration, temperature, etc.) is applied to a system in a state of dynamic equilibrium, the equilibrium shifts in the direction that minimizes the effect of that stress.*

Figure 16.1 A dynamic equilibrium exists between reactant NO_2 molecules and product N_2O_4 molecules.

*Numerous online websites discuss LeChâtelier's Principle.

This is **LeChâtelier's principle,** proposed by Henri Louis LeChâtelier in 1888.

Often the equilibrium concentrations of all species in the system can be determined. From this information an equilibrium constant can be calculated; its magnitude indicates the relative position of the equilibrium. This constant is determined in Experiments 22, 26, and 34.

Two factors affecting equilibrium position are studied in this experiment: changes in concentration and changes in temperature.

Changes in Concentration

Metal–Ammonia Ions. Aqueous solutions of copper ions and nickel ions appear sky blue and green, respectively. The colors of the solutions change, however, in the presence of added ammonia, NH_3. Because the metal–ammonia bond is stronger than the metal–water bond, ammonia substitution occurs and the following equilibria shift *right,* forming the metal–ammonia **complex ions:**[1]

Complex ion: a metal ion bonded to a number of Lewis bases. The complex ion is generally identified by its enclosure with brackets, [ ].

$$[Cu(H_2O)_4]^{2+}(aq) + 4\,NH_3(aq) \rightleftharpoons [Cu(NH_3)_4]^{2+}(aq) + 4\,H_2O(l) \qquad (16.2)$$

$$[Ni(H_2O)_6]^{2+}(aq) + 6\,NH_3(aq) \rightleftharpoons [Ni(NH_3)_6]^{2+}(aq) + 6\,H_2O(l) \qquad (16.3)$$

Addition of strong acid, H^+, affects these equilibria by its reaction with ammonia (a base) on the left side of the equations.

$$NH_3(aq) + H^+(aq) \rightarrow NH_4^+(aq) \qquad (16.4)$$

The ammonia is removed from the equilibria, and the reactions shift *left* to relieve the stress caused by the removal of the ammonia, re-forming the aqueous Cu^{2+} (sky blue) and Ni^{2+} (green) solutions. For copper ions, this equilibrium shift may be represented as

$$[Cu(H_2O)_4]^{2+}(aq) + 4\,NH_3(aq) \overset{\leftarrow}{\rightleftharpoons} [Cu(NH_3)_4]^{2+}(aq) + 4\,H_2O(l) \qquad (16.5)$$
$$\quad\quad\quad | \quad 4\,H^+(aq)$$
$$\quad\quad\quad \rightarrow 4\,NH_4^+(aq)$$

$[Cu(H_2O)_4]^{2+}$ is a sky-blue color (left), but $[Cu(NH_3)_4]^{2+}$ is a deep-blue color (right).

Multiple Equilibria with the Silver Ion

Many salts are only slightly soluble in water. Silver ion, Ag^+, forms a number of these salts. Several equilibria involving the relative solubilities of the silver salts of the carbonate, CO_3^{2-}, chloride, Cl^-, iodide, I^-, and sulfide, S^{2-}, anions are investigated in this experiment.

Silver Carbonate Equilibrium. The first of the silver salt equilibria is a saturated solution of silver carbonate, Ag_2CO_3, in dynamic equilibrium with its silver and carbonate ions in solution.

$$Ag_2CO_3(s) \rightleftharpoons 2\,Ag^+(aq) + CO_3^{2-}(aq) \qquad (16.6)$$

Nitric acid, HNO_3, dissolves silver carbonate: H^+ ions react with (and remove) the CO_3^{2-} ions on the right; the system, in trying to replace the CO_3^{2-} ions, shifts to the *right.* The Ag_2CO_3 dissolves, and carbonic acid, H_2CO_3, forms.

$$Ag_2CO_3(s) \overset{\rightarrow}{\rightleftharpoons} 2\,Ag^+(aq) + CO_3^{2-}(aq) \qquad (16.7)$$
$$\quad\quad\quad | \quad 2\,H^+(aq)$$
$$\quad\quad\quad \rightarrow H_2CO_3(aq) \rightarrow H_2O(l) + CO_2(g) \quad (16.8)$$

The carbonic acid, being unstable at room temperature and pressure, decomposes to water and carbon dioxide. The silver ion and nitrate ion (from HNO_3) remain in solution.

---

[1] A further explanation of complex ions appears in Experiment 36.

Silver Chloride Equilibrium. Chloride ion precipitates silver ion as AgCl. Addition of chloride ion (from HCl) to the above solution, containing Ag^+, causes the formation of a silver chloride, AgCl, precipitate, now in dynamic equilibrium with its Ag^+ and Cl^- ions (Figure 16.2).

$$Ag^+(aq) + Cl^-(aq) \rightleftharpoons AgCl(s) \qquad (16.9)$$

Aqueous ammonia, NH_3, "ties up" (i.e., it forms a complex ion with) silver ion, producing the soluble diamminesilver(I) ion, $[Ag(NH_3)_2]^+$. The addition of NH_3 removes silver ion from the equilibrium in Equation 16.9, shifting its equilibrium position to the *left* and causing AgCl to dissolve.

$$\overset{\longleftarrow}{Ag^+(aq) + Cl^-(aq) \rightleftharpoons AgCl(s)} \qquad (16.10)$$
$$\text{2 } NH_3(aq)$$
$$[Ag(NH_3)_2]^+(aq)$$

Figure 16.2 Solid AgCl quickly forms when solutions containing Ag^+ and Cl^- are mixed.

Adding acid, H^+, to the solution again frees silver ion to recombine with chloride ion and re-forms solid silver chloride. This occurs because H^+ reacts with the NH_3 (see Equation 16.4) in Equation 16.10, restoring the system to that described by Equation 16.9.

$$\overset{\longrightarrow}{Ag^+(aq) + Cl^-(aq) \rightleftharpoons AgCl(s)} \qquad (16.11)$$
$$2\ NH_3(aq) + 2\ H^+(aq) \rightarrow 2\ NH_4^+(aq)$$
$$[Ag(NH_3)_2]^+(aq)$$

Silver Iodide Equilibrium. Iodide ion, I^- (from KI), added to the $Ag^+(aq) + 2\ NH_3(aq) \rightleftharpoons Ag(NH_3)_2^+ (aq)$ equilibrium in Equation 16.10 results in the formation of solid silver iodide, AgI.

$$\overset{\longleftarrow}{Ag^+(aq) + 2\ NH_3(aq) \rightleftharpoons [Ag(NH_3)_2]^+(aq)} \qquad (16.12)$$
$$I^-(aq)$$
$$AgI(s)$$

The iodide ion removes the silver ion, causing a dissociation of the $[Ag(NH_3)_2]^+$ ion and a shift to the *left*.

Silver Sulfide Equilibrium. Silver sulfide, Ag_2S, is less soluble than silver iodide, AgI. Therefore an addition of sulfide ion (from Na_2S) to the $AgI(s) \rightleftharpoons Ag^+(aq) + I^-(aq)$ dynamic equilibrium in Equation 16.12 removes silver ion; AgI dissolves but solid silver sulfide forms.

$$\overset{\longrightarrow}{AgI(s) \rightleftharpoons Ag^+(aq) + I^- (aq)} \qquad (16.13)$$
$$\tfrac{1}{2}\ S^{2-}(aq)$$
$$\tfrac{1}{2}\ Ag_2S(s)$$

Buffers

In many areas of research, chemists need an aqueous solution that resists a pH change when small amounts of acid or base are added. Biologists often grow cultures that are very susceptible to changes in pH and therefore a buffered medium is required (Figure 16.3).

A buffer solution must be able to consume small additions of H_3O^+ and OH^- without undergoing large pH changes. Therefore, it must have present a basic component that can react with added H_3O^+ *and* an acidic component that can react with added OH^-. Such a buffer solution consists of a weak acid and its conjugate base (or weak base and its conjugate acid). This experiment shows that an acetic acid–acetate buffer system can minimize large pH changes.

$$CH_3COOH(aq) + H_2O(l) \rightleftharpoons H_3O^+(aq) + CH_3CO_2^-(aq) \qquad (16.14)$$

The addition of OH^- shifts the buffer equilibrium, according to LeChâtelier's principle, to the *right* because of its reaction with H_3O^+, forming H_2O. The shift right is by an amount that is essentially equal to the moles of OH^- added to the buffer system. Thus, the amount of $CH_3CO_2^-$ increases and the amount of CH_3COOH decreases by an amount equal to the moles of OH^- added.

$$CH_3COOH(aq) + H_2O(l) \overset{\rightarrow}{\rightleftharpoons} H_3O^+(aq) + CH_3CO_2^-(aq) \qquad (16.15)$$
$$\begin{array}{c} | \; OH^-(aq) \\ \downarrow 2\,H_2O(l) \end{array}$$

Conversely, the addition of H_3O^+ from a strong acid to the buffer system causes the equilibrium to shift *left:* the H_3O^+ combines with the acetate ion (a base) to form more acetic acid, an amount (moles) equal to the amount of H_3O^+ added to the system.

$$CH_3COOH(aq) + H_2O(l) \overset{\leftarrow}{\rightleftharpoons} H_3O^+(aq) + CH_3CO_2^-(aq) \qquad (16.16)$$
$$\uparrow \; H_3O^+(aq)$$

As a consequence of the addition of strong acid, the amount of CH_3COOH increases and the amount of $CH_3CO_2^-$ decreases by an amount equal to the moles of strong acid added to the buffer system.

This experiment compares the pH changes of a buffered solution to those of an unbuffered solution when varying amounts of strong acid or base are added to each.

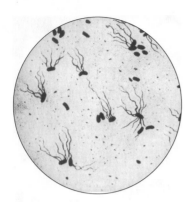

Figure 16.3 Bacteria cultures survive in media that exist over a narrow pH range. Buffers are used to control large changes in pH.

Common-Ion Effect

The effect of adding an ion or ions common to those already present in a system at a state of dynamic equilibrium is called the **common-ion effect.** The effect is observed in this experiment for the following equilibrium:

$$4\,Cl^-(aq) + [Co(H_2O)_6]^{2+}(aq) \rightleftharpoons [CoCl_4]^{2-}(aq) + 6\,H_2O(l) \qquad (16.17)$$

Ligand: a Lewis base that donates a lone pair of electrons to a metal ion, generally a transition metal ion (see Experiment 36).

Equation 16.17 represents an equilibrium of the **ligands** Cl^- and H_2O bonded to the cobalt(II) ion—the equilibrium is shifted because of a change in the concentrations of the chloride ion and water.

Changes in Temperature

Referring again to Equation 16.1,

$$2\,NO_2(g) \rightleftharpoons N_2O_4(g) + 58\ kJ \qquad \text{(repeat of 16.1)}$$

Exothermic: characterized by energy release from the system to the surroundings

The reaction for the formation of colorless N_2O_4 is **exothermic** by 58 kJ. To favor the formation of N_2O_4, the reaction vessel should be kept cool (Figure 16.4 right); removing heat from the system causes the equilibrium to replace the removed heat and the equilibrium therefore shifts *right.* Added heat shifts the equilibrium in the direction that absorbs heat; for this reaction, a shift to the left occurs with addition of heat.

Coordination sphere: all ligands of the complex ion (collectively with the metal ion they are enclosed in square brackets when writing the formula of the complex ion). See Experiment 36.

This experiment examines the effect of temperature on the system described by Equation 16.17. This system involves an equilibrium between the **coordination spheres,** the water versus the Cl^- about the cobalt(II) ion; the equilibrium is concentration *and* temperature dependent. The tetrachlorocobaltate(II) ion, $[CoCl_4]^{2-}$ is more stable at higher temperatures.

EXPERIMENTAL PROCEDURE

Procedure Overview: A large number of qualitative tests and observations are performed. The effects that concentration changes and temperature changes have on a system at equilibrium are observed and interpreted using LeChâtelier's principle. The functioning of a buffer system and the effect of a common ion on equilibria are observed.

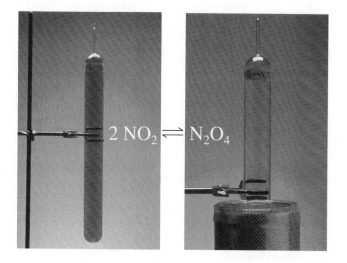

$$2\,NO_2 \rightleftharpoons N_2O_4$$

Figure 16.4 NO_2, a red-brown gas (left), is favored at higher temperatures; N_2O_4, a colorless gas (right), is favored at lower temperatures.

Perform this experiment with a partner. At each circled, superscript ①–㉑ in the procedure, *stop,* and record your observations on the Report Sheet. Discuss your observations with your lab partner and instructor. Account for the changes in appearance of the solution after each addition in terms of LeChâtelier's principle.

Ask your instructor which parts of the Experimental Procedure are to be completed. Prepare a hot water bath for Part E.

1. **Formation of Metal–Ammonia Ions.** Place 1 mL (<20 drops) of 0.1 *M* $CuSO_4$ (or 0.1 *M* $NiCl_2$) in a small, clean test tube.① Add drops of "conc" NH_3 (**Caution:** *strong odor, do not inhale*) until a color change occurs and the solution is clear (*not* colorless).②

2. **Shift of Equilibrium.** Add drops of 1 *M* HCl until the color again changes.③

A. Metal-Ammonia Ions

1. **Silver Carbonate Equilibrium.** In a 150-mm test tube (Figure 16.5) add ½ mL (≤10 drops) of 0.01 *M* $AgNO_3$ to ½ mL of 0.1 *M* Na_2CO_3.④ Add drops of 6 *M* HNO_3 (**Caution:** *6 M HNO₃ reacts with the skin!*) to the precipitate until evidence of a chemical change occurs.⑤

2. **Silver Chloride Equilibrium.** To the "clear" solution from Part B.1, add ~ 5 drops of 0.1 *M* HCl.⑥ Add drops of conc NH_3 (**Caution!**) *avoid breathing vapors and avoid skin contact*) until evidence of a chemical change.*⑦ Reacidify the solution with 6 *M* HNO_3 (**Caution!**) and record your observations.⑧ What happens if excess conc NH_3 is again added? Try it.⑨

3. **Silver Iodide Equilibrium.** After "trying it," add drops of 0.1 *M* KI.⑩

4. **Silver Sulfide Equilibrium.** To the mixture from Part B.3, add drops of 0.1 *M* Na_2S† until evidence of chemical change has occurred.⑪

B. Multiple Equilibria with the Silver Ion

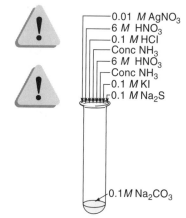

- 0.01 *M* $AgNO_3$
- 6 *M* HNO_3
- 0.1 *M* HCl
- Conc NH_3
- 6 *M* HNO_3
- Conc NH_3
- 0.1 *M* KI
- 0.1 *M* Na_2S

0.1*M* Na_2CO_3

Figure 16.5 Sequence of added reagents for the study of silver ion equilibria.

*At this point, the solution should be "clear and colorless."
†The Na_2S solution should be freshly prepared.

Disposal: Dispose of the waste silver salt solutions in the "Waste Silver Salts" container.

CLEANUP: Rinse the test tube twice with tap water and discard in the "Waste Silver Salts" container. Rinse twice with deionized water and discard in the sink.

C. A Buffer System

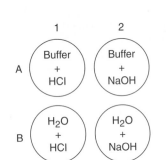

The use of a well plate is recommended. Appropriately labeled 75-mm test tubes are equally useful for performing the experiments.

1. **Preparation of Buffered and Unbuffered Systems.** Transfer 10 drops of 0.10 *M* CH₃COOH to wells A1 and A2 of a 24-well plate, add 3 drops of universal indicator,† and note the color.⑫ Compare the color of the solution with the pH color chart for the universal indicator.⑫ Now add 10 drops of 0.10 *M* NaCH₃CO₂ to each well.⑬ Place 20 drops of deionized water into wells B1 and B2 and add 3 drops of universal indicator.⑭

2. **Effect of Strong Acid.** Add 5–6 drops of 0.10 *M* HCl to wells A1 and B1, estimate the pH, and record each pH *change*.⑮

3. **Effect of Strong Base.** Add 5–6 drops of 0.10 *M* NaOH to wells A2 and B2, estimate the pH, and record each pH *change*.⑯

4. **Effect of a Buffer System.** Explain the observed pH change for a buffered system (as compared with an unbuffered system) when a strong acid or strong base is added to it.⑰

D. [Co(H₂O)₆]²⁺, [CoCl₄]²⁻ Equilibrium (Common-Ion Effect)

⚠️

1. **Effect of Concentrated HCl.** Place about 10 drops of 1.0 *M* CoCl₂ in a 75-mm test tube.⑱ Add drops of conc HCl (**Caution:** *avoid inhalation and skin contact*) until a color change occurs.⑲ Slowly add water to the system and stir.⑳

E. [Co(H₂O)₆]²⁺, [CoCl₄]²⁻ Equilibrium (Temperature Effect)

1. **What Does Heat Do?** Place about 1.0 mL of 1.0 *M* CoCl₂ in a 75-mm test tube into the boiling water bath. Compare the color of the "hot" solution with that of the original "cool" solution.㉑

Disposal for Parts A, C, D, and E: Dispose of the waste solutions in the "Waste Salt Solutions" container.

CLEANUP: Rinse the test tubes and 24-well plate twice with tap water and discard in the "Waste Salt Solutions" container. Do two final rinses with deionized water and discard in the sink.

The Next Step

Buffers are vital to biochemical systems. (1) What is the pH of blood and what are the blood buffers that maintain that pH? (2) Natural waters (rivers, oceans, etc.) are buffered for the existence of plant and animal life (Experiment 20). What are those buffers? Experimentally, see how they resist pH changes with the additions of strong acid and/or strong base. (3) Review Prelaboratory Question #6, equilibria which also account for the existence of hard waters (Experiment 21).

---

†pH indicator paper may be substituted for the universal indicator to measure the pH of the solutions.

LeChâtelier's Principle; Buffers

Date _____ Lab Sec. _____ Name _____ Desk No. _____

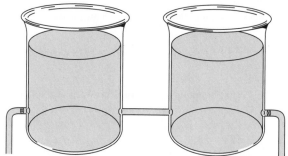

1. a. Describe the dynamic equilibrium that exists in the two
       water tanks at right.

    b. Explain how LeChâtelier's principle applies when the
       faucet on the right tank is opened.

    c. Explain how LeChâtelier's principle applies when water is added to the right tank.

2. a. Experimental Procedure. Cite the reason for each of the four cautions in the experiment.

    b. Experimental Procedure. How is "bumping" avoided in the preparation of a hot water bath?

3. The following chemical equilibria are studied in this experiment. To become familiar with their behavior, indicate the
    direction, left or right, of the equilibrium shift when the accompanying stress is applied to the system.

    a. $NH_3(aq)$ is added to $Ag^+(aq) + Cl^-(aq) \rightleftharpoons AgCl(s)$ _____

    b. $HNO_3(aq)$ is added to $Ag_2CO_3(s) \rightleftharpoons Ag^+(aq) + CO_3^{2-}(aq)$ _____

    c. $KI(aq)$ is added to $Ag^+(aq) + 2\,NH_3(aq) \rightleftharpoons [Ag(NH_3)_2]^+(aq)$ _____

    d. $Na_2S(aq)$ is added to $AgI(s) \rightleftharpoons Ag^+(aq) + I^-(aq)$ _____

    e. $KOH(aq)$ is added to $CH_3COOH(aq) + H_2O(l) \rightleftharpoons H_3O^+(aq) + CH_3CO_2^-(aq)$ _____

    f. $HCl(aq)$ is added to $4\,Cl^-(aq) + Co(H_2O)_6^{2+}(aq) \rightleftharpoons CoCl_4^{2-}(aq) + 6\,H_2O(l)$ _____

4. Experimental Procedure, Part C. Explain how the addition of $NaCH_3CO_2$ affects the pH of a CH_3COOH solution.

5. A state of dynamic equilibrium, $Ag_2CO_3(s) \rightleftharpoons 2Ag^+(aq) + CO_3^{2-}(aq)$, exists in solution.

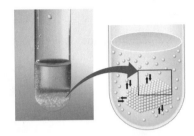

   a. What happens to the equilibrium if more $Ag_2CO_3(s)$ is added to the system?

   b. What happens to the equilibrium if $AgNO_3(aq)$ is added to the system?

   *c. What happens to the equilibrium if $HCl(aq)$ is added to the system?

6. a. The carbon dioxide of the atmosphere, being a nonmetallic oxide and acid anhydride, has a low solubility in rainwater but produces a slightly acidic solution. Write an equilibrium equation for the dissolution of carbon dioxide in water and identify the acid that is formed.

   b. Calcium carbonate, the major component of limestone, is slightly soluble in water. Write an equilibrium equation showing the slight solubility of calcium carbonate in water.

   c. Write a Brønsted equation for the weak acid formed in part a, showing how it produces hydronium ion in aqueous solution.

   d. Using LeChâtelier's principle, explain how the weak acid formed in rainwater causes the dissolution of calcium carbonate. The consequence of this reaction contributes to the degree of hardness in underground and surface water supplies and also accounts for the formation of stalagmites and stalactites.

LeChâtelier's Principle; Buffers

Date _____ Lab Sec. _____ Name _____ Desk No. _____

A. Metal–Ammonia Ions

| $CuSO_4(aq)$ or $NiCl_2(aq)$ | $[Cu(NH_3)_4]^{2+}$ or $[Ni(NH_3)_6]^{2+}$ | HCl Addition |

Color① _____ ② _____ ③ _____

Account for the effects of $NH_3(aq)$ and $HCl(aq)$ on the $CuSO_4$ or $NiCl_2$ solution.

B. Multiple Equilibria with the Silver Ion

④Observation and net ionic equation for reaction.

⑤Account for the observed chemical change from HNO_3 addition.

⑥Observation from HCl addition and net ionic equation for the reaction.

⑦Effect of conc NH₃. Explain.

⑧What does the HNO₃ do? Explain.

⑨What result does excess NH₃ produce? Explain.

⑩Effect of added KI. Explain.

⑪Effect of Na₂S and net ionic equation for the reaction.

C. A Buffer System

⑫Write the Brønsted acid equation for $CH_3COOH(aq)$.

Color of universal indicator in CH_3COOH _____ pH _____

⑬Color of universal indicator after addition of $NaCH_3CO_2$ _____ pH _____

Effect of $NaCH_3CO_2$ on the equilibrium.

⑭Color of universal indicator in water _____ pH _____

	Buffer System		Water	
	Well A1 (or test tube)	**Well A2** (or test tube)	**Well B1** (or test tube)	**Well B2** (or test tube)
⑮Color after 0.10 M HCl addition	_____	...	_____	...
Approximate pH	_____	...	_____	...
Approximate ΔpH	_____	...	_____	...
⑯Color after 0.10 M NaOH addition	...	_____	...	_____
Approximate pH	...	_____	...	_____
Approximate ΔpH	...	_____	...	_____

⑰Discuss in detail the magnitude of the changes in pH that are observed in wells A1 and A2 relative to those observed in wells B1 and B2.

D. $[Co(H_2O)_6]^{2+}$, $[CoCl_4]^{2-}$ Equilibrium (Common-Ion Effect)

⑱Color of $CoCl_2(aq)$

⑲Observation from conc HCl addition and net ionic equation for the reaction.

⑳Account for the observation resulting from the addition of water.

E. [Co(H₂O)₆]²⁺, [CoCl₄]²⁻ Equilibrium (Temperature Effect)

$\text{E. } [Co(H_2O)_6]^{2+}, [CoCl_4]^{2-}$ **Equilibrium (Temperature Effect)**

㉑ Effect of heat. What happens to the equilibrium? Explain.

Laboratory Questions

Circle the questions that have been assigned.

1. Part A.1. NH_3 is a weak base; NaOH is a strong base. Predict what would appear in the solution if NaOH were been added to the $CuSO_4$ solution instead of the NH_3. (*Hint:* See Appendix G.)

2. Part B.1a. HNO_3, a strong acid, is added to shift the Ag_2CO_3 equilibrium (Equation 16.6) to the right. Explain why the shift occurs.
   b. What would have been observed if HCl (also a strong acid) had been added instead of the HNO_3?

3. Part B.4. Silver bromide, a pale yellow precipitate, is *more* soluble than silver iodide. Predict what would happen if 0.1 *M* NaBr had been added to the solution in Part B.3 instead of the Na_2S solution. Explain.

4. Write an equation that shows the pH dependence on the chromate, CrO_4^{2-}/dichromate, $Cr_2O_7^{2-}$, equilibrium system (see opening photo of the experiment).

5. Part C. HCl(*aq*) is a much stronger acid that $CH_3COOH(aq)$. However, when 5 drops of 0.10 *M* HCl(*aq*) is added to 20 drops of a buffer solution that is 0.10 *M* CH_3COOH and 0.10 *M* $CH_3CO_2^-$ only a very small change in pH occurs. Explain.

6. Part C. Explain why equal volumes of 0.1 *M* CH_3COOH and 0.1 *M* $NaCH_3CO_2$ function as a buffer solution, but equal volumes of 0.1 *M* HCl and 0.1 *M* NaOH do not.

*7. Part C. At what point is a buffer solution no longer effective in resisting a pH change when a strong acid is added?

8. Part E. Consider the following endothermic equilibrium reaction system in aqueous solution:

$$4\,Cl^-(aq) + [Co(H_2O)_6]^{2+}(aq) \rightleftharpoons [CoCl_4]^{2-}(aq) + 6\,H_2O(l)$$

If the equilibrium system were stored in a vessel (with no heat transfer into or out of the vessel from the surroundings), predict what would happen to the temperature reading on a thermometer placed in the solution when hydrochloric acid is added. Explain.

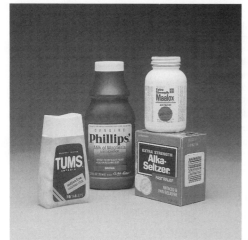

All antacids, as weak bases, reduce the acidity of the stomach.

- To determine the neutralizing effectiveness per gram of a commercial **antacid**

OBJECTIVE

The following techniques are used in the Experimental Procedure

TECHNIQUES

INTRODUCTION

Various commercial antacids claim to be the "most effective" for relieving acid indigestion. All antacids, regardless of their claims or effectiveness, have one purpose—to neutralize the *excess* hydrogen ion in the stomach to relieve acid indigestion.

The **pH** of the "gastric juice" in the stomach ranges from 1.0 to 2.0. This acid, primarily hydrochloric acid, is necessary for the digestion of foods. Acid is continually secreted while eating; consequently, overeating may lead to an excess of stomach acid, leading to acid indigestion and a pH less than 1. An excess of acid can, on occasion, cause an irritation of the stomach lining, particularly the upper intestinal tract, causing "heartburn." An antacid reacts with the hydronium ion to relieve the symptoms. Excessive use of antacids can cause the stomach to have a pH greater than 2, which stimulates the stomach to excrete additional acid, a potentially dangerous condition.

The most common bases used for over-the-counter antacids are:

aluminum hydroxide, $Al(OH)_3$ magnesium hydroxide, $Mg(OH)_2$
calcium carbonate, $CaCO_3$ sodium bicarbonate, $NaHCO_3$
magnesium carbonate, $MgCO_3$ potassium bicarbonate, $KHCO_3$

Milk of magnesia (Figure 17.1), an aqueous suspension of magnesium hydroxide, $Mg(OH)_2$, and sodium bicarbonate, $NaHCO_3$, commonly called baking soda, are simple antacids (and thus, bases) that neutralize hydronium ion, H_3O^+:

$$Mg(OH)_2(s) + 2\,H_3O^+(aq) \rightarrow Mg^{2+}(aq) + 4\,H_2O(l) \qquad (17.1)$$

$$NaHCO_3(aq) + H_3O^+(aq) \rightarrow Na^+(aq) + CO_2(g) + 2\,H_2O(l) \qquad (17.2)$$

The release of carbon dioxide gas from the action of sodium bicarbonate on hydronium ion (Equation 17.2) causes one to "belch."

To decrease the possibility of the stomach becoming too basic from the antacid, **buffers** are often added as part of the formulation of some antacids. The more common, "faster relief" commercial antacids that buffer the pH of the stomach are those

Antacid: dissolved in water, it forms a basic solution

pH: negative logarithm of the molar concentration of hydronium ion, $-\log [H_3O^+]$ (see Experiment 6)

Appendix D

Figure 17.1 Milk of magnesia is an aqueous suspension of slightly soluble magnesium hydroxide.

Buffers: substances in an aqueous system that are present for the purpose of resisting changes in acidity or basicity

Table 17.1 Common Antacids

Principal Active Ingredient(s)	Formulation	Commercial Antacid
$CaCO_3$	Tablet	Tums®, Titralac®, Chooz®, Maalox®
$CaCO_3$, $Mg(OH)_2$	Tablet	Rolaids®, Di-Gel®, Mylanta®
$MgCO_3$, $Al(OH)_3$	Tablet	Gaviscon® Extra Strength
$Mg(OH)_2$, $Al(OH)_3$	Tablet	Gelasil®, Tempo®
$NaHCO_3$, citric acid, aspirin	Tablet	Alka-Seltzer®
$Mg(OH)_2$	Tablet	Phillips'® Milk of Magnesia
$Mg(OH)_2$	Liquid	Phillips'® Milk of Magnesia
$Mg(OH)_2$, $Al(OH)_3$	Liquid	Maalox®, Mylanta® Extra Strength
$MgCO_3$, $Al(OH)_3$	Liquid	Gaviscon® Extra Strength

containing calcium carbonate, $CaCO_3$, and/or sodium bicarbonate. A HCO_3^-/CO_3^{2-} buffer system[1] is established in the stomach with these antacids.

$$CO_3^{2-}(aq) + H_3O^+(aq) \rightarrow HCO_3^-(aq) + H_2O(l) \tag{17.3}$$

$$HCO_3^-(aq) + H_3O^+(aq) \rightarrow CO_2(g) + 2\,H_2O(l) \tag{17.4}$$

Rolaids® is an antacid that consists of a combination of $Mg(OH)_2$ and $CaCO_3$ in a mass ratio of $1:5$, thus providing the effectiveness of the hydroxide base and the carbonate/bicarbonate buffer. Some of the more common over-the-counter antacids and their major active antacid ingredient(s) are listed in Table 17.1.

In this experiment, the "neutralizing power" of several antacids is determined using a strong acid–strong base titration. To obtain the quantitative data for the analysis, which requires a well-defined **endpoint** in the titration, the buffer action is eliminated.

Endpoint: the point in the titration when an indicator changes color

The buffering component of the antacid is eliminated when an *excess* of standardized hydrochloric acid, HCl, is added to the antacid solution; this addition drives the HCO_3^-/CO_3^{2-} reactions in Equations 17.3 and 17.4 far to the right. The solution is then heated to remove carbon dioxide. At this point *all* moles of base in the antacid (whether or not a buffer is present) have reacted with the standardized HCl solution.

The *unreacted* HCl is then titrated with a standardized sodium hydroxide, NaOH, solution.[2] This analytical technique is referred to as a **back titration.**

Back titration: an analytical procedure by which the analyte is "swamped" with an excess of a standardized neutralizing agent; the excess neutralizing agent is, in return, neutralized to a final stoichiometric point

The number of moles of base in the antacid of the commercial sample *plus* the number of moles of NaOH used in the titration equals the number of moles of HCl added to the original antacid sample:

$$\text{moles}_{\text{base, antacid}} + \text{moles}_{\text{NaOH}} = \text{moles}_{\text{HCl}} \tag{17.5}$$

A rearrangement of the equation provides the moles of base in the antacid in the sample:

$$\text{moles}_{\text{base, antacid}} = \text{moles}_{\text{HCl}} - \text{moles}_{\text{NaOH}} \tag{17.6}$$

The moles of base in the antacid per gram of antacid provide the data required for a comparison of the antacid effectiveness of commercial antacids. If purchase prices for the antacids are available, a final cost analysis of various antacids can be made.

EXPERIMENTAL PROCEDURE

Procedure Overview: The amount of base in an antacid sample is determined. The sample is dissolved, and the buffer components of the antacid are eliminated with the addition of an excess of standardized HCl solution. The unreacted HCl is back titrated with a standardized NaOH solution.

---

[1] A buffer system resists large changes in the acidity of a solution. To analyze for the amount of antacid in this experiment, we want to *remove* this buffering property to determine the total effectiveness of the antacid.

[2] A standardized NaOH solution is one in which the concentration of NaOH has been very carefully determined.

At least two analyses should be completed per antacid if two antacids are to be analyzed to compare their neutralizing powers. If only one antacid is to be analyzed, complete three trials.

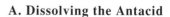

A. Dissolving the Antacid

1. **Determine the Mass of Antacid for Analysis.** If your antacid is a tablet, pulverize and/or grind the antacid tablet with a mortar and pestle. Measure and record the mass (± 0.001 g) of a 250-mL Erlenmeyer flask. Add no more than 0.2 g of the pulverized commercial antacid (or 0.2 g of a liquid antacid) to the flask and measure and record the combined mass (± 0.001 g).

2. **Prepare the Antacid for Analysis.** Pipet 25.0 mL of a standardized 0.1 *M* HCl solution (stomach acid equivalent) into the flask and swirl.[3] Record the actual molar concentration of the HCl on the Report Sheet. Warm the solution to a very *gentle* boil and maintain the heat for 1 minute to remove dissolved CO_2 . . . using a hot plate (Figure 17.2a) or a direct flame and a gentle swirl (Figure 17.2b). Add 4–8 drops of bromophenol blue indicator.[4] If the solution is blue, pipet an additional 10.0 mL of 0.1 *M* HCl into the solution and boil again. Repeat as often as necessary. Record the *total* volume of HCl that is added to the antacid.

B. Analyzing the Antacid Sample

Read Technique 16c closely

Read the buret to the correct number of significant figures

Obtain about 75 mL of a standardized 0.1 *M* NaOH solution. The solution may have been previously prepared by the stockroom personnel. If not, prepare a standardized 0.1 *M* NaOH solution, as described in Experiment 9. Consult with your laboratory instructor.

1. **Prepare the Buret for Titration.** Prepare a *clean* buret. Rinse the clean buret with two 3- to 5-mL portions of the standardized NaOH solution and drain through the buret tip. Record the actual molar concentration of the NaOH on the Report Sheet. Fill the buret with the NaOH solution; be sure no air bubbles are in the buret tip. Wait for 10–15 seconds, then read and record its initial volume, "using all certain digits *plus* one uncertain digit."

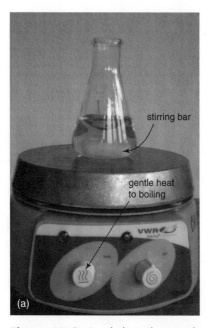
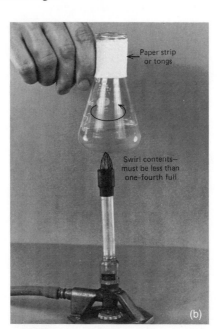

Figure 17.2 Gently heat the sample to remove CO_2 gas.

[3]If the sample is a tablet, swirl to dissolve. Some of the *inert* ingredients—fillers and binding agents used in the formulation of the antacid tablet—may not dissolve.
[4]Bromophenol blue is yellow at a pH less than 3.0 and blue at a pH greater than 4.6.

Be constantly aware of the use of significant figures that reflect the precision of your measuring instrument

2. **Titrate the Sample.** Once the antacid solution has cooled, titrate the sample with the NaOH solution to a faint blue endpoint. Watch closely; the endpoint may appear after only a few milliliters of titrant, depending on the concentration of the antacid in the sample. When a single drop (or half-drop) of NaOH solution changes the sample solution from yellow to blue, **stop.** Wait for 10–15 seconds and then read and record the final volume of NaOH solution in the buret.

3. **Repeat the Titration of the Same Antacid.** Refill the buret and repeat the experiment, starting at Part A.1.

4. **Analyze Another Antacid.** Perform the experiment, in duplicate, for another antacid. Record all data on the Report Sheet.

> *Disposal:* Dispose of the test solutions as directed by your instructor.

CLEANUP: Discard the remaining NaOH titrant as directed by your instructor. Flush the buret several times with tap water and dispense through the buret tip, followed by several portions of deionized water. Dispose of all buret washings in the sink.

C. Calculations

1. Determine the number of moles of HCl added to the antacid sample.

2. How many moles of NaOH titrant were required to neutralize the *unreacted* acid?

3. Calculate the number of moles of base in the antacid sample.

4. Calculate the number of moles of base in the antacid sample *per gram* of sample.

5. (Optional) If the store-bought antacid and its purchase price are available, calculate its cost per gram. Complete a cost analysis—determine the best buy!

Antacid Analysis

Date _____ Lab Sec. _____ Name _____ Desk No. _____

1. Write balanced equations for the reactions of the active ingredients in Gaviscon® Extra Strength with excess acid. See Table 17.1.

2. Identify the two most common anions present in antacids.

3. If the antacid for analysis (Part A.2) is known to be Phillips'® Milk of Magnesia, the solution does not need to be heated, but if the sample is Maalox®, the solution must be heated. Explain the difference in experimental procedures.

4. a. How much time should be allowed for the titrant to drain from the buret wall before a reading is made?

   b. What criterion is followed in reading and recording the volume of titrant of a buret?

   c. Bromophenol blue is the indicator used in detecting the endpoint for the antacid analysis in this experiment. What is the expected color change at the endpoint?

5. A 0.187-g sample of a CO_3^{2-} antacid is dissolved with 25.0 mL of 0.0984 M HCl. The hydrochloric acid that is *not* neutralized by the antacid is titrated to a bromophenol blue endpoint with 5.85 mL of 0.0911 M NaOH.

a. Assuming the active ingredient in the antacid sample is $CaCO_3$, calculate the mass of $CaCO_3$ in the sample.

b. What is the percent active ingredient in the antacid sample?

6. a. How many moles of stomach acid would be neutralized by one tablet of Regular Strength Maalox® that contains 600 mg of calcium carbonate?

b. Assuming the volume of the stomach to be 1.0 L, what will be the pH change of the stomach acid resulting from the ingestion of one Regular Strength Maalox® tablet?

7. One tablet of Regular Strength Maalox® claims to contain 600 mg $CaCO_3$. If 7.25 mL of 0.100 M NaOH titrant is used to back titrate the excess 0.100 M HCl from the analysis of one-third of a Maalox® tablet, how many milliliters of 0.100 M HCl must have been initially added to the 200 mg of $CaCO_3$ in the Maalox® sample?

Antacid Analysis

Date _____ Lab Sec. _____ Name _____ Desk No. _____

A. Dissolving the Antacid	*Trial 1*	*Trial 2*	*Trial 1*	*Trial 2*
1. Mass of flask (*g*)	_____	_____	_____	_____
2. Mass of flask + antacid sample (*g*)	_____	_____	_____	_____
3. Mass (or tared mass) of antacid sample (*g*)	▨▨▨▨	▨▨▨▨	▨▨▨▨	▨▨▨▨
4. Volume of HCl added (*mL*)	_____	_____	_____	_____
5. Molar concentration of HCl (*mol/L*)	_____		_____	

B. Analyzing the Antacid Sample

	Trial 1	*Trial 2*	*Trial 1*	*Trial 2*
1. Molar concentration of NaOH (*mol/L*)	_____		_____	
2. Buret reading, *initial* (*mL*)	_____	_____	_____	_____
3. Buret reading, *final* (*mL*)	_____	_____	_____	_____
4. Volume of NaOH (*mL*)	▨▨▨▨	▨▨▨▨	▨▨▨▨	▨▨▨▨

C. Calculations

	Trial 1	*Trial 2*	*Trial 1*	*Trial 2*
1. Moles of HCl added, *total* (*mol*)	▨▨▨▨	▨▨▨▨	▨▨▨▨	▨▨▨▨
2. Moles of NaOH added (*mol*)	▨▨▨▨	▨▨▨▨	▨▨▨▨	▨▨▨▨
3. Moles of base in antacid sample (*mol*)	▨▨▨▨	▨▨▨▨	▨▨▨▨	▨▨▨▨
4. $\dfrac{\text{mol base in antacid}}{\text{mass of antacid sample}}$ (*mol/g*)	▨▨▨▨ *	▨▨▨▨	▨▨▨▨ *	▨▨▨▨
5. Average $\dfrac{\text{mol base in antacid}}{\text{mass of antacid sample}}$ (*mol/g*)	▨▨▨▨		▨▨▨▨	
6. Cost per gram of antacid (*cents/g*), from bottle (if available)	▨▨▨▨		▨▨▨▨	
7. Antacid effectiveness $\left(\dfrac{\text{mol base in antacid}}{\text{cent}}\right)$	▨▨▨▨		▨▨▨▨	
8. Best buy		▨▨▨▨		

*Show calculation(s) for trial 1 on the next page.

Calculations:

Laboratory Questions

Circle the questions that have been assigned.

1. Part A.1. The antacid tablet for analysis was not finely pulverized before its reaction with hydrochloric acid. How might this technique error affect the reported amount of antacid in the sample? Explain.

2. Part A.2. The HCl(aq) solution has a lower concentration than what is indicated on the reagent bottle. Will this result indicate the presence of more or less moles of base in the antacid? Explain.

3. Part A.2. All of the CO_2 is not removed by gentle boiling after the addition of HCl. Will the reported amount of antacid in the sample be too high, too low, or unaffected? Explain.

4. Part B.1. An air bubble was initially trapped in the buret but was dispensed during the back titration of the unreacted HCl (Part B.2). As a result of this technique error, will the reported amount of antacid in the sample be too high or too low? Explain.

5. Part B.2. The bromophenol blue indicator changed from yellow to blue as desired in the back titration of the excess HCl. Upon leaving the solution in the flask until the end of the laboratory period however, the solution may return to a yellow color. Explain why this might be possible.

6. Part B.2. The bromophenol blue endpoint is surpassed in the back titration of the excess HCl with the sodium hydroxide titrant. As a result of this technique error, will the reported amount of antacid in the sample be too high or too low? Explain.

*7. A few of the "newer" antacids contain sodium citrate, $Na_3C_6H_5O_7$, as the effective, but more mild antacid ingredient.
   a. Write a balanced equation representing the antacid effect of the citrate ion, $C_6H_5O_7^{3-}$.
   b. Will 500 mg of $Na_3C_6H_5O_7$ (258.1 g/mol) or 500 mg of $Mg(OH)_2$ (58.32 g/mol) neutralize more moles of hydronium ion? Show calculations. Assume that both the $C_6H_5O_7^{3-}$ and the OH^- ions become fully protonated.

Potentiometric Analyses

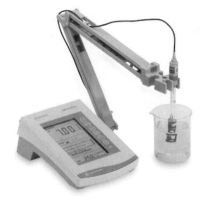

A modern pH meter with a combination electrode.

- To operate a pH meter
- To graphically determine a stoichiometric point
- To determine the molar concentration of a weak acid solution
- To determine the molar mass of a solid weak acid
- To determine the pK_a of a weak acid

The following techniques are used in the Experimental Procedure

INTRODUCTION

A "probe" connected to an instrument that provides a direct reading of the concentration of a particular substance in an aqueous system is a convenient form of analysis. Such convenience is particularly advantageous when the analysis for a large number of samples needs to be performed. The probe, or electrode, senses a difference in concentrations between the substance in solution and the substance in the probe itself. The concentration difference causes a voltage (or potential difference), which is recorded by the instrument, called a **potentiometer.**

A potentiometric chemical analysis is a powerful, convenient method for determining the concentrations of various ions in solution. To list only a few, the molar concentrations of the cations H^+, Li^+, Na^+, K^+, Ag^+, Ca^{2+}, Cu^{2+}, Pb^{2+}; the anions F^-, Cl^-, Br^-, I^-, CN^-, SO_4^{2-}; and the gases O_2, CO_2, NH_3, SO_2, H_2S, NO_x can be measured directly using an electrode specifically designed for its measurement (a specific selective electrode).

The H^+ concentration of a solution is measured with a potentiometer called a **pH meter,** an instrument that measures a potential difference (or voltage) caused by a difference in the hydrogen concentration of the test solution relative to that of the 0.1 M HCl contained within the electrode. The electrode, called a combination electrode, is shown in Figure 18.1.

The measured voltage recorded by the potentiometer, E_{cell}, is a function of the pH of the solution at 25°C by the equation

$$E_{cell} = E' + 0.0592 \, pH \qquad (18.1)$$

E' is a cell constant, an internal parameter that is characteristic of the pH meter and its electrode.

Potentiometer: an instrument that measures a potential difference— often called a voltmeter. See Experiment 32.

pH meter: an instrument that measures the pH of a solution

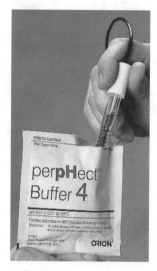

Buffer solutions are used to calibrate pH meters.

Buffer solution: a solution that maintains a relatively constant, reproducible pH

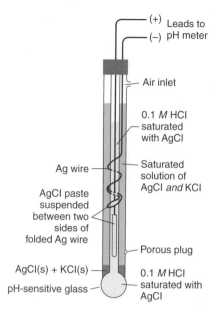

Figure 18.1 A combination electrode for measuring pH.

Before any pH measurements are made, the pH meter is calibrated (that is, E' is set). The electrode is placed into a **buffer solution** of known pH (see margin photo), and the E_{cell}, the potential difference between the $[H^+]$ of the glass electrode and the $[H^+]$ of the buffer, is manually adjusted to read the pH of the buffer—this adjustment sets E'.

The readout of the pH meter is E_{cell}, expressed in volts, but since E_{cell} is directly proportional to pH (Equation 18.1), the meter for the readout is expressed directly in pH units.

Indicators and pH of a Weak Acid Solution

The selection of an indicator for the titration of a strong acid with a strong base is relatively easy in that the color change at the stoichiometric point always occurs at a pH of 7 (at 25°C). Usually phenolphthalein can be used, because its color changes at a pH close to 7. However, when a weak acid is titrated with a strong base, the stoichiometric point is at a pH greater than 7 and a different indicator may need to be selected.[1] If the weak acid is an unknown acid, then the proper indicator cannot be selected because the pH at the stoichiometric point cannot be predetermined. The color change of a selected indicator may *not* occur at (or even near) the pH of the stoichiometric point for the titration. To better detect a stoichiometric point for the titration of an unknown weak acid, a pH meter is more reliable.

Molar Concentration of a Weak Acid Solution

Titrimetric analysis: a titration procedure that is chosen for an analysis

Titration curve: a data plot of pH versus volume of titrant

In Part A of this experiment, a **titrimetric analysis** is used to determine the molar concentration of a weak acid solution. A pH meter is used to detect the stoichiometric point of the titration. An acid–base indicator will *not* be used. A standardized sodium hydroxide solution is used as the titrant.[2]

The pH of a weak acid solution increases as the standardized NaOH solution is added. A plot of the pH of the weak acid solution as the strong base is being added, pH versus V_{NaOH}, is called the **titration curve** (Figure 18.2) for the reaction. The inflection

---

[1]The pH is greater than 7 at the stoichiometric point for the titration of a weak monoprotic acid because of the basicity of the conjugate base, A^-, of the weak acid, HA:

$$A^-(aq) + H_2O(l) \rightarrow HA(aq) + OH^-(aq)$$

[2]The procedure for preparing a standardized NaOH solution is described in Experiment 9.

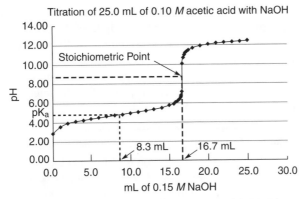

Titration of 25.0 mL of 0.10 *M* acetic acid with NaOH

Figure 18.2 Titration curve for 25.0 mL of 0.10 *M* CH₃COOH with 0.15 *M* NaOH

point in the sharp vertical portion of the plot (about midway on the vertical rise) is the **stoichiometric point.**

Stoichiometric point: also called the equivalence point

The moles of NaOH used for the analysis equals the volume of NaOH, dispensed from the buret, times its molar concentration.

$$\text{moles NaOH (mol)} = \text{volume (L)} \times \text{molar concentration (mol/L)} \qquad (18.2)$$

For a **monoprotic acid,** HA, one mole of OH⁻, neutralizes one mole of acid:

Monoprotic acid: a substance capable of donating a single proton, H⁺

$$\text{HA}(aq) + \text{OH}^-(aq) \rightarrow \text{H}_2\text{O}(l) + \text{A}^-(aq) \qquad (18.3)$$

The molar concentration of the acid is determined by dividing its number of moles in solution by its measured volume in liters:

$$\text{molar concentration HA (mol/L)} = \frac{\text{mol HA}}{\text{volume HA(L)}} \qquad (18.4)$$

For a **diprotic acid,** H₂X, 2 mol of OH⁻ neutralizes 1 mol of acid:

Diprotic acid: a substance capable of donating two protons

$$\text{H}_2\text{X}(aq) + 2\,\text{OH}^-(aq) \rightarrow 2\,\text{H}_2\text{O}(l) + \text{X}^{2-}(aq) \qquad (18.5)$$

Molar Mass of a Weak Acid

In Part B, the molar mass and the pK_a of an unknown *solid* weak acid are determined. The standardized NaOH solution is used to titrate a carefully measured mass of the *dissolved* acid to the stoichiometric point. A plot of pH versus V_{NaOH} is required to define the stoichiometric point.

The moles of acid is determined as described in Equations 18.2 and 18.3.

The molar mass of the acid is calculated from the moles of the solid acid neutralized at the stoichiometric point and its measured mass.

$$\text{molar mass (g/mol)} = \frac{\text{mass of solid acid(g)}}{\text{moles of solid acid}} \qquad (18.6)$$

pK_a of a Weak Acid

A weak acid, HA, in water undergoes only partial ionization,

$$\text{HA}(aq) + \text{H}_2\text{O}(l) \rightleftharpoons \text{H}_3\text{O}^+(aq) + \text{A}^-(aq) \qquad (18.7)$$

At equilibrium conditions, the mass action expression for the weak acid system equals the equilibrium constant.

$$K_a = \frac{[\text{H}_3\text{O}^+][\text{A}^-]}{[\text{HA}]} \qquad (18.8)$$

When one-half of the weak acid is neutralized by the NaOH titrant in a titration, mol HA = mol A⁻ and also [HA] = [A⁻]. Since [HA] = [A⁻] at this point in the titration, then $K_a = [H_3O^+]$. If one takes the negative logarithm of both sides of this equality, then $pK_a = pH$. As pH is recorded directly from the pH meter, the pK_a of the weak acid is readily obtained at the "halfway point" (halfway to the stoichiometric point) in the titration (Figure 18.2).

The stoichiometric point is again determined from the complete titration curve of pH versus V_{NaOH}. If the weak acid is diprotic and if both stoichiometric points are detected, then pK_{a1} and pK_{a2} can be determined.

EXPERIMENTAL PROCEDURE

Procedure Overview: The pH meter is used in conjunction with a titration apparatus and a standardized sodium hydroxide solution to determine the molar concentration of a weak acid solution and the molar mass and pK_a of a solid, weak acid. Plots of pH versus volume of NaOH are used to determine the stoichiometric point of each titration.

The number of pH meters in the laboratory is limited. You may need to share one with a partner or with a larger group. Ask your instructor for details of the arrangement. Consult with your instructor for directions on the proper care and use of the pH meter. Also inquire about the calibration of the pH meter.

Because of time and equipment constraints, it may be impossible to do all parts of the experiment in one laboratory period. Time is required not only to collect and graph the data, but also to interpret the data and complete the calculations. Discuss the expectations from the experiment with your instructor.

The pH versus V_{NaOH} curves to be plotted in Parts A.6 and B.3 can be established by using a pH probe that is connected directly to either a calculator or computer with the appropriate software. If this pH sensing/recording apparatus is available in the laboratory, consult with your instructor for its use and adaptation to the experiment. The probe merely replaces the pH electrode in Figure 18.3. However, volume readings from the buret will still need to be recorded.

A. Molar Concentration of a Weak Acid Solution

Obtain about 90 mL of an acid solution with an unknown concentration from your instructor. Your instructor will advise you as to whether your acid is monoprotic or diprotic. Record the sample number on the Report Sheet. Clean three 250-mL beakers.

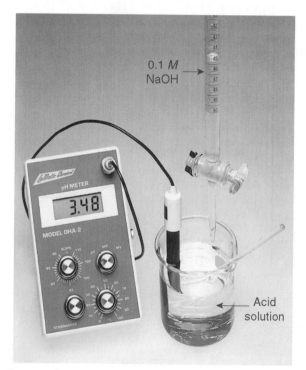

Figure 18.3 The setup for a potentiometric titration. Use a stirring rod or magnetic stirrer to stir the solution.

1. **Obtain a Standardized NaOH Solution.** A standardized (~0.1 *M*) NaOH solution was prepared in Experiment 9. If that solution was saved, it is to be used for this experiment. If the solution was not saved, you must either again prepare and standardize the solution (Experiment 9, Part A) or obtain about 200 mL of a standardized NaOH solution prepared by stockroom personnel. Record the *exact* molar concentration of the NaOH solution. Your instructor will advise you.

2. **Prepare the Buret with the Standardized NaOH Solution.** Properly clean a buret; rinse twice with tap water, twice with deionized water, and finally with three 5-mL portions of the standardized 0.1 *M* NaOH. Drain each rinse through the tip of the buret. Fill the buret with the 0.1 *M* NaOH. After 10–15 seconds, properly read[3] and **record** the volume of solution, "using all certain digits *plus* one uncertain digit."

Record the volume to the correct number of significant figures

3. **Prepare the Weak Acid Solution for Analysis.** Pipet 25 mL of the unknown weak acid solution into each of three *labeled* 250-mL beakers; add 50 mL of deionized water to each.

Set up the titration apparatus as shown in Figure 18.3. Remove the electrode from the deionized water and touch-dry the electrode with lint-free paper (Kimwipes). Immerse the electrode about one-half inch deep into the solution of Beaker 1. Swirl or stir the solution and read and record the initial pH.[4]

4. **Titrate the Weak Acid Solution.** Add the NaOH titrant, initially in 1- to 2-mL increments, and swirl or stir the solution. After each addition, allow the pH meter to stabilize; read and record the pH and buret readings on a *self-designed* data sheet. Repeat the additions until the stoichiometric point is near,[5] then slow the addition. When the stoichiometric point is imminent, add the NaOH titrant drop-wise.[6] Use a minimum volume of deionized water from a wash bottle to rinse the wall of the beaker or to add half-drop volumes of NaOH. Dilution affects pH readings.

5. **Titrate Beyond the Stoichiometric Point.** After reaching the stoichiometric point, first add drops of NaOH, then 1 mL, and finally 2- to 3-mL **aliquots** until at least 10 mL of NaOH solution has been added beyond the stoichiometric point. Read and record the pH and buret readings after each addition.

Aliquot: an undefined, generally small, volume of a solution

6. **Plot the Data.** Use appropriate software, such as Excel, to plot the data for the titration curve, pH versus V_{NaOH}. Draw a smooth curve through the data points (do not "follow the dots!"). Properly label your graph and obtain your instructor's approval.

Appendix C

From the plotted data, determine the volume of NaOH titrant added to reach the stoichiometric point.

7. **Repeat the Analysis.** Repeat the titration of the samples of weak acid in Beakers 2 and 3. Determine the average molar concentration of the acid.

Three samples of the solid weak acid are to be analyzed. Prepare three clean 250-mL beakers for this determination. Obtain an unknown solid acid from your instructor and record the sample number; your instructor will advise you as to whether your unknown acid is monoprotic or diprotic.

B. Molar Mass and the pK_a of a Solid Weak Acid

---

[3]Remember to read the bottom of the meniscus with the aid of a black mark drawn on a white card.
[4]A magnetic stirrer and magnetic stirring bar may be used to swirl the solution during the addition of the titrant. Ask your instructor.
[5]The stoichiometric point is near when larger changes in pH occur with smaller additions of the NaOH titrant.
[6]Suggestion: It may save time to quickly titrate a test sample to determine an approximate volume to reach the stoichiometric point.

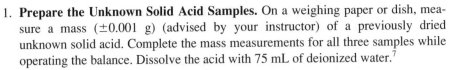

1. **Prepare the Unknown Solid Acid Samples.** On a weighing paper or dish, measure a mass (± 0.001 g) (advised by your instructor) of a previously dried unknown solid acid. Complete the mass measurements for all three samples while operating the balance. Dissolve the acid with 75 mL of deionized water.[7]

2. **Fill the Buret and Titrate.** Refill the buret with the standardized NaOH solution and, after 10–15 seconds, read and record the initial volume and the initial pH. Refer to Parts A.4 and A.5. Titrate each sample to 10 mL beyond the stoichiometric point.

Appendix C

3. **Plot and Interpret the Data.** Use appropriate software, such as Excel, to plot the data for a titration curve, pH versus V_{NaOH}. From the plot, determine the volume of NaOH used to reach the stoichiometric point of the titration. Obtain your instructor's approval.

4. **Calculate the Molar Mass *and* the pK_a of the Weak Acid.** a. Calculate the molar mass of the weak acid.

   b. Note the volume of NaOH titrant required to reach the stoichiometric point. Determine the pH (and therefore pK_a of the weak acid) at the point where one-half of the acid was neutralized.

5. **Repeat.** Similarly titrate the other unknown solid acid samples and handle the data accordingly.

Disposal: Dispose of all test solutions as directed by your instructor.

CLEANUP: Discard the sodium hydroxide solution remaining in the buret as directed by your instructor. Rinse the buret twice with tap water and twice with deionized water, discarding each rinse through the buret tip into the sink.

Appendix B

6. **Collect the Data.** Obtain the pK_a for the same sample number from other student chemists in the laboratory. Calculate the standard deviation and the relative standard deviation (%RSD) for the pK_a measurement for the acid.

The Next Step

While most common for the determination of hydrogen ion concentrations (and pH), potentiometric titrations are also utilized for the determination of any ion's concentration where a specific ion electrode is available (see Introduction). Develop a plan/procedure for determining the concentration of an ion in solution potentiometrically, using a specific ion electrode.

For example, a chloride specific ion electrode would read pCl directly. What would be the x-axis label in the titration curve?

---

[7]The solid acid may be relatively insoluble, but with the addition of the NaOH solution from the buret, it will gradually dissolve and react. The addition of 10 mL of ethanol may be necessary to dissolve the acid. Consult with your instructor.

Potentiometric Analyses

Date _____ Lab Sec. _____ Name _____ Desk No. _____

1. a. What does it mean when the titration of an acid with a base has reached the stoichiometric point? Be specific.

   b. For a weak acid (e.g., CH_3COOH) that is titrated with a strong base (e.g., NaOH), what species (ions/molecules) are present in the solution at the stoichiometric point?

   c. For a weak acid (e.g., CH_3COOH) that is titrated with a strong base (e.g., NaOH), what species (ions/molecules) are present in the solution at the halfway point in the titration toward the stoichiometric point?

   d. Explain why the pH at the stoichiometric point of a weak acid titrated with a strong base is always greater than 7, whereas that for a strong acid/strong base equals 7 at 25°C. Hint: See Question 1.b.

2. Briefly explain how the pK_a for a weak acid is determined in this experiment.

3. A pH meter is often used to measure the pH of an existing solution. An old bottle of hydrochloric acid was found in the stockroom with an unknown concentration (the label fell off), and it is desirable to dispose of the solution. A pH meter was used to determine that the pH of the hydrochloric acid solution is 0.94.
   a. How many moles of hydronium ion are present in 50.0 mL of the sample?

b. How many milliliters of 0.091 M NaOH would be required to neutralize (i.e., increase its pH to 7.0) a 50.0-mL sample of the hydrochloric acid solution so that it can be discarded?

4. Data in the following table were obtained for the titration of 25.00 mL of a weak acid with a 0.150 M KOH solution. Plot (at right) pH (ordinate) versus V_{KOH} (abscissa).

V_{KOH} added (mL)	pH
0.00	1.96
2.00	2.22
4.00	2.46
7.00	2.77
10.00	3.06
12.00	3.29
14.00	3.60
16.00	4.26
17.00	11.08
18.00	11.67
20.00	12.05
25.00	12.40

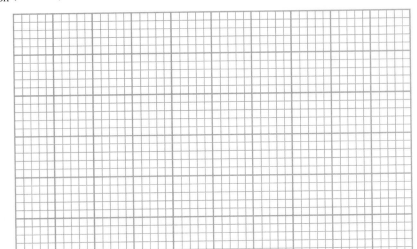

a. What volume of the KOH solution is required to reach the stoichiometric point?

b. What is the pH at the stoichiometric point?

c. What is the pK_a of the weak acid?

d. What is the molar concentration of the weak acid?

Potentiometric Analyses

Date _____ Lab Sec. _____ Name _____ Desk No. _____

A. Molar Concentration of a Weak Acid Solution

Sample No. _____ Monoprotic or diprotic acid? _____

	Trial 1	*Trial 2*	*Trial 3*
1. Molar concentration of NaOH (*mol/L*)		_____	
2. Volume of weak acid (*mL*)	_____	_____	_____
3. Buret readng of NaOH, *initial* (*mL*)	_____	_____	_____
4. Buret reading NaOH at stoichiometric point, *final* (*mL*)	_____	_____	_____
5. Volume of NaOH dispensed (*mL*)			
6. Instructor's approval of pH vs. V_{NaOH} graph	_____	_____	_____
7. Moles of NaOH to stoichiometric point (*mol*)			
8. Moles of acid (*mol*)			
9. Molar concentration of acid (*mol/L*)			
10. Average molar concentration of acid (*mol/L*)			

B. Molar Mass and the pK_a of a Solid Weak Acid

Sample No. _____ Monoprotic or diprotic acid? _____ Suggested mass _____

	Trial 1	*Trial 2*	*Trial 3*
1. Mass of dry, solid acid (*g*)	_____	_____	_____
2. Molar concentration of NaOH (*mol/L*)		_____	
3. Buret readng of NaOH, *initial* (*mL*)	_____	_____	_____
4. Buret reading NaOH at stoichiometric point, *final* (*mL*)	_____	_____	_____
5. Volume of NaOH dispensed (*mL*)			
6. Instructor's approval of pH vs. V_{NaOH} graph	_____	_____	_____
7. Moles of NaOH to stoichiometric point (*mol*)			
8. Moles of acid (*mol*)			
9. Molar mass of acid (*g/mol*)		*	
10. Average molar mass of acid (*g/mol*)			
11. Volume of NaOH halfway to stoichiometric point (*mL*)	_____	_____	_____
12. pK_{a1} of weak acid (from graph)	_____	_____	_____
13. Average pK_{a1}			

*Show calculations for Trial 1 on next page.

*Calculations for Trial 1.

Class Data/Group	1	2	3	4	5	6
pK_a for Sample No. _____						

Standard deviation and the relative standard deviation (%RSD) for the pK_a of an acid from the class data.

Laboratory Questions

Circle the questions that have been assigned.

1. The pH meter was calibrated with a buffer solution that reads 1.0 pH unit higher than what the label indicates.
   a. Part A. How will this miscalibration affect the determination of the molar concentration of the weak acid? Explain.
   b. Part B. Is the determined pK_a of the weak acid too high, too low, or unaffected by the miscalibration? Explain.

2. a. Part A.4. The pH reading is taken before the pH meter stabilizes. As a result the pH reading may be too low. Explain.
   b. Part A.4. Explain why it is good technique to slow the addition of NaOH titrant near the stoichiometric point.
   c. Part A.5. While not absolutely necessary, why is it good technique to add NaOH titrant beyond the stoichiometric point?

3. Part B.1. The solid acid is dissolved in 100 mL of deionized water, followed by 10 mL of ethanol. How does this added volume affect the reported molar mass of the weak acid . . . too high, too low, or unaffected? Explain.

4. Part B.4. As a result of adding the NaOH titrant too rapidly and an unwillingness to allow the pH meter to equilibrate before reading its pH, the stoichiometric point is ill-defined. As a result of this technique error,
   a. will the molar mass of the weak acid be reported too high or too low? Explain.
   b. will the pK_a of the weak acid be reported too high or too low? Explain.

5. Ideally how many stoichiometric points would be observed on a pH titration curve for a diprotic acid? Sketch the appearance of its titration curve and explain.

6. For polyprotic acids the successive conjugate bases increase in strength. Explain why the "last" stoichiometric point for a polyprotic acid (e.g., H_3PO_4) may be ill-defined or undefined on a pH titration curve?

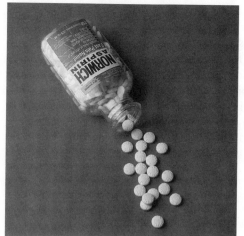

Aspirin Synthesis and Analysis

Aspirin is a leading commercial pain reliever, first synthesized in a pure and stable form by Felix Hoffman in 1897.

OBJECTIVES

- To synthesize aspirin
- To determine the purity of the synthesized aspirin or a commercial aspirin tablet

TECHNIQUES

The following techniques are used in the Experimental Procedure

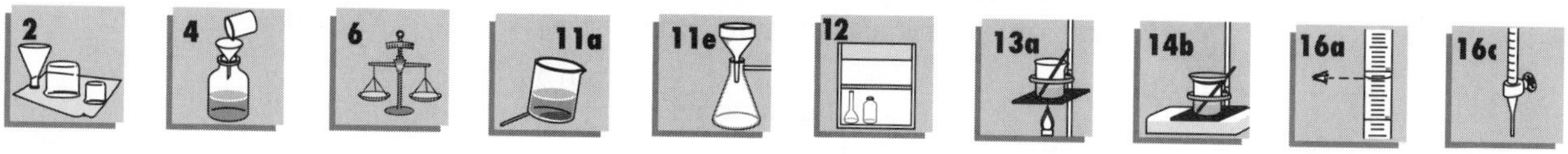

INTRODUCTION

Pure aspirin, chemically called acetylsalicylic acid, is both an organic **ester** and an **organic acid.** It is used extensively as a painkiller (analgesic) and as a fever-reducing drug (antipyretic). When ingested, acetylsalicylic acid remains intact in the acidic stomach, but in the basic medium of the upper intestinal tract, it forms the salicylate and acetate ions.

Ester: a [ester structure] *grouping of atoms in a molecule*

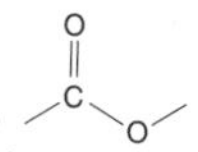

Organic acid: a [acid structure] *grouping of atoms in a molecule*

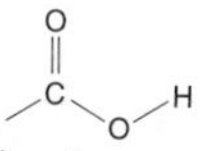

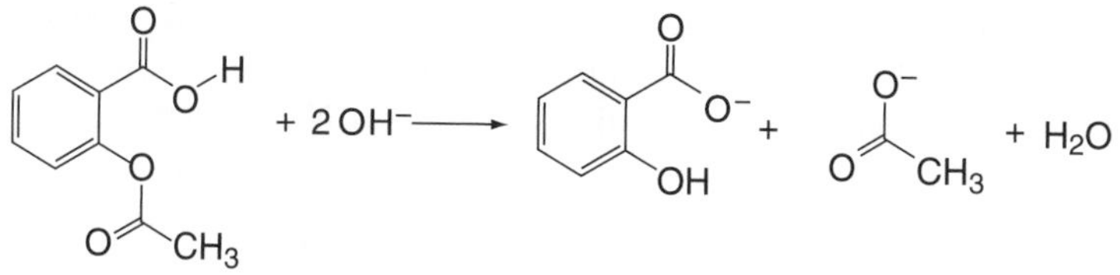

[reaction diagram]

aspirin (acetylsalicylic acid)   salicylate ion   acetate ion

$$+ 2\,OH^- \longrightarrow \qquad + \qquad + H_2O \qquad (19.1)$$

The analgesic action of aspirin is undoubtedly due to the salicylate ion; however, its additional physiological effects and biochemical reactions are still not thoroughly understood. It is known that **salicylic acid** has the same therapeutic effects as aspirin; however, due in part to the fact that it is an acid, salicylic acid causes a more severe upset stomach than does aspirin.

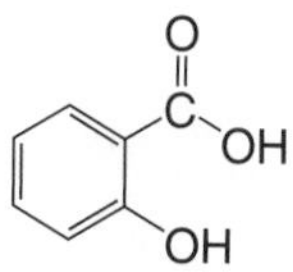

Aspirin (molar mass of 180.2 g/mol) is prepared by reacting salicylic acid (molar mass of 138.1 g/mol) with acetic anhydride (molar mass of 102.1 g/mol). Aspirin, like many other organic acids, is a weak **monoprotic acid.**

[salicylic acid structure]

salicylic acid

Monoprotic acid: a molecule that provides one proton for neutralization

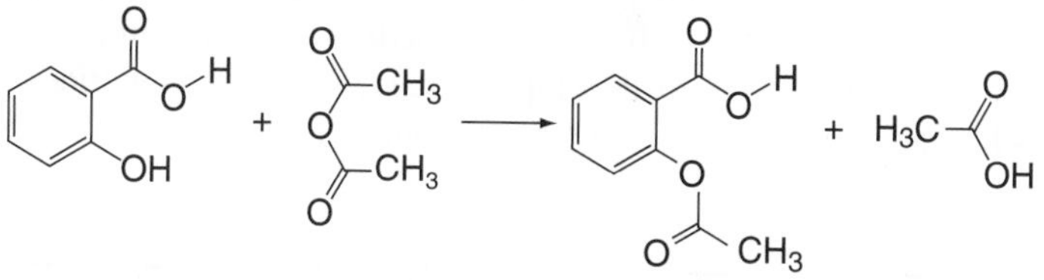

[reaction diagram]

salicylic acid   acetic anhydride   aspirin (acetylsalicylic acid)   acetic acid

$$(19.2)$$

Qualitatively, the purity of an aspirin sample can be determined from its melting point. The melting point of a substance is essentially independent of atmospheric pressure, but it is always lowered by the presence of impurities (a colligative property of pure substances; see Experiment 14). The degree of lowering of the melting point depends on the nature and the concentration of the impurities.

Quantitatively, the purity of an aspirin sample can be determined by a simple acid–base titration. The acetylsalicylic acid reacts with hydroxide ion, from a standardized sodium hydroxide solution, accordingly.

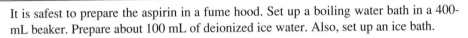

$$+ \; OH^- \longrightarrow \qquad\qquad + \; H_2O \qquad\qquad (19.3)$$

Phenolphthalein: an acid-base indicator that is colorless at a pH less than 8.2 and pink at a pH greater than 10.0

A standardized NaOH solution titrates the acetylsalicylic acid to the **phenolphthalein** endpoint, where

$$\text{volume of NaOH (L)} \times \text{molar concentration of NaOH (mol/L)} = \text{mol NaOH} \quad (19.4)$$

According to Equation 19.3, one mole of OH^- reacts with one mole of acetylsalicylic acid; thus, the moles and mass of acetylsalicylic acid in the prepared sample are calculated. Knowing the calculated mass of the acid and the measured mass of the aspirin sample, the percent purity of the aspirin sample can be calculated:

$$\text{mol acetylsalicylic acid} \times \frac{180.2 \text{ g}}{\text{mol}} = \text{g acetylsalicylic acid} \qquad (19.5)$$

$$\% \text{ purity} = \frac{\text{g acetylsalicylic acid}}{\text{g aspirin sample}} \times 100 \qquad (19.6)$$

In Part C, the analysis for the percent acetylsalicylic acid in an aspirin sample can be performed on the aspirin prepared in Part A *or* on a commercial aspirin tablet.

EXPERIMENTAL PROCEDURE

Procedure Overview: Crystalline aspirin is synthesized and then purified by the procedure of recrystallization. The melting point and the percent purity of the aspirin are determined, the latter by titration with a standardized NaOH solution.

A. Preparation of Aspirin

It is safest to prepare the aspirin in a fume hood. Set up a boiling water bath in a 400-mL beaker. Prepare about 100 mL of deionized ice water. Also, set up an ice bath.

1. **Mix the Starting Materials and Heat.** Measure about 2 g (± 0.01 g) of salicylic acid (**Caution:** *this is a skin irritant*) in a *dry* 125-mL Erlenmeyer flask. Cover the crystals with 4–5 mL of acetic anhydride. (**Caution:** *Acetic anhydride is a severe eye irritant—avoid skin and eye contact.*) Swirl the flask to wet the salicylic acid crystals. Add 5 drops of conc H_2SO_4 (**Caution:** H_2SO_4 *causes severe skin burns.*) to the mixture and gently heat the flask in a boiling water bath (Figure 19.1[1]) for 5–10 minutes.

2. **Cool to Crystallize the Aspirin.** Remove the flask from the hot water bath and, to the reaction mixture, add 10 mL of deionized *ice* water to decompose any excess acetic anhydride. Chill the solution in an ice bath until crystals of aspirin no longer form, stirring occasionally to decompose residual acetic anhydride. *If* an "oil" appears instead of a solid, reheat the flask in the hot water bath until the oil disappears and again cool.

---

[1] A Bunsen flame may be substituted for the hot plate.

3. **Separate the Solid Aspirin from the Solution.** Set up a vacuum filtration apparatus and "turn it on." Seal the filter paper with water in the Büchner funnel. *Decant* the liquid onto the filter paper; minimize any transfer of the solid aspirin. Some aspirin, however, may be inadvertently transferred to the filter; that's O.K.

4. **Filter, Wash, and Transfer the Aspirin.** Add 15 mL of *ice* water to the flask, swirl, chill briefly, and decant onto the filter. Repeat until the transfer of the crystals to the vacuum filter is complete; maintain the vacuum to dry the crystals as best possible. Wash the aspirin crystals on the filter paper with 10 mL of ice water. Keep all of the filtrate until the aspirin has been transferred to the filter.

   If aspirin forms in the filtrate, transfer this filtrate and aspirin to a beaker, chill in an ice bath, and vacuum filter as before, using a new piece of filter paper.

Disposal: Dispose of the "final" filtrate as directed by your laboratory instructor.

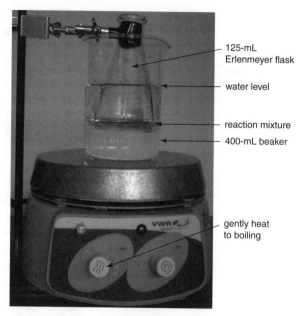

Figure 19.1 Boiling water bath for the dissolution of the acetylsalicylic acid crystals.

125-mL Erlenmeyer flask

water level

reaction mixture

400-mL beaker

gently heat to boiling

5. **Recrystallize the Aspirin.** Transfer the crystals from the filter paper(s) to a 100-mL beaker. Add repetitive small volumes of ethanol (e.g., 3-mL volumes) to the aspirin until the crystals *just* dissolve ($\leq$20 mL is required). Warm the mixture in a 60°C water bath (**Caution:** *no flame*, use a hot plate or a hot water bath). Pour 50 mL of ~60°C water into the solution. If a solid forms, continue warming until the solid dissolves.

   Cover the beaker with a watchglass, remove it from the heat, and set it aside to cool slowly to room temperature. Then set the beaker in an ice bath. Beautiful needlelike crystals of acetylsalicylic acid form.

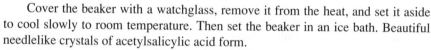

6. **How Much Did You Prepare?** Vacuum filter the crystals on filter paper, the mass of which has been previously measured ($\pm$0.01 g). Wash the crystals with two 10-mL volumes of *ice* water. Place the filter paper and aspirin sample on a watchglass and allow them to air-dry. The time for air-drying the sample may require that it be left in your lab drawer until the next laboratory period.

   Determine the mass of the dry filter paper and sample. Dispose of the filtrate as directed by your laboratory instructor.

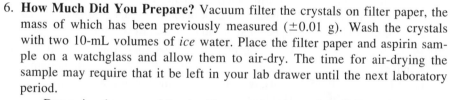

7. **Correct for Residual Solubility.** The solubility of acetylsalicylic acid is 0.25 g per 100 mL of water. Correcting for this inherent loss of product due to the wash water in Part A.6, calculate the percent yield.

8. **What Do You Do with It?** Don't use it for a headache! Place the sample in a properly labeled test tube, stopper, and submit it along with your Report Sheet to your laboratory instructor at the conclusion of the experiment.

The melting point of the aspirin sample can be determined with either a commercial melting point apparatus (Figure 15.5) or with the apparatus shown in Figure 19.2 and described in Part B.1. Consult with your instructor.

B. Melting Point of the Aspirin Sample

1. **Prepare the Sample.** Fill a capillary melting point tube to a depth of 1 cm with the recrystalized aspirin prepared in Part A.6. See Figures 15.3 and 15.4. Attach the tube to a 360°C glass or digital thermometer with a rubber band (or band of rubber tubing).

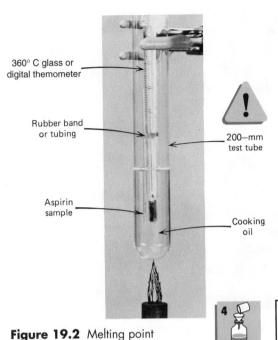

360° C glass or digital thermometer

Rubber band or tubing

200—mm test tube

Aspirin sample

Cooking oil

Figure 19.2 Melting point apparatus for aspirin.

Place the sample alongside the thermometer bulb (Figure 19.2) or thermal sensor. As the melting point for aspirin is greater than 100°C, a cooking oil must be used for the heating bath.

2. **Determine the Melting Point.** *Slowly* and *gently* heat the oil bath at a rate of ~5°C per minute until the aspirin melts. (**Caution:** *The oil bath is at a temperature greater than 100°C—do not touch!*) Cool the bath and aspirin to just below this approximate melting point until the aspirin in the tube solidifies; at a slower ~1°C per minute rate, heat again until it melts; this is the melting point of your prepared aspirin.

3. **A Purity Check of the Sample.** If the melting point of your prepared aspirin sample is less than 130°C, repeat Part A.5 to recrystallize the sample for the purpose of increasing its purity. After the recrystallization, repeat Parts B.1 and B.2.

4. **Repeat the Melting Point Measurement.** Again, cool the bath and aspirin to just below the melting point until the aspirin in the tube solidifies; at a 1°C per minute rate, heat again until it melts.

Disposal: Ask your instructor about the proper disposal of the oil. Be sure the oil is cool when handling it. Dispose of the capillary tube in the "Waste Glass" container.

C. Percent Acetylsalicylic Acid in the Aspirin Sample

Three trials are to be completed in the analysis of the aspirin. Prepare three clean 125- or 250-mL Erlenmeyer flasks and determine the mass of three aspirin samples while occupying the balance. Obtain a 50-mL buret.

1. **Prepare the Aspirin Sample for Analysis.** *Assuming* 100% purity of your aspirin sample, calculate the mass of aspirin that requires 20 mL of 0.1 M NaOH to reach the stoichiometric point. Show the calculation on the Report Sheet. On weighing paper measure the calculated mass (± 0.001 g) of the aspirin you have just prepared (or a crushed commercial aspirin tablet) and transfer it to the flask. Add 10 mL of 95% ethanol, followed by about 50 mL of deionized water, and swirl to dissolve the aspirin. Add 2 drops of phenolphthalein indicator. Repeat for trials 2 and 3.

2. **Prepare the Buret for Titration.** Prepare a clean buret, rinse, and fill it with a standardized 0.1 M NaOH solution.[2] Be sure that no air bubbles are present in the buret tip. After 10–15 seconds, read and record the volume, and the *actual* molar concentration of the NaOH solution.

3. **Titrate the Sample.** Slowly add the NaOH solution from the buret to the dissolved aspirin sample until the endpoint is reached. The endpoint in the titration should be within one-half drop of a faint pink color. The color should persist for 30 seconds. Read and record the final volume of NaOH in the buret.

Disposal: Discard the test solution in the "Waste Acids" container or as advised by your instructor.

CLEANUP: Discard the NaOH titrant into a properly labeled bottle; rinse the buret with several 5-mL volumes of tap water, followed by two 5-mL volumes of deionized water.

The Next Step

The purity of an aspirin sample can also be determined spectrophotometrically. Research the Internet for the procedure and refer to Experiment 35 for details.

---

[2]You may need to prepare the 0.1 M NaOH solution using the procedure in Experiment 9, or the stockroom personnel may have it already prepared.

Aspirin Synthesis and Analysis

Date _____ Lab Sec. _____ Name _____ Desk No. _____

1. A 0.421-g sample of aspirin prepared in the laboratory is dissolved in 95% ethanol, diluted with water, and titrated to the phenolphthalein endpoint with 17.3 mL of 0.114 M NaOH.

    a. How many moles of acetylsalicylic acid (molar mass = 180.2 g/mol) are present in the sample?

    b. Calculate the percent purity of acetylsalicylic acid in the aspirin sample.

2. Experimental Procedure, Part A.1. In the experiment 2.00 g of salicylic acid (molar mass = 138.1 g/mol) reacts with an excess amount of acetic anhydride. Calculate the theoretical yield of acetylsalicylic acid (molar mass = 180.2 g/mol) for this synthesis.

3. Experimental Procedure, Part C.1. Determine the number of grams of acetylsalicylic acid that will react with 20.0 mL of 0.100 M NaOH. Show calculation here and on the Report Sheet.

4. Experimental Procedure, Part A.5. What is the purpose of recrystallizing the aspirin?

5. Experimental Procedure, Part B.3. The melting point of the prepared aspirin in this experiment will most likely be less than (but not greater than) that of pure aspirin. Explain. See Experiment 14.

6. Where and why is Technique 14b used in this experiment?

7. Describe the procedure for "seating" the filter paper in the funnel for a vacuum filtration.

8. Identify the five **cautions** cited in the Experimental Procedure for this experiment.

Aspirin Synthesis and Analysis

Date _____ Lab Sec. _____ Name _____ Desk No. _____

A. Preparation of Aspirin

1. Mass of salicylic acid (*g*) _____

2. Theoretical yield of aspirin (*g*) _____

3. Experimental yield of aspirin (*g*) _____

4. *Total* volume of solutions in contact with aspirin in Part A.6 (*mL*) _____

5. Experimental yield, corrected for solubility (*g*) _____

6. Percent yield (%) _____

B. Melting Point of the Aspirin Sample

1. Melting point measurements (°*C*) _____ _____ _____

2. Average melting point of aspirin (°*C*) _____

C. Percent Acetylsalicylic Acid in the Aspirin Sample

Calculation for the mass of aspirin for the titrimetric analysis.

	Trial 1	*Trial 2*	*Trial 3*
1. Mass of weighing paper (*g*)	_____	_____	_____
2. Mass of weighing paper plus aspirin sample (*g*)	_____	_____	_____
3. Mass of aspirin sample (*g*)	_____	_____	_____
4. Molar concentration of the NaOH solution (*mol/L*)	_____		
5. Buret reading, initial (*mol/L*)	_____	_____	_____

	Trial 1	Trial 2	Trial 3
6. Buret reading, *final* (mL)			
7. Volume of NaOH used (mL)			
8. Moles of NaOH added (mol)			
9. Moles of acetylsalicylic acid (mol)			
10. Mass of acetylsalicylic acid (g)			
11. Percent purity of aspirin sample (%)			
12. Average percent purity of aspirin sample (%)			

Class Data/Group	1	2	3	4	5	6
Average percent acid in sample						

Calculate the standard deviation for the percent acetylsalicylic acid in aspirin from the class data. See Appendix B.

Calculate the relative standard deviation (%RSD).

Laboratory Questions

Circle the questions that have been assigned.

1. Part A.1. According to LeChâtelier's principle, explain why it is necessary to add the conc H_2SO_4 during the preparation of the acetylsalicylic acid. Also see Equation 19.1.

2. Part A.1. "Anhydride" means without water. Suppose 1 M H_2SO_4 were substituted for the conc H_2SO_4. Would the yield of acetylsalicylic acid be increased, decreased, or unaffected by the substitution? Explain.

3. Part A.2. Deionized ice water is added to decompose excess acetic anhydride. What is the product for the hydrolysis of acetic anhydride? Write a balanced equation.

4. Part A.4. Some of the aspirin passed through the filter into the filtrate. How does the aspirin in the filtrate differ from that collected on the filter paper?

5. Part A.5. The product crystals are dissolved in a minimum volume of ethanol. Is acetylsalicylic acid more soluble in ethanol or water? Explain.

6. Part B.2. Would the product isolated after Part A.4 have a higher or lower melting point than that isolated after Part A.6? Explain.

7. Part C.2. The molar concentration of the NaOH solution is recorded as being 0.1 M instead of the actual molar concentration of 0.151 M. If the recorded concentration is used to calculate the purity of the aspirin sample, will the percent purity be reported too high or too low? Explain.

Experiment 20

Alkalinity of a Water Resource

Icebergs have low levels of alkalinity.

OBJECTIVES

- To properly obtain a water sample
- To determine two levels of alkalinity in a water sample and to explain their differences
- To learn the chemistry used in determining the alkalinity of a water sample

TECHNIQUES

The following techniques are used in the Experimental Procedure.

INTRODUCTION

Chemists and chemical engineers who have the responsibility of monitoring water that must pass through a water treatment plant prior to its final use are keenly aware of the alkalinity of the water source. Water with a high alkalinity in conjunction with high hardness leads to pipe corrosion and failure of water conduits and distribution systems. Conversely water with low alkalinity may lead to an acidic water supply that may also be corrosive to water distribution systems. Water with high alkalinity generally exists where the water is or has been in contact with rocks and minerals high in carbonates and phosphates. Low alkalinity in water can be attributed to discharge of industrial acid wastes or from heavy rainfall runoff.

Alkalinity is a measure of the "buffering capacity" of the water. Most of the buffering action in natural waters is due *primarily* to the presence of carbonates and bicarbonates, a buffering system with pH values between 5.0 and 8.0, a pH range over which most marine organisms readily survive. Therefore, water that supports aquatic life in this pH range must have an adequate buffering capacity. Other weak acid/conjugate base buffering systems may also contribute to the alkalinity of the water.

The source of alkalinity is primarily from water passing over (surface waters) or through (underground aquifers) limestone (carbonate) deposits. The established equilibria that result from this water movement is the carbonic acid, bicarbonate, carbonate system expressed as

$$H_2CO_3(aq) + H_2O(l) \leftrightarrow H_3O^+(aq) + HCO_3^-(aq) \quad K_{a1} = 4.3 \times 10^{-7} \quad (20.1)$$

$$HCO_3^-(aq) + H_2O(l) \leftrightarrow H_3O^+(aq) + CO_3^{2-}(aq) \quad K_{a2} = 4.8 \times 10^{-11} \quad (20.2)$$

The dominance of carbonic acid, bicarbonate, or carbonate species in the water sample is determined by the pH. According to Figure 20.1, the fraction of the HCO_3^- ion is highest in the 5.0–9.0 pH range. Carbonic acid, (also noted as $CO_2(aq)$) is dominate at pH values less than 3; carbonate ion is dominate at pH values greater than 11. Therefore, HCO_3^- plays the primary role of the buffer—buffering bases to produce CO_3^{2-} or buffering acids to produce carbonic acid, $CO_2(aq)$, in the natural environment.

Other "alkaline-contributing" ions may include, for example, phosphates, PO_4^{3-} and silicates, SiO_3^{2-}.

Alkalinity concentrations are expressed as **ppm** $CaCO_3$ (see below). While alkalinity concentrations below 100 ppm $CaCO_3$ are desirable for municipal and industrial water supplies, drinking water alkalinity ranges from 30 to 400 ppm $CaCO_3$. Moderately alkaline concentrations of ~350 ppm $CaCO_3$ tend to inhibit pipe corrosion, but alkalinities above 500 ppm $CaCO_3$ are generally quite basic and tend to promote corrosion especially in hot water piping. Alkalinities below ~100 ppm $CaCO_3$ are generally acidic and also tend to be corrosive to piping.

ppm: parts per million by mass; 1 g per 1 × 10⁶ g or 1 mg per 1 kg. Since 1 kg ~ 1L of solution, also 1 mg/L.

Analysis for Alkalinity

The alkalinity of a water sample is determined by its titration with a standard strong acid (typically HCl(aq)) solution. However, alkalinity may have two reported values depending upon the indicator that is selected for the titration. Two stoichiometric points occur for the titration of a bicarbonate/carbonate system with a strong acid. The first occurs at a pH near the endpoint for phenolphthalein (pH = 8.2) at which point the stronger bases of the water sample have been mono-protonated (often called the "P" or *carbonate alkalinity*). The "alkaline-contributing" ions are primarily carbonate, CO_3^{2-}, hydroxide, and phosphate, PO_4^{3-}. From Figure 20.1, the fraction of HCO_3^- ≈ 1 at the phenolphthalein endpoint (colorless at pH = 8.2).

The "P" alkalinity is a measure of the corrosiveness of the water sample, measuring the concentrations of the stronger bases in the water sample, e.g., OH^-, CO_3^{2-}, PO_4^{3-}.

The second stoichiometric point occurs at a pH near the endpoint for methyl orange where essentially all bases contributing to alkalinity including HCO_3^- and HPO_4^{2-} have been fully protonated (often called the "T" or *bicarbonate alkalinity*)

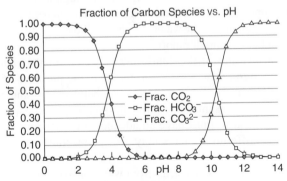

Figure 20.1 The fraction of carbon species present in water as a function of pH.

providing a "total" alkalinity value. Note from Figure 20.1 that the fraction of $CO_2(aq)$ ≈ 0.9 at the methyl orange endpoint (red-orange at pH = 3.4).

The "T" alkalinity is a measure of the total buffering capacity of the water sample, measuring the total concentration of proton-accepting species in the sample.

Figure 20.2 is a titration curve for the carbonate species titrated with hydrochloric acid showing the changes in pH at the phenolphthalein (pH = 8.2) and methyl orange (pH = 3.4) endpoints.

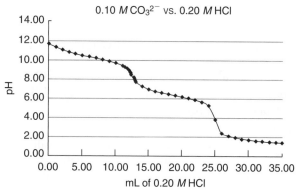

0.10 M CO_3^{2-} vs. 0.20 M HCl

Figure 20.2 A pH curve for the titration 25.0 mL of 0.10 M CO_3^{2-} with 0.20 M HCl.

Alkalinity Calculation

To express alkalinity in ppm $CaCO_3$ (mg $CaCO_3$/L sample), titrating with a standard HCl solution:

1. The volume and concentration of the HCl solution measure the moles of "alkaline-contributing" ions in the water sample at the respective endpoint ("P" or "T").

2. Two mole of $H_3O^+(aq)$ reacts with one mole $CO_3^{2-}(aq)$ to produce one mole $H_2CO_3(aq)$, often written $CO_2(aq)$:

$$2\,H_3O^+(aq) + CO_3^{2-}(aq)$$
$$\rightarrow H_2CO_3(aq) + 2\,H_2O(l) \rightarrow CO_2(aq) + 3\,H_2O(l) \quad (20.3)$$

One mole CO_3^{2-} is equivalent to one mole of $CaCO_3$.

3. Conversion of moles of $CaCO_3$ (100.1 g/mol) to milligrams divided by the volume of the titrated sample (in liters) produces an alkalinity value in ppm $CaCO_3$.

If only carbonate and bicarbonate ions are present in the water sample, then the "T" alkalinity would be twice the "P" alkalinity (see Prelaboratory Assignment) because twice the volume of HCl titrant would be needed to convert from CO_3^{2-} to H_2CO_3 at the methyl orange endpoint as would be needed to convert CO_3^{2-} to HCO_3^- at the phenolphthalein endpoint.

Most reported alkalinity values appearing in the literature or reported to state and national government water monitoring agencies are "T" alkalinities.

EXPERIMENTAL PROCEDURE

Procedure Overview. A water sample is obtained from either a "natural" or a potable source of water. At least three samples (filtered if necessary) of the water are titrated to the phenolphthalein endpoint and, again, to the methyl orange endpoint to determine the "P" and "T" alkalinities respectively for a total of six (minimum) titrations. Calculations with the data are to present the results in units of ppm $CaCO_3$. An interpretation of the results and a conclusion will be presented.

A. A Standard 0.015 *M* HCl Solution

Record the reading on the buret to the correct number of significant figures.

Ask your instructor if a standard solution of HCl is available. If so, proceed to Part B of the Experimental Procedure.

Create and *design your own Report Sheet* for this part of the experiment.

1. **Preparation of an HCl Solution.** Prepare 500 mL of a ~0.015 *M* HCl solution, starting with 1 *M* HCl. Show the calculations of your Report Sheet.

2. **Samples of $Na_2CO_3(s)$, a Primary Standard.** Calculate the mass of Na_2CO_3 (molar mass = 105.99 g/mol) that is neutralized to the methyl orange endpoint with about 25 mL of 0.015 *M* HCl (show the calculation on your Report Sheet . . . this is a *small* mass!). Use weighing paper to measure the tared mass (± 0.001 g) of at least three samples of anhydrous Na_2CO_3 (previously dried at 110°C)[1] based upon your calculations.

   Transfer each sample to a clean, dry 125- or 250-mL Erlenmeyer flask, dissolve with deionized water, and add several drops of methyl orange indicator.

   Place a white sheet of paper beneath the Erlenmeyer flask.

3. **Prepare a Buret.** Prepare a *clean,* 50-mL buret for titration. Rinse the buret with 3–5 mL of the HCl solution, roll the buret to rinse the wall, and drain through the buret tip. Fill the buret with the HCl solution, drain the tip of air bubbles, and after 10–15 seconds, read and record the volume "using all certain digits plus one uncertain digit."

4. **Titrate to Standardize the HCl Solution.** Slowly add the HCl titrant to the Na_2CO_3 solution. As the color change slows with the addition of titrant, *gently warm* the Erlenmeyer flask on a hot plate (Figure 20.3a) or over an open flame (Figure 20.3b) for 1–2 minutes to evolve any of the $CO_2(aq)$ that has formed during the titration.

   Continue to titrate the solution to the methyl orange endpoint.[2] Read and record the volume of the HCl solution in the buret.

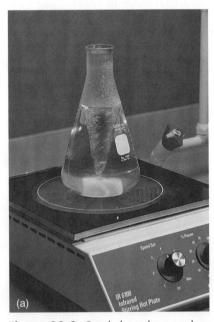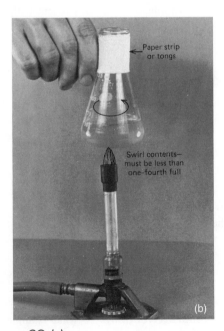

Figure 20.3 Gently heat the sample to remove $CO_2(g)$.

---

[1]The Na_2CO_3 should have been dried prior to your coming to the laboratory. Ask your instructor.
[2]The methyl orange color change, yellow (base) to red-orange (acid), requires a watchful eye. The color change is subtle and requires some technique for reproducibility.

242 Alkalinity of a Water Resource

5. **Additional trials.** Repeat the analysis the two remaining samples.

6. **Calculations.** Calculate the molar concentration of HCl for each sample and the average molar concentration. Show a sample calculation on your Report Sheet.

> *Disposal:* Dispose of the test solutions as directed by your instructor.

---

1. **Select and prepare the water sample.** Select a site for securing a water sample—one recommended by the instructor or by the student water chemist. Use a clean 500-mL Erlenmeyer flask to obtain a water sample. If the sample is from a "natural" source (i.e., stream, lake, river, etc.), gravity filter the sample before proceeding to Parts C and D.

   Write a brief description characterizing your water sample.

B. Preparation of Water Sample

---

1. **Prepare the sample for titration.** Label three, clean 125-mL Erlenmeyer flasks for the "P" alkalinity analysis. Pipet 25.0 mL of the water sample (filtered if necessary) into each of the flasks and add 2–3 drops of phenolphthalein. [3]

   Place a white sheet of paper beneath the receiving flask.

2. **Prepare the titrant.** Prepare a clean buret. Add 3–5 mL of the standard HCl solution to the buret, roll the solution to wet the wall of the buret and dispense through the buret tip and discard. Use a clean funnel to fill the buret . . . dispense a small portion through the buret tip to remove air bubbles. Read and record the volume of HCl solution in the buret according to Technique 16A.2, "using all certain digits plus one uncertain digit."

3. **Titrate for "P" alkalinity.** Slowly dispense the HCl titrant into the water sample. Swirl the flask as titrant is added (Technique 16C.4). As the color fade of the phenolphthalein slowly decreases and the return to pink slows, add the HCl titrant dropwise. When one drop (ideally half-drop) results in the disappearance of the slightly pink color, stop the titration and again (after ~15 seconds) read and record the volume of titrant in the buret.

4. **Additional trials.** Repeat the "P" analysis on at least two more samples.

5. **Calculations.** Calculate the "P" alkalinity for each sample and the average "P" alkalinity, expressed in ppm $CaCO_3$.

C. Determination of "P" Alkalinity

> *Disposal:* Dispose of the test solutions as directed by your instructor.

---

1. **Prepare the sample for titration.** Repeat Part C.1, substituting methyl orange indicator for phenolphthalein.

2. **Analysis for "T" alkalinity.** Repeat Part C.2 if necessary. Slowly add the HCl titrant to the water sample. As the rate of the color change decreases with the addition of titrant, *gently warm* the Erlenmeyer flask on a hot plate or over an open flame (Figure 20.3) for 1–2 minutes to evolve any of the $CO_2(aq)$ that has formed during the titration.[4] Continue to titrate the solution to the methyl

D. Determination of "T" Alkalinity

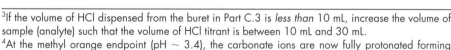

---

[3]If the volume of HCl dispensed from the buret in Part C.3 is *less than* 10 mL, increase the volume of sample (analyte) such that the volume of HCl titrant is between 10 mL and 30 mL.

[4]At the methyl orange endpoint (pH ~ 3.4), the carbonate ions are now fully protonated forming H_2CO_3 (or $CO_2(aq)$ and H_2O). The dissolved carbon dioxide not only produces a slight acidity but also produces a slight buffering action: $CO_2(aq) + H_2O(l) \leftrightarrow HCO_3^-(aq)$. Since $CO_2(aq)$ has a low water solubility, heating removes the CO_2 as a gas and shifts the buffer equilibrium, which results in a sharper methyl orange endpoint.

orange endpoint (see footnote #2). Read and record the volume of the HCl solution in the buret.

3. **Additional trials.** Repeat the "T" analysis on at least two more samples.

4. **Calculations.** Calculate the "T" alkalinity for each sample and the average "T" alkalinity, expressed in ppm $CaCO_3$.

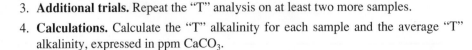

Disposal: Dispose of the test solutions as directed by your instructor.

The Next Step

All natural water samples have alkalinity. Design a systematic study of the alkalinity changes of water as a function of, e.g., sampling sites along a river, drinking water sources of your area (surface and/or subsurface sources), wastewater, or storm water discharge sites of the city, ...

Experiment 20 *Prelaboratory Assignment*

Alkalinity of a Water Resource

Date ________ Lab Sec. ______ Name ______________________________ Desk No. ________

1. a. Define alkalinity of a water sample.

   b. What is/are the principle ions contributing to the alkalinity of a water sample?

2. "P" and "T" alkalinities are determined for a water sample in this experiment.
   a. Differentiate between the Experimental Procedures of the two determinations.

   b. What is the color change at the "P" endpoint?

   What is the color change at the "T" endpoint?

   c. Explain why the "T" alkalinity is generally greater than the "P" alkalinity for a water sample.

   d. Suppose hydroxide ion, OH^-, is the only anion contributing to the alkalinity of the sample. How will the "P" and "T" alkalinity values compare?

3. The "P" alkalinity represents a value at which the bases dissolved in the water sample have been mono-protonated while "T" alkalinity represents a value at which all bases have been fully protonated by the strong acid in the titration. What is the chemical formula of the following at the "P" and "T" endpoints?

	"P" endpoint	**"T" endpoint**
a. CO_3^{2-}	________	________
b. OH^-	________	________
c. PO_4^{3-}	________	________
d. SiO_3^{2-}	________	________

4. A hot plate or Bunsen burner is required as part of the experiment. State the purpose in the Experimental Procedure. Explain.

5. Speculate as to how the alkalinity levels might compare between surface water and ground (well) water.

6. A chemist chooses to standardize an approximately 0.015 M HCl solution to the methyl orange endpoint using $Na_2CO_3(s)$ as a primary standard. If 22.7 mL of the 0.015 M HCl neutralizes 0.0144 grams of Na_2CO_3 (molar mass = 105.99 g/mol), what is the standardized molar concentration of the HCl solution?

7. A chemist titrates a 50.0 mL water sample to the phenolphthalein endpoint with a 17.7 mL of a 0.015 M HCl standard solution.
   a. How many moles of "alkaline-contributing" ions are present in the sample?

   b. What is the "P" alkalinity of the solution expressed in ppm $CaCO_3$? (Assume density = 1.00 g/mL)

8. A chemist titrates a 50.0 mL water sample to the methyl orange endpoint with a 25.4 mL of a 0.0120 M HCl standard solution.
   a. How many moles of "alkaline-contributing" ions are present in the sample?

   b. What is the "T" alkalinity of the solution expressed in ppm $CaCO_3$? (Assume density = 1.00 g/mL)

Experiment 20 *Report Sheet*

Alkalinity of a Water Resource

Date __________ Lab Sec. ______ Name ______________________________ Desk No. __________

A. A Standard 0.015 *M* HCl Solution

Molar concentration of HCl from self-designed (and attached) Report Sheet (*mol/L*) ______________

Calculation of molar concentration of HCl

B Preparation of Water Sample

Write a brief description of your water sample.

C. Determination of "P" Alkalinity

	Trial 1	*Trial 2*	*Trial 3*
1. Sample volume (*mL*)			
2. Buret reading, *initial* (*mL*)			
3. Buret reading, *final* (*mL*)			
4. Volume HCl dispensed (*mL*)			
5. Moles of HCl dispensed (*mol*)			
6. Moles of equivalent $CaCO_3$ (*mol*)			
7. mg $CaCO_3$ (*mg*)			
8. "P" alkalinity, ppm $CaCO_3$ (*mg/L*)			
9. Average "P" alkalinity, ppm $CaCO_3$			

D. Determination of "T" Alkalinity

	Trial 1	Trial 2	Trial 3
1. Sample volume (*mL*)			
2. Buret reading, *initial* (*mL*)			
3. Buret reading, *final* (*mL*)			
4. Volume HCl dispensed (*mL*)			
5. Moles of HCl dispensed (*mol*)			
6. Moles of equivalent $CaCO_3$ (*mol*)			
7. mg $CaCO_3$ (*mg*)			
8. "T" alkalinity, ppm $CaCO_3$ (*mg/L*)			
9. Average "T" alkalinity, ppm $CaCO_3$			

Write a short summary based upon an interpretation of your analytical data.

Laboratory Questions

Circle the questions that have been assigned.

1. a. Explain what may have contributed to the measured "P" alkalinity being greater than one-half the "T" alkalinity in your water sample. Be specific as to the greater than expected "P" alkalinity.
    b. Explain what may have contributed to the measured "P" alkalinity being less than one-half the "T" alkalinity in a water sample. Be specific as to the greater than expected "T" alkalinity.

2. *If* the phenolphthalein endpoint had been surpassed in the analysis, would the alkalinity of the sample been reported unchanged, too high, or too low? Explain.

3. *If* an air bubble in the buret containing the standard hydrochloric acid is discharged during the analysis, would the reported alkalinity remain unchanged, too high, or too low? Explain.

4. Suppose the analysis of the water sample had indicated a very low alkalinity. Since alkalinity represents a "buffering action" of the water, how might this low value affect the quality of the marine life?

5. The alkalinity of a stream or small lake (pond) is generally lower after a rainfall than before. Explain why this may be so.

6. Trisodium phosphate (TSP), formerly used as a detergent additive, is used as cleaning agent and degreaser to prepare surfaces for painting. TSP also reacts with the ferric ion of rust to form a $FePO_4$ impermeable coating to which paint will bind. How would the excessive use of TSP affect the alkalinity of the wastewater from such an operation? Explain.

7. Assuming a "P" alkalinity of 300 ppm $CaCO_3$, calculate the predicted volume of 0.0150 *M* HCl required to reach the phenolphthalein endpoint for a 25.0-mL sample.

Experiment 21

Hard Water Analysis

Deposits of hardening ions (generally calcium carbonate deposits) can reduce the flow of water in plumbing.

OBJECTIVES

OBJECTIVES

- To learn the cause and effects of hard water
- To determine the hardness of a water sample

TECHNIQUES

The following techniques are used in the Experimental Procedure

INTRODUCTION

Hardening ions present in natural waters are the result of slightly acidic rainwater flowing over mineral deposits of varying compositions; the acidic rainwater[1] reacts with the *very* slightly soluble carbonate salts of calcium and magnesium and with various iron-containing rocks. A partial dissolution of these salts releases the ions into the water supply, which may be **surface water** or ground water.

$$CO_2(aq) + H_2O(l) + CaCO_3(s) \rightarrow Ca^{2+}(aq) + 2\,HCO_3^-(aq) \qquad (21.1)$$

Surface water: water that is collected from a watershed, e.g., lakes, rivers, and streams

Hardening ions, such as Ca^{2+}, Mg^{2+}, and Fe^{2+} (and other divalent, 2^+, ions), form insoluble compounds with soaps and cause many detergents to be less effective. Soaps, which are sodium salts of fatty acids such as sodium stearate, $C_{17}H_{35}CO_2^-Na^+$, are very effective cleansing agents so long as they remain soluble; the presence of the hardening ions however causes the formation of a gray, insoluble soap scum such as $(C_{17}H_{35}CO_2)_2Ca$.

$$2\,C_{17}H_{35}CO_2^-Na^+(aq) + Ca^{2+}(aq) \rightarrow (C_{17}H_{35}CO_2)_2Ca(s) + 2\,Na^+(aq) \quad (21.2)$$

This gray precipitate appears as a "bathtub ring" and it also clings to clothes, causing white clothes to appear gray. Dishes and glasses may have spots, shower stalls and lavatories may have a sticky film, clothes may feel rough and scratchy, hair may be dull and unmanageable, and your skin may be irritated and sticky because of hard water.

Hard water is also responsible for the appearance and undesirable formation of "boiler scale" on tea kettles and pots used for heating water. The boiler scale is a poor conductor of heat and thus reduces the efficiency of transferring heat. Boiler scale also builds on the inside of hot water pipes causing a decrease in the flow of water (see opening photo); in extreme cases, this buildup causes the pipe to burst.

Boiler scale consists primarily of the carbonate salts of the hardening ions and is formed according to

$$Ca^{2+}(aq) + 2\,HCO_3^-(aq) \overset{\Delta}{\rightarrow} CaCO_3(s) + CO_2(g) + H_2O(l) \qquad (21.3)$$

---

[1]CO_2 dissolved in rainwater makes rainwater slightly acidic.

$$CO_2(g) + 2\,H_2O(l) \rightarrow H_3O^+(aq) + HCO_3^-(aq)$$

The greater the $CO_2(g)$ levels in the atmosphere due to fossil fuel combustion, the more acidic will be the rainwater.

Figure 21.1 Stalactite and stalagmite formations are present in regions having large deposits of limestone, a major contributor of hardening ions. Colored formations are often due to trace amounts of Fe^{2+}, Mn^{2+}, or Sr^{2+}, also hardening ions.

Table 21.1 Hardness Classification of Water*

Hardness (*ppm CaCO$_3$*)	Classification
<17.1 ppm	soft water
17.1 ppm–60 ppm	slightly hard water
60 ppm–120 ppm	moderately hard water
120 ppm–180 ppm	hard water
>180 ppm	very hard water

*U.S. Department of Interior and the Water Quality Association

Notice that this reaction is just the reverse of the reaction for the formation of hard water (Equation 21.1). The same two reactions are also key to the formation of stalactites and stalagmites for caves located in regions with large limestone deposits (Figure 21.1).

Because of the relatively large natural abundance of limestone deposits and other calcium minerals, such as gypsum, $CaSO_4 \cdot 2H_2O$, it is not surprising that Ca^{2+} ion, in conjunction with Mg^{2+}, is a major component of the dissolved solids in hard water.

Hard water, however, is not a health hazard. In fact, the presence of Ca^{2+} and Mg^{2+} in hard water can be considered dietary supplements to the point of providing their daily recommended allowance (RDA). Some research studies (though disputed) have also indicated a positive correlation between water hardness and decreased heart disease.

The concentration of the hardening ions in a water sample is commonly expressed as though the hardness is due exclusively to $CaCO_3$. Hardness is commonly expressed as mg $CaCO_3$/L, which is also ppm $CaCO_3$,[2] ... or grains per gallon, gpg $CaCO_3$, where 1 gpg $CaCO_3$ = 17.1 mg $CaCO_3$/L. A general classification of hard waters is listed in Table 21.1.

Theory of Analysis

Complex ion: generally a cation of a metal ion to which is bonded a number of molecules or anions (see Experiment 36)

Titrant: the solution placed in the buret in a titrimetric analysis

Analyte: the solution containing the substance being analyzed, generally in the receiving flask in a titration setup

In this experiment a titration technique is used to measure the combined hardening divalent ion concentrations (primarily Ca^{2+} and Mg^{2+}) in a water sample. The titrant is the disodium salt of ethylenediaminetetraacetic acid (abbreviated Na_2H_2Y).[3]

In aqueous solution Na_2H_2Y dissociates into Na^+ and H_2Y^{2-} ions. The H_2Y^{2-} ion reacts with the hardening ions, Ca^{2+} and Mg^{2+}, to form very stable **complex ions,** especially in a solution buffered at a pH of about 10. An ammonia–ammonium ion buffer is often used for this pH adjustment in the analysis.

As H_2Y^{2-} **titrant** is added to the **analyte,** it complexes with the "free" Ca^{2+} and Mg^{2+} of the water sample to form the respective complex ions:

$$Ca^{2+}(aq) + H_2Y^{2-}(aq) \rightarrow [CaY]^{2-}(aq) + 2\,H^+(aq) \qquad (21.4a)$$

$$Mg^{2+}(aq) + H_2Y^{2-}(aq) \rightarrow [MgY]^{2-}(aq) + 2\,H^+(aq) \qquad (21.4b)$$

From the balanced equations, it is apparent that once the molar concentration of the Na_2H_2Y solution is known, the moles of hardening ions in a water sample can be calculated, a 1:1 stoichiometric ratio.

$$\text{volume } H_2Y^{2-} \times \text{molar concentration of } H_2Y^{2-} = \text{moles } H_2Y^{2-}$$
$$= \text{moles hardening ions} \quad (21.5)$$

The hardening ions, for reporting purposes, are assumed to be exclusively Ca^{2+} from the dissolving of $CaCO_3$. Since one mole of Ca^{2+} forms from one mole of $CaCO_3$, the hardness of the water sample expressed as mg $CaCO_3$ per liter of sample is

$$\text{moles hardening ions} = \text{moles } Ca^{2+} = \text{moles of } CaCO_3 \quad (21.6)$$

$$\text{ppm } CaCO_3 \left(\frac{\text{mg } CaCO_3}{\text{L sample}} \right) = \frac{\text{mol } CaCO_3}{\text{L sample}} \times \frac{100.1 \text{ g } CaCO_3}{\text{mol}} \times \frac{\text{mg}}{10^{-3}\text{g}} \quad (21.7)$$

$$
\begin{array}{c}
Na^+\,^-O \qquad\qquad O\,^-Na^+ \\
| \qquad\qquad\qquad | \\
O{=}C \qquad\qquad C{=}O \\
H_2C \qquad\quad CH_2 \\
\diagdown \quad N \quad \diagup \\
| \\
CH_2 \\
| \\
CH_2 \\
| \\
H_2C \qquad\quad CH_2 \\
\diagup \quad N \quad \diagdown \\
O{=}C \qquad\qquad C{=}O \\
| \qquad\qquad\qquad | \\
OH \qquad\qquad OH \\
Na_2H_2Y
\end{array}
$$

[2] ppm means "parts per million"—1 mg of $CaCO_3$ in 1 000 000 mg (or 1 kg) solution is 1 ppm $CaCO_3$. Assuming the density of the solution is 1 g/mL (or 1 kg/L), then 1 000 000 mg solution = 1 L solution. Therefore, 1 mg/L is an expression of ppm.

[3] **E**thylene**d**iamine**t**etra**a**cetic acid is often simply referred to as EDTA with an abbreviated formula of H_4Y.

A special indicator is used to detect the endpoint in the titration. Called Eriochrome Black T (EBT),[4] it forms complex ions with the Ca^{2+} and Mg^{2+} ions, but binds more strongly to Mg^{2+} ions. Because only a small amount of EBT is added, only Mg^{2+} complexes; no Ca^{2+} ion complexes to EBT—therefore, most all of the hardening ions remain "free" in solution. The EBT indicator is sky-blue in solution but forms a wine-red complex with Mg^{2+}.

$$Mg^{2+}(aq) + EBT(aq) \rightleftharpoons [Mg\text{-}EBT]^{2+}(aq) \qquad (21.8)$$
$$\text{sky-blue} \qquad \text{wine-red}$$

Therefore before any H_2Y^{2-} titrant is added for the analysis, the analyte is wine-red because of the $[Mg\text{-}EBT]^{2+}$ complex ion.

As the H_2Y^{2-} titrant is added, all of the "free" Ca^{2+} and Mg^{2+} ions in the water sample become complexed just prior to the endpoint; thereafter, the H_2Y^{2-} removes the trace amount of Mg^{2+} from the wine-red $[Mg\text{-}EBT]^{2+}$ complex. At this point, the solution changes from the wine-red color back to the original sky-blue color of the EBT indicator to reach the endpoint. All hardening ions have been complexed with H_2Y^{2-}.

$$[Mg^{2+}\text{-}EBT]^{2+}(aq) + H_2Y^{2-}(aq) \rightarrow [MgY]^{2-}(aq) + 2\,H^+(aq) + EBT(aq) \quad (21.9)$$
$$\text{wine-red} \qquad\qquad\qquad\qquad\qquad\qquad \text{sky-blue}$$

Therefore, the presence of Mg^{2+} in the sample is a must in order for the color change from wine-red to sky-blue to be observed. To ensure the appearance of the endpoint, oftentimes a small amount of Mg^{2+} as $[MgY]^{2-}$ is initially added to the analyte along with the EBT indicator to form the wine-red color of $[Mg\text{-}EBT]^{2+}$.

Eriochrome Black T

The mechanism for the process of adding both $[MgY]^{2-}$ and EBT is: the $[MgY]^{2-}$ dissociates in the analyte because the Y^{4-} (as H_2Y^{2-} in water) is more strongly bonded to the Ca^{2+} of the sample; the "freed" Mg^{2+} then combines with the EBT to form the wine-red color (Equation 21.8). The complexing of the "free" Ca^{2+} and Mg^{2+} with the H_2Y^{2-} titrant continues until both are depleted. At that point, the H_2Y^{2-} reacts with the $[Mg\text{-}EBT]^{2+}$ in the sample until the endpoint is reached (Equation 21.9).

Because Mg^{2+} and Y^{4-} (as H_2Y^{2-}) are freed initially from the added $[MgY]^{2-}$, but later consumed at the endpoint, no additional H_2Y^{2-} titrant is required for the analysis of hardness in the water sample.

The standardization of a Na_2H_2Y solution is determined by its reaction with a known amount of calcium ion in a (primary) standard Ca^{2+} solution (Equation 21.4a). The measured aliquot of the standard Ca^{2+} solution is buffered to a pH of 10 and titrated with the Na_2H_2Y solution to the Eriochrome Black T *sky-blue* endpoint (Equation 21.9). To achieve the endpoint a small amount of Mg^{2+} in the form of $[MgY]^{2-}$ is added to the standard Ca^{2+} solution.

Note that the standardization of the Na_2H_2Y solution with a standard Ca^{2+} solution in Part A is reversed in Part B where the (now) standardized Na_2H_2Y solution is used to determine the concentration of Ca^{2+} (and other hardening ions) in a sample.

Procedure Overview: A (primary) standard solution of Ca^{2+} is used to standardize a prepared $\sim0.01\,M$ Na_2H_2Y solution. The (secondary) standardized Na_2H_2Y solution is subsequently used to titrate the hardening ions of a water sample to the Eriochrome Black T (or Calmagite) indicator endpoint.

The standardized Na_2H_2Y solution may have already been prepared by stockroom personnel. If so, obtain 100 mL of the solution and proceed to Part B. Consult with your instructor.

Three trials are to be completed for the standardization of the $\sim0.01\,M$ Na_2H_2Y solution. Initially prepare three, clean 125-mL Erlenmeyer flasks for Part A.3.

---

[4]Calmagite may be substituted for Eriochrome Black T as an indicator. The same wine-red to sky-blue endpoint is observed. Ask your instructor.

Read and record the volume in the buret to the correct number of significant figures.

1. **Measure of the Mass of the Na₂H₂Y Solution.** Calculate the mass of $Na_2H_2Y \cdot 2H_2O$ (molar mass = 372.24 g/mol) required to prepare 250 mL of a 0.01 M Na_2H_2Y solution. Show this calculation on the Report Sheet. Measure this mass on weighing paper, transfer it to a 250-mL volumetric flask containing 100 mL of deionized water, swirl to dissolve, and dilute to the "mark" (slight heating may be required).

2. **Prepare a Buret for Titration.** Rinse a *clean* buret with the Na_2H_2Y solution several times and then fill. Record the volume of the titrant "using all certain digits plus one uncertain digit."

3. **Prepare the Standard Ca²⁺ Solution.** Obtain ~80 mL of a standard Ca^{2+} solution and record its exact molar concentration (~0.01 M). Pipet 25.0 mL of the standard Ca^{2+} solution into a 125-mL Erlenmeyer flask, add 1 mL of buffer (pH = 10) solution, and 2 drops of EBT indicator (containing a small amount of $[MgY]^{2-}$).

4. **Titrate the Standard Ca²⁺ Solution.** Titrate the standard Ca^{2+} solution with the Na_2H_2Y titrant; swirl continuously. Near the endpoint, slow the rate of addition to drops; the last few drops should be added at 3–5 s intervals. The solution changes from wine-red to purple to sky-blue—no tinge of the wine-red color should remain; the solution is *blue* at the endpoint. Record the final volume in the buret.

5. **Repeat the Titration of the Standard Ca²⁺ Solution.** Repeat the titrations on the remaining two samples. Calculate the molar concentration of the Na_2H_2Y solution. Save the standard Ca^{2+} solution for Part B.

B. Analysis of Water Sample

Complete three trials for your analysis. The first trial is an indication of the hardness of your water sample. You may want to adjust the volume of water for the analysis of the second and third trials.

1. **Obtain the Water Sample for Analysis.** Obtain about 100 mL of a water sample from your instructor. You may use your own water sample or simply the tap water in the laboratory.

    If the water sample is from a lake, stream, or ocean, you will need to gravity filter the sample before the analysis.

    If your sample is acidic, add 1 M NH_3 until it is basic to litmus (or pH paper).

2. **Prepare the Water Sample for Analysis.** Pipet 25.0 mL of your (filtered, if necessary) water sample[5] into a 125-mL Erlenmeyer flask, add 1 mL of the buffer (pH = 10) solution, and 2 drops of EBT indicator.

3. **Titrate the Water Sample.** Titrate the water sample with the standardized Na_2H_2Y until the *blue* endpoint appears (as described in Part A.4). Repeat (twice) the analysis of the water sample to determine its hardness.

Disposal: Dispose of the analyzed solutions in the "Waste EDTA" container.

The Next Step

(1) Because hardness of a water source varies with temperature, rainfall, seasons, water treatment, etc., design a systematic study of the hardness of a water source as a function of one or more variables. (2) Compare the incoming vs. the outgoing water hardness of a continuous water supply. (3) Compare the water hardness of drinking water for adjacent city/county water supplies and account for the differences.

---

[5]If your water is known to have a high hardness, decrease the volume of the water proportionally until it takes about 15 mL of Na_2H_2Y titrant for your second and third trials. Similarly if your water sample is known to have a low hardness, increase the volume of the water proportionally.

Date _____ Lab Sec. _____ Name _____ Desk No. _____

1. What cations are responsible for water hardness?

2. Experimental Procedure, Part A.1. Calculate the mass of disodium ethylenediaminetetraacetate (molar mass = 372.24 g/mol) required to prepare 250 mL of a 0.010 *M* solution. Show the calculation here and on the Report Sheet.

3. Experimental Procedure, Part A.3. A 25.7-mL volume of a prepared Na_2H_2Y solution titrates 25.0 mL of a standard 0.0107 *M* Ca^{2+} solution to the Eriochrome Black T endpoint. What is the molar concentration of the Na_2H_2Y solution?

4. a. Which hardening ion, Ca^{2+} or Mg^{2+}, binds more tightly to (forms a stronger complex ion with) the Eriochrome Black T indicator used for today's analysis?

   b. What is the color change at the endpoint?

5. A 50.0-mL water sample requires 16.33 mL of 0.0109 M Na$_2$H$_2$Y to reach the Eriochrome Black T endpoint.

   a. Calculate the moles of hardening ions in the water sample.

   b. Assuming the hardness is due exclusively to CaCO$_3$, express the hardness concentration in mg CaCO$_3$/L sample.

   c. What is this hardness concentration expressed in ppm CaCO$_3$?

   d. Classify the hardness of this water according to Table 21.1.

6. The hardness of a water sample is known to be 320 ppm CaCO$_3$. In an analysis of a 50-mL water sample, what volume of 0.0100 M Na$_2$H$_2$Y is needed to reach the Eriochrome Black T endpoint?

Experiment 21 *Report Sheet*

Hard Water Analysis

Date ________ Lab Sec. ______ Name ______________________________ Desk No. ________

A. A Standard 0.01 *M* Disodium Ethylenediaminetetraacetate, Na_2H_2Y, Solution

Calculate the mass of $Na_2H_2Y \cdot 2H_2O$ required to prepare 250 mL of a 0.01 *M* Na_2H_2Y solution.

	Trial 1	*Trial 2*	*Trial 3*
1. Volume of standard Ca^{2+} solution (*mL*)	25.0	25.0	25.0
2. Concentration of standard Ca^{2+} solution			
3. Mol Ca^{2+} = mol Na_2H_2Y (*mol*)			
4. Buret reading, *initial* (*mL*)			
5. Buret reading, *final* (*mL*)			
6. Volume of Na_2H_2Y titrant (*mL*)			
7. Molar concentration of Na_2H_2Y solution (*mol/L*)			
8. Average molar concentration of Na_2H_2Y solution (*mol/L*)			

B. Analysis of Water Sample

	Trial 1	*Trial 2*	*Trial 3*
1. Volume of water sample (*mL*)			
2. Buret reading, *initial* (*mL*)			
3. Buret reading, *final* (*mL*)			
4. Volume of Na_2H_2Y titrant (*mL*)			

5. Mol Na_2H_2Y = mol hardening ions,
   Ca^{2+} and Mg^{2+} (*mol*)

   _____ _____ _____

6. Mass of equivalent $CaCO_3$ (*g*)

   _____ _____ _____

7. ppm $CaCO_3$ (*mg $CaCO_3$/L sample*)

   _____ _____ _____

8. Average ppm $CaCO_3$

   _____

9. Average gpg $CaCO_3$

   _____

10. Standard deviation of ppm $CaCO_3$

    _____

11. Relative standard deviation of ppm $CaCO_3$ (%RSD)

    _____

Laboratory Questions

Circle the questions that have been assigned.

1. Part A.3. State the purpose for the 1 mL of buffer (pH = 10) being added to the standard Ca^{2+} solution.

2. Part A.3. The Eriochrome Black T indicator is mistakenly omitted. What is the color of the analyte (standard Ca^{2+} solution)? Describe the appearance of the analyte with the continued addition of the Na_2H_2Y solution. Explain.

*3. Part A.3. The buffer solution is omitted from the titration procedure, the Eriochrome Black T indicator and a small amount of Mg^{2+} are added, and the standard Ca^{2+} solution is acidic.
   a. What is the color of the solution? Explain.
   b. The Na_2H_2Y solution is dispensed from the buret. What color changes are observed? Explain.

4. Part A.4. Deionized water from the wash bottle is used to wash the side of the Erlenmeyer flask. How does this affect the reported molar concentration of the Na_2H_2Y solution . . . too high, too low, or unaffected? Explain.

5. Part A.4. The dispensing of the Na_2H_2Y solution from the buret is discontinued when the solution turns purple. Because of this technique error, will the reported molar concentration of the Na_2H_2Y solution be too high, too low, or unaffected? Explain.

6. Part B.3. The dispensing of the Na_2H_2Y solution from the buret is discontinued when the solution turns purple. Because of this technique error, will the reported hardness of the water sample be too high, too low, or unaffected? Explain.

7. Part A.4 and Part B.3. The dispensing of the Na_2H_2Y solution from the buret is discontinued when the solution turns purple. However in Part B.3, the standardized Na_2H_2Y solution is then used to titrate a water sample to the (correct) *blue* endpoint. Will the reported hardness of the water sample be too high, too low, or unaffected? Explain.

*8. Washing soda, $Na_2CO_3 \cdot 10H_2O$ (molar mass = 286 g/mol), is often used to "soften" hard water, i.e., to remove hardening ions. Assuming hardness is due to Ca^{2+}, the CO_3^{2-} ion precipitates the Ca^{2+}.

$$Ca^{2+}(aq) + CO_3^{2-}(aq) \rightarrow CaCO_3(s)$$

How many grams and pounds of washing soda are needed to remove the hardness from 500 gallons of water having a hardness of 200 ppm $CaCO_3$ (see Appendix A for conversion factors)?

Experiment 22

Molar Solubility, Common-Ion Effect

Silver oxide forms a brown mudlike precipitate from a mixture of silver nitrate and sodium hydroxide solutions.

OBJECTIVES

- To determine the molar solubility and the solubility constant of calcium hydroxide
- To study the effect of a common ion on the molar solubility of calcium hydroxide

TECHNIQUES

The following techniques are used in the Experimental Procedure

INTRODUCTION

Salts that have a very limited solubility in water are called **slightly soluble** (or "insoluble") **salts.** A saturated solution of a slightly soluble salt is a result of a dynamic equilibrium between the solid salt and its ions in solution; however, because the salt is only slightly soluble, the concentrations of the ions in solution are low. For example, in a saturated silver sulfate, Ag_2SO_4, solution, the **dynamic equilibrium** between solid Ag_2SO_4 and the Ag^+ and SO_4^{2-} ions in solution lies far to the *left* because of the low solubility of silver sulfate.

Slightly soluble salt: a qualitative term that reflects the very low solubility of a salt

Dynamic equilibrium: the rate of the forward reaction equals the rate of the reverse reaction

$$Ag_2SO_4(s) \rightleftharpoons 2\,Ag^+(aq) + SO_4^{2-}(aq) \qquad (22.1)$$

The mass action expression for this system is

$$[Ag^+]^2[SO_4^{2-}] \qquad (22.2)$$

As Ag_2SO_4 is a solid, its concentration is constant and, therefore, does not appear in the mass action expression. At equilibrium, the mass action expression equals K_{sp}, called the **solubility product** or, more simply, the equilibrium constant for this slightly soluble salt.

The **molar solubility** of Ag_2SO_4, determined experimentally, is 1.4×10^{-2} mol/L. This means that in 1.0 L of a saturated Ag_2SO_4 solution, only 1.4×10^{-2} mol of silver sulfate dissolves, forming 2.8×10^{-2} mol of Ag^+ and 1.4×10^{-2} mol of SO_4^{2-}. The solubility product of silver sulfate equals the product of the molar concentrations of the ions, each raised to the power of its coefficient in the balanced equation:

Molar solubility: the number of moles of salt that dissolve per liter of (aqueous) solution

$$K_{sp} = [Ag^+]^2[SO_4^{2-}] = [2.8 \times 10^{-2}]^2[1.4 \times 10^{-2}] = 1.1 \times 10^{-5} \qquad (22.3)$$

What happens to the molar solubility of a salt when an ion, common to the salt, is added to the saturated solution? According to LeChâtelier's principle (Experiment 16),

Figure 22.1 The addition of conc HCl to a saturated NaCl solution results in the formation of solid NaCl.

Figure 22.2 The solid $Ca(OH)_2$ in a saturated $Ca(OH)_2$ solution is slow to settle.

the equilibrium for the salt shifts to compensate for the added ions; that is, it shifts *left* to favor the formation of more of the solid salt. This effect, caused by the addition of an ion common to an existing equilibrium, is called the **common-ion effect.** As a result of the common-ion addition and the corresponding shift in the equilibrium, fewer moles of the salt dissolve in solution, lowering the molar solubility of the salt.

While molar solubility is often associated with slightly soluble salts, soluble salts are also affected by the addition of a common-ion to the equilibrium. For example, consider the equilibrium for a saturated solution of sodium chloride:

$$NaCl(s) \rightleftharpoons Na^+(aq) + Cl^-(aq) \tag{22.4}$$

The addition of chloride ion, a common-ion in the equilibrium, shifts the equilibrium left to cause the formation of solid sodium chloride (Figure 22.1).

In Part A this experiment, you will determine the molar solubility and the solubility product for calcium hydroxide, $Ca(OH)_2$. A saturated $Ca(OH)_2$ solution[1] (Figure 22.2) is prepared; after an equilibrium is established between the solid $Ca(OH)_2$ and the Ca^{2+} and OH^- ions in solution, the supernatant solution is analyzed. The hydroxide ion, OH^-, in the supernatant solution is titrated with a standardized HCl solution to determine its molar concentration.

According to the equation

$$Ca(OH)_2(s) \rightleftharpoons Ca^{2+}(aq) + 2\,OH^-(aq) \tag{22.5}$$

for each mole of $Ca(OH)_2$ that dissolves, 1 mol of Ca^{2+} and 2 mol of OH^- are present in solution. Thus, by determining the molar concentration of hydroxide ion, the $[Ca^{2+}]$, the K_{sp}, and the molar solubility of $Ca(OH)_2$ can be calculated.

$$[Ca^{2+}] = \tfrac{1}{2}[OH^-]$$
$$K_{sp} = [Ca^{2+}][OH^-]^2 = \tfrac{1}{2}[OH^-]^3 \tag{22.6}$$
$$\text{molar solubility of } Ca(OH)_2 = [Ca^{2+}] = \tfrac{1}{2}[OH^-]$$

Likewise, the same procedure is used in Part B to determine the molar solubility of $Ca(OH)_2$ in the presence of added calcium ion, an ion common to the slightly soluble salt equilibrium.

---

[1] A saturated solution of calcium hydroxide is called **limewater.**

Procedure Overview: The supernatant from a saturated calcium hydroxide solution is titrated with a standardized hydrochloric acid solution to the methyl orange endpoint. An analysis of the data results in the determination of the molar solubility and solubility product of calcium hydroxide. The procedure is repeated on the supernatant from a saturated calcium hydroxide solution containing added calcium ion.

EXPERIMENTAL PROCEDURE

Three analyses are to be completed. To hasten the analyses, prepare three, *clean* labeled 125- or 250-mL Erlenmeyer flasks. Obtain no more than 50 mL of standardized 0.05 *M* HCl for use in Part A.5.

Ask your laboratory instructor about the status of the saturated Ca(OH)₂ solution: if you are to prepare the solution, then omit Part A.3; if the stockroom personnel has prepared the solution, then omit Parts A.1 and A.2.

A. Molar Solubility and Solubility Product of Calcium Hydroxide

1. **Prepare the Stock Calcium Hydroxide Solution.** Prepare a saturated Ca(OH)₂ solution 1 week before the experiment by adding approximately 3 g of Ca(OH)₂ to 120 mL of boiled, deionized water in a 125-mL Erlenmeyer flask. Stir the solution and stopper the flask.

2. **Transfer the Saturated Calcium Hydroxide Solution.** Allow the solid Ca(OH)₂ to remain settled (from Part A.1). *Carefully* [do not disturb the finely divided Ca(OH)₂ solid] decant about 90 mL of the saturated Ca(OH)₂ solution into a second 125-mL flask. Proceed to Part A.4.

3. **Obtain a Saturated Calcium Hydroxide Solution (Alternate).** Submit a clean, dry 150-mL beaker to your laboratory instructor (or stockroom) for the purpose of obtaining ~90 mL of supernatant from a saturated Ca(OH)₂ solution for analysis.

4. **Prepare a Sample for Analysis.** Rinse a 25-mL pipet at least twice with 1- to 2-mL portions of the saturated Ca(OH)₂ solution and discard. Pipet 25 mL of the saturated Ca(OH)₂ solution into a clean 125-mL flask and add 2 drops of methyl orange indicator.[2]

5. **Set Up the Titration Apparatus.** Prepare a clean, 50-mL buret for titration. Rinse the clean buret and tip with three 5-mL portions of the standardized 0.05 *M* HCl solution and discard. Fill the buret with standardized 0.05 *M* HCl, remove the air bubbles in the buret tip, and, after 10–15 seconds, read and record the initial volume in the buret to the correct number of significant figures. Record the *actual* concentration of the 0.05 *M* HCl on the Report Sheet. Place a sheet of white paper beneath the receiving flask.

6. **Titrate.** Titrate the Ca(OH)₂ solution with the standardized HCl solution to the methyl orange endpoint, where the color changes from yellow to a faint red-orange. Remember the addition of HCl should stop within *one-half drop* of the endpoint. After 10–15 seconds of the persistent endpoint, read and record the final volume of standard HCl in the buret.

7. **Repeat.** Titrate two additional samples of the saturated Ca(OH)₂ solution until 1% reproducibility is achieved.

8. **Do the Calculations.** Complete your calculations as outlined on the Report Sheet. The reported values for the K_{sp} of Ca(OH)₂ will vary from chemist to chemist.

Three analyses are to be completed. Clean and label three 125- or 250-mL Erlenmeyer flasks.

Again, as in Part A, ask your instructor about the procedure by which you are to obtain the saturated Ca(OH)₂ solution with the added CaCl₂ for analysis. If you are

B. Molar Solubility of Calcium Hydroxide in the Presence of a Common Ion

---

[2]Methyl orange changes from red-orange to yellow in the pH range of from 3.2 to 4.4.

to prepare the solution, then complete Parts B.1 and A.2; if the stockroom personnel prepared the "spiked" saturated $Ca(OH)_2$ solution, then repeat Part A.3.

In either case, complete Part B.2 in its entirety for the analysis of the saturated $Ca(OH)_2$ solution with the added $CaCl_2$.

1. **Prepare the Stock Solution.** Mix 3 g of $Ca(OH)_2$ and 1 g of $CaCl_2 \cdot 2H_2O$ with 120 mL of boiled, deionized water in a 125-mL flask 1 week before the experiment. Stir and stopper the flask.

2. **Prepare a Buret for Analysis, Prepare the Sample, and Titrate.** Repeat Parts A.4–A.8.

> *Disposal:* Discard all of the reaction mixtures as advised by your instructor.

CLEANUP: Discard the HCl solution in the buret as advised by your instructor. Rinse the buret twice with tap water and twice with deionized water.

The Next Step

(1) Design an experiment to determine the molar solubility of a salt without using a titration procedure. (2) How does the molar solubility of the hydroxide salts vary within a group and/or within a period of the periodic table? (3) Does the amount of $CaCl_2$ added to saturated $Ca(OH)_2$ solution produce a linear correlation to its molar solubility? Try additional sample preparations.

Molar Solubility, Common-Ion Effect

Date _____ Lab Sec. _____ Name _____ Desk No. _____

1. The molar solubility of PbI_2 is 1.5×10^{-3} mol/L (see photo).
   a. What is the molar concentration of iodide ion in a saturated PbI_2 solution?

   b. Determine the solubility constant, K_{sp}, for lead(II) iodide.

Lead iodide precipitate.

2. Experimental Procedure, Part A.4. What is the purpose of rinsing the pipet twice with aliquots of the saturated $Ca(OH)_2$ solution?

3. Experimental Procedure, Part A.4. a. What is the indicator used to detect the endpoint in the titration for this experiment?

   b. What is the expected color change at the endpoint in this experiment?

4. Experimental Procedure, Part A.5. The directions are to "read and record the initial volume of the buret to the correct number of significant figures." Explain what this means.

5. Experimental Procedure, Part A.6 vs. Part B.2. Would you expect more or less standard 0.05 *M* HCl to be used to reach the methyl orange endpoint in Part B.2? Explain.

6. A saturated solution of strontium hydroxide is prepared and the excess solid strontium hydroxide is allowed to settle. A 25.0-mL aliquot of the saturated solution is withdrawn and transferred to an Erlenmeyer flask, and two drops of methyl orange indicator are added. A 0.101 *M* HCl solution (titrant) is dispensed from a buret into the solution (analyte). The solution turns from yellow to a very faint red-orange after the addition of 21.3 mL.
   a. How many moles of hydroxide ion are neutralized in the analysis?

   b. What is the molar concentration of the hydroxide ion in the saturated solution?

   c. What is the molar solubility of strontium hydroxide?

   d. What is the solubility product, K_{sp}, for strontium hydroxide?

*7. Phenolphthalein has a color change over the pH range of 8.2 to 10.0; methyl orange has a color change over the pH range of 3.2 to 4.4. Although the phenolphthalein indicator is commonly used for neutralization reactions, why instead is the methyl orange indicator recommended for this experiment?

Molar Solubility, Common-Ion Effect

Date _____ Lab Sec. _____ Name _____ Desk No. _____

A. Molar Solubility and Solubility Product of Calcium Hydroxide

	Trial 1	*Trial 2*	*Trial 3*
1. Volume of sat'd $Ca(OH)_2$ solution (*mL*)	25.0	25.0	25.0
2. Concentration of standardized HCl solution (*mol/L*)			
3. Buret reading, *initial* (*mL*)			
4. Buret reading, *final* (*mL*)			
5. Volume of HCl added (*mL*)			
6. Moles of HCl added (*mol*)			
7. Moles of OH^- in sat'd solution (*mol*)			
8. $[OH^-]$, equilibrium (*mol/L*)			
9. $[Ca^{2+}]$, equilibrium (*mol/L*)			
10. Molar solubility of $Ca(OH)_2$ (*mol/L*)	*		
11. Average molar solubility of $Ca(OH)_2$ (*mol/L*)			
12. K_{sp} of $Ca(OH)_2$	*		
13. Average K_{sp}			
14. Standard deviation of K_{sp}			
15. Relative standard deviation of K_{sp} (%RSD)			

*Calculations for Trial 1.

B. Molar Solubility of Calcium Hydroxide in the Presence of a Common Ion

	Trial 1	Trial 2	Trial 3
1. Volume of sat'd $Ca(OH)_2$ with added $CaCl_2$ solution (mL)	25.0	25.0	25.0
2. Concentration of standardized HCl solution (mol/L)			
3. Buret reading, *initial* (mL)			
4. Buret reading, *final* (mL)			
5. Volume of HCl added (mL)			
6. Moles of HCl added (mol)			
7. Moles of OH^- in sat'd solution (mol)			
8. $[OH^-]$, equilibrium (mol/L)			
9. Molar solubility of $Ca(OH)_2$ with added $CaCl_2$ (mol/L)			
10. Average molar solubility of $Ca(OH)_2$ with added $CaCl_2$ (mol/L)			

Account for the different molar solubilities in Part A.11 and Part B.10 (Report Sheet).

Laboratory Questions

Circle the questions that are to be answered.

1. Part A.2. Suppose some of the solid calcium hydroxide is inadvertently transferred along with the supernatant liquid for analysis.
   a. Will more, less, or the same amount of hydrochloric acid titrant be used for the analysis in Part A.6? Explain.
   b. How will this inadvertent transfer affect the calculated solubility product for calcium hydroxide? Explain.
   c. How will this inadvertent transfer affect the calculated molar solubility for calcium hydroxide? Explain.

2. Part A.6. Does adding boiled, deionized water to the titrating flask to wash the wall of the Erlenmeyer flask and the buret tip affect the K_{sp} value of the $Ca(OH)_2$? Explain.

*3. How will tap water instead of boiled, deionized water affect the K_{sp} value of $Ca(OH)_2$ in Part A? *Hint:* How will the minerals in the water affect the solubility of $Ca(OH)_2$?

4. Part A.8. Jerry forgot to record the actual molar concentration of the standard HCl solution (which was actually 0.044 M). However, to complete the calculations quickly, the ~0.05 M concentration was used. Will the reported molar solubility of $Ca(OH)_2$ be too high or too low? Explain.

5. Refer to the Prelaboratory Assignment, Question 7. If phenolphthalein had been used in this experiment instead of the methyl orange, would the reported solubility product for calcium hydroxide be too high, too low, or unaffected? Explain.

6. The ethylenediaminetetraacetate ion, H_2Y^{2-}, forms a strong complex with the calcium ion. How does the addition of H_2Y^{2-} affect the molar solubility of $Ca(OH)_2$? Explain. See Experiment 21 for the reaction of Ca^{2+} with H_2Y^{2-}.

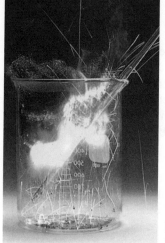

Experiment 23

Factors Affecting Reaction Rates

Iron reacts slowly in air to form iron(III) oxide, commonly called rust. When finely-divided pure iron is heated and thrust into pure oxygen, the reaction is rapid.

• To study the various factors that affect the rates of chemical reactions

OBJECTIVE

The following techniques are used in the Experimental Procedure

TECHNIQUES

Chemical kinetics is the study of chemical reaction rates, how reaction rates are controlled, and the pathway or mechanism by which a reaction proceeds from its reactants to its products.

Reaction rates vary from the very fast, in which the reaction, such as the explosion of a hydrogen/oxygen mixture, is essentially complete in microseconds or even **nanoseconds,** to the very slow, in which the reaction, such as the setting of concrete, requires years to complete.

The rate of a chemical reaction may be expressed as a *change* in the concentration of a reactant (or product) as a function of time (e.g., per second)—the greater the change in the concentration per unit of time, the faster the rate of the reaction. Other parameters that can follow the change in concentration of a **species** as a function of time in a chemical reaction are color (expressed as absorbance, Figure 23.1), temperature, pH, odor, and conductivity. The parameter chosen for following the rate of a particular reaction depends on the nature of the reaction and the species of the reaction.

We will investigate four of five factors that can be controlled to affect the rate of a chemical reaction. The first four factors listed below are systematically studied in this experiment:

- Nature of the reactants
- Temperature of the chemical system
- Presence of a catalyst
- Concentration of the reactants
- Surface area of the reactants

INTRODUCTION

Nanosecond: 1×10^{-9} second

Figure 23.1 The higher concentration of light-absorbing species, the more intense is the color of the solution.

Species: any atom, molecule, or ion that may be a reactant or product of a chemical reaction

---

Some substances are naturally more reactive than others and, therefore, undergo rapid chemical changes. For example, the reaction of **sodium metal and water** is a very rapid, exothermic reaction (see Experiment 11, Part F), whereas the corrosion of iron is much slower. Plastics, reinforced with fibers such as carbon or glass, are now being substituted for iron and steel in specialized applications where corrosion has historically been a problem.

Nature of the Reactants

Sodium metal and water: the reaction releases $H_2(g)$ which ignites with the oxygen in the air to produce a yellow/blue flame, the yellow resulting from the presence of Na^+ in the flame

Temperature of the Chemical System

Internal energy: the energy contained within the molecules/ions when they collide

As a rule of thumb, a 10°C rise in temperature doubles (increases by a factor of 2) the rate of a chemical reaction. The added heat not only increases the number of collisions[1] between reactant molecules, but also, and more importantly, increases their kinetic energy. On collision of the reactant molecules, this kinetic energy is converted into an **internal energy** that is distributed throughout the collision system. This increased internal energy increases the probability for the weaker bonds to be broken and the new bonds to be formed.

Presence of a Catalyst

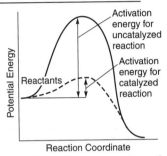

Figure 23.2 Reaction profiles of an uncatalyzed and a catalyzed reaction.

A **catalyst** increases the rate of a chemical reaction without undergoing any *net* chemical change. Some catalysts increase the rate of only one specific chemical reaction without affecting similar reactions. Other catalysts are more general and affect an entire set of similar reactions. Catalysts generally reroute the pathway of a chemical reaction so that this "alternate" path, although perhaps more circuitous, has a lower activation energy for reaction than the uncatalyzed reaction (Figure 23.2).

Concentration of the Reactants

An increase in the concentration of a reactant generally increases the reaction rate. See the opening photo. The larger concentration of reactant molecules increases the probability of an "effective" collision between reacting molecules for the formation of product. On occasion, such an increase may have no effect or may even decrease the reaction rate. A quantitative investigation on the affect of concentration changes on reaction rate is undertaken in Experiment 24.

Surface Area of the Reactants

Generally speaking, the greater the exposed surface area of the reactant, the greater the reaction rate. Again, see opening photo. For example, a large piece of coal burns very slowly, but coal *dust* burns rapidly, a consequence of which can lead to a disastrous coal mine explosion; solid potassium iodide reacts very slowly with solid lead nitrate, but when both are dissolved in solution, the formation of lead iodide is instantaneous.

EXPERIMENTAL PROCEDURE

Procedure Overview: A series of qualitative experiments are conducted to determine how various factors affect the rate of a chemical reaction.

Caution: *A number of strong acids are used in the experiment. Handle with care; do not allow them to touch the skin or clothing.*

Perform the experiment with a partner. At each circled superscript ①–⑲ in the procedure, *stop,* and record your observation on the Report Sheet. Discuss your observations with your lab partner and your instructor.

Ask your instructor which parts of the Experimental Procedure you are to complete. Prepare the hot water bath needed for Parts B.3 and C.3,4. Use a 250-mL beaker.

A. Nature of the Reactants

1. **Different Acids Affect Reaction Rates.** Half-fill a set of four, labeled small test tubes (Figure 23.3) with 3 M H_2SO_4, 6 M HCl, 6 M CH_3COOH, and 6 M H_3PO_4, respectively. (**Caution:** *Avoid skin contact with the acids.*) Submerge a 1-cm strip of magnesium ribbon into each test tube. Compare the reaction rates and record your observations.①

2. **Different Metals Affect Reaction Rates.** Half-fill a set of three, labeled small test tubes (Figure 23.4) with 6 M HCl. Submerge 1-cm strips of zinc, magnesium, and copper separately into the test tubes. Compare the reaction rates of each metal in HCl and record your observations.② Match the relative reactivity of the metals with the photos in Figure 23.5.③

---

[1] A 10°C temperature rise only increases the collision frequency between reactant molecules by a factor of 1.02—nowhere near the factor of 2 that is normally experienced in a reaction rate.

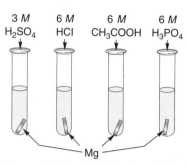

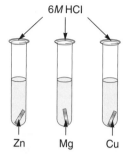

Figure 23.3 Setup for the effect of acid type on reaction rate.

Figure 23.4 Setup for the effect of metal type on reaction rate.

Figure 23.5 Zinc, copper, and magnesium react at different rates with 6 M HCl. Identify the metals in the photo according to their reactivity.[3]

Ask your instructor to determine if *both* Parts B and C are to be completed. You should perform the experiment with a partner; as one student combines the test solutions, the other notes the time.

The oxidation–reduction reaction that occurs between hydrochloric acid and sodium thiosulfate, $Na_2S_2O_3$, produces insoluble sulfur as a product.

$$2\ HCl(aq) + Na_2S_2O_3(aq) \rightarrow S(s) + SO_2(g) + 2\ NaCl(aq) + H_2O(l) \quad (23.1)$$

The time required for the cloudiness of sulfur to appear is a measure of the reaction rate. Measure each volume of reactant with separate graduated pipets.

1. **Prepare the Solutions.** Pipet 2 mL of 0.1 *M* $Na_2S_2O_3$ into each of a set of three 150-mm, *clean* test tubes. Into a second set of three 150-mm test tubes pipet 2 mL of 0.1 *M* HCl. Label each set of test tubes.

2. **Record the Time for Reaction at the "Lower" Temperature.** Place a $Na_2S_2O_3$–HCl pair of test tubes in an ice water bath until thermal equilibrium is established (~5 minutes). Pour the HCl solution into the $Na_2S_2O_3$ solution, START TIME, agitate the mixture for several seconds, and return the reaction mixture to the ice bath. STOP TIME when the cloudiness of the sulfur appears. Record the time lapse for the reaction and the temperature of the bath, "using all certain digits *plus* one uncertain digit."[4]

3. **Record the Time for Reaction at the "Higher" Temperature.** Place a second $Na_2S_2O_3$–HCl pair of test tubes in a warm water (<60°C) bath until thermal equilibrium is established (~5 minutes). Pour the HCl solution into the $Na_2S_2O_3$ solution,

B. Temperature of the Reaction: Hydrochloric Acid–Sodium Thiosulfate Reaction System

START TIME, agitate the mixture for several seconds, and return the reaction mixture to the warm water bath. STOP TIME when the cloudiness of the sulfur appears. Record the temperature of the bath.[5]

4. **Record the Time for Reaction at "Room" Temperature.** Combine the remaining set of $Na_2S_2O_3$–HCl test solutions at room temperature and proceed as in Parts B.2 and B.3. Record the appropriate data.[6] Repeat any of the above reactions as necessary.

Appendix C

5. **Plot the Data.** Plot temperature (y axis) versus time (x axis) on one-half of a sheet of linear graph paper or by using appropriate software for the three data points. Have the instructor approve your graph.[7] Further interpret your data as suggested on the Report Sheet.

Disposal: Dispose of the reaction solutions in the "Waste Inorganic Test Solutions" container.

C. Temperature of the Reaction: Oxalic Acid–Potassium Permanganate Reaction System

The reaction rate for the oxidation–reduction reaction between oxalic acid, $H_2C_2O_4$, and potassium permanganate, $KMnO_4$, is measured by recording the time elapsed for the (purple) color of the permanganate ion, MnO_4^-, to *disappear* in the reaction:

$$5\ H_2C_2O_4(aq) + 2\ KMnO_4(aq) + 3\ H_2SO_4(aq) \rightarrow$$
$$10\ CO_2(g) + 2\ MnSO_4(aq) + K_2SO_4(aq) + 8\ H_2O(l) \quad (23.2)$$

Measure the volume of each solution with separate, *clean* graduated pipets. As one student pours the test solutions together, the other notes the time.

1. **Prepare the Solutions.** Into a set of three, clean 150-mm test tubes, pipet 1 mL of $0.01\ M\ KMnO_4$ (in $3\ M\ H_2SO_4$) and 4 mL of $3\ M\ H_2SO_4$. (**Caution:** *$KMnO_4$ is a strong oxidant and causes brown skin stains; H_2SO_4 is a severe skin irritant and is corrosive. Do not allow either chemical to make skin contact.*) Into a second set of three, clean 150-mm test tubes pipet 5 mL of $0.33\ M\ H_2C_2O_4$.

2. **Record the Time for Reaction at "Room" Temperature.** Select a $KMnO_4$–$H_2C_2O_4$ pair of test tubes. Pour the $H_2C_2O_4$ solution into the $KMnO_4$ solution. START TIME. Agitate the mixture. Record the time for the purple color of the permanganate ion to disappear. Record "room" temperature "using all certain digits *plus* one uncertain digit."[8]

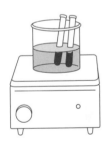

3. **Record the Time for Reaction at the "Higher" Temperature.** Place a second $KMnO_4$–$H_2C_2O_4$ pair of test tubes in a warm water (~40°C) bath until thermal equilibrium is established (~5 minutes). Pour the $H_2C_2O_4$ solution into the $KMnO_4$ solution, START TIME, agitate the mixture for several seconds, and return the reaction mixture to the warm water bath. Record the time for the disappearance of the purple color. Record the temperature of the bath.[9]

4. **Record the Time for Reaction at the "Highest" Temperature.** Repeat Part C.3, but increase the temperature of the bath to ~60°C. Record the appropriate data.[10] Repeat any of the preceding reactions as necessary.

Appendix C

5. **Plot the Data.** Plot temperature (y axis) versus time (x axis) on one-half of a sheet of linear graph paper or by using apropriate software for the three data points. Have the instructor approve your graph.[11]

Disposal: Dispose of the reaction solutions in the "Waste Inorganic Test Solutions" container.

Hydrogen peroxide is relatively stable, but it readily decomposes in the presence of a catalyst.

D. Presence of a Catalyst

1. **Add a Catalyst.** Place approximately 2 mL of a 3% H_2O_2 solution in a small test tube. Add 1 or 2 crystals of MnO_2 to the solution and observe. Note its instability.[12]

Ask your instructor for advice in completing *both* Parts E and F.

E. Concentration of Reactants: Magnesium–Hydrochloric Acid System

1. **Prepare the Reactants.** Into a set of four, clean, labeled test tubes, pipet 5 mL of 6 *M* HCl, 4 *M* HCl, 3 *M* HCl, and 1 *M* HCl, respectively (Figure 23.6).[2] Determine the mass (± 0.001 g)—separately (for identification)—of four 1-cm strips of *polished* (with steel wool or sand paper) magnesium. Calculate the number of moles of magnesium in each strip.[13]

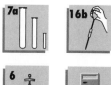

6 *M* HCl   4 *M* HCl   3 *M* HCl   1 *M* HCl

Figure 23.6 Setup for the effect of acid concentration on reaction rate.

2. **Record the Time for Completion of the Reaction.** Add the first magnesium strip to the 6 *M* HCl solution, START TIME, and record the time for all traces of the magnesium strip to disappear. Repeat the experiment with the remaining three magnesium strips and the 4 *M* HCl, 3 *M* HCl, and 1 *M* HCl, solutions.[14]

3. **Plot the Data.** Plot $\dfrac{\text{mol HCl}}{\text{mol Mg}}$ (*y* axis) versus time in seconds (*x* axis) for the four tests on one-half of a sheet of linear graph paper. Have the instructor approve your graph.[15]

Appendix C

Disposal: Dispose of the reaction solutions in the test tubes in the "Waste Inorganic Test Solutions" container.

CLEANUP: Rinse the test tubes twice with tap water and twice with deionized water. Discard each rinse in the sink; flush the sink with water.

A series of interrelated oxidation–reduction reactions occur between iodic acid, HIO_3, and sulfurous acid, H_2SO_3, that ultimately lead to the formation of triiodide ion, I_3^-, and sulfuric acid, H_2SO_4, as the final products.

F. Concentration of Reactants: Iodic Acid–Sulfurous Acid System

$$3\ HIO_3(aq) + 8\ H_2SO_3(aq) \rightarrow H^+(aq) + I_3^-(aq) + 8\ H_2SO_4(aq) + H_2O(l) \quad (23.3)$$

The triodide ion, I_3^- ($[I_2 \bullet I]^-$), appears *only* after all of the sulfurous acid is consumed in the reaction. Once the I_3^- forms, its presence is detected by its reaction with starch, forming a deep-blue complex.

$$I_3^-(aq) + \text{starch}(aq) \rightarrow I_3^- \bullet \text{starch}(aq)\ (\text{deep-blue}) \quad (23.4)$$

1. **Prepare the Test Solutions.** Review the preparation of the test solutions in Table 23.1. Set up five, clean and labeled test tubes (Figure 23.7). Measure the volumes of the 0.01 *M* HIO_3, starch, and water with dropping (or Beral) pipets.[3] Calibrate the

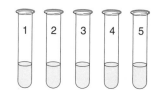

Figure 23.7 Setup for changes in HIO_3 concentration on reaction rate.

---

[2]Remember to properly rinse the pipet with the appropriate solution before dispensing it into the test tube.

[3]Be careful! Do not intermix the dropping pipets between solutions. This error in technique causes a significant error in the data.

Table 23.1 Reactant Concentration and Reaction Rate

| Test Tube | Solution in Test Tube | | | Add to Test Tube |
	0.01 M HIO$_3$	Starch	H$_2$O	0.01 M H$_2$SO$_3$
1	3 drops	1 drop	17 drops	1.0 mL
2	6 drops	1 drop	14 drops	1.0 mL
3	12 drops	1 drop	8 drops	1.0 mL
4	15 drops	1 drop	5 drops	1.0 mL
5	20 drops	1 drop	0 drops	1.0 mL

HIO$_3$ dropping pipet to determine the volume (mL) per drop.⑯ Calibrate a second dropping (or Beral) pipet with water to determine the number of milliliters per drop.⑰

Calibrate a third dropping (or Beral) pipet for the 0.01 M H$_2$SO$_3$ solution that delivers 1 mL; mark the level on the pipet so that quick delivery of 1 mL of the H$_2$SO$_3$ solution to each test tube can be made. Alternatively, use a calibrated 1-mL Beral pipet.

2. **Record the Time for the Reaction.** Place a sheet of white paper beside the test tube (Figure 23.8). As one student quickly transfers 1.0 mL of the 0.01 M H$_2$SO$_3$ to the respective test tube, the other notes the time. *Immediately* agitate the test tube; record the time lapse (seconds) for the deep-blue I$_3^-$•starch complex to appear.[4]

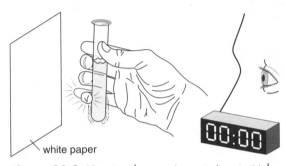

white paper

Figure 23.8 Viewing the reaction rate in a test tube.

3. **Complete Remaining Reactions.** Repeat Part F.2. for the remaining reaction mixtures in Table 23.1. Repeat any of the trials as necessary.⑱

Appendix C

4. **Plot the Data.** On one-half of a sheet of linear graph paper or by using appropriate software, plot for each solution the initial concentration of iodic acid,[5] [HIO$_3$]$_0$ (y axis), versus the time in seconds (x axis) for the reaction.⑲

Disposal: Dispose of all test solutions in the "Waste Inorganic Test Solutions" container.

CLEANUP: Rinse the test tubes twice with tap water and discard each into the "Waste Inorganic Test Solutions" container. Two final rinses with deionized water can be discarded in the sink.

The Next Step

(1) The dissolution of dissolved gases, such as CO$_2$(aq) in carbonated beverages, changes significantly with temperature changes. Study the kinetics of the dissolution of dissolved gases, such as CO$_2$(aq) or O$_2$(g) using such things as the candy, mentos™, salt, rust, etc. The study may be qualitative or quantitative. For the dissolution of O$_2$(g), refer to Experiment 31 in this manual. (2) Corrosion of iron in deionized water, tap water, boiled deionized/tap water, salt water (varying concentrations), etc. all affect the economy.

---

[4]Be ready! The appearance of the deep-blue solution is sudden.
[5]Remember that in calculating [HIO$_3$]$_0$, its total volume is the *sum* of the volumes of the two solutions.

Factors Affecting Reaction Rates

Date _____ Lab Sec. _____ Name _____ Desk No. _____

1. Identify the major factor affecting reaction rates that accounts for the following observations:

   a. Tadpoles grow more rapidly near the cooling water discharge from a power plant.

   b. Enzymes accelerate certain biochemical reactions, but are not consumed.

   c. Campfires are started wth twigs rather than logs of wood.

   d. Iron and steel corrode more rapidly near the coast of an ocean than in the desert.

2. Assuming that the rate of a chemical reaction doubles for every 10°C temperature increase, by what factor would a chemical reaction increase if the temperature were increased from −5°C (a *very* cold winter morning) to 25°C (room temperature)?

3. Experimental Procedure, Part B. a. Identify the visual evidence used for timing the reaction.

   b. A data plot is used to predict reaction rates at other conditions. What are the coordinates of the data plot?

   *c. How could the data be manipulated to reflect a concentration expression along the *y*-axis instead of the expression suggested in the experiment?

4. Experimental Procedure, Part E.3. a. A 15-mg strip of magnesium metal reacts in 5.0 mL of 1.0 M HCl over a given time period. Evaluate the $\dfrac{\text{mol HCl}}{\text{mol Mg}}$ ratio for the reaction.

b. What are the correct labelings of the axes for the data plot?

5. Experimental Procedure, Part F. A 1.0-mL volume of 0.010 M H_2SO_3 is added to a mixture of 12 drops of 0.01 M HIO_3, 8 drops of deionized water, and 1 drop of starch solution. A color change in the reaction mixture occurred in 40 seconds.
   a. Assuming 25 drops per milliliter, determine the initial molar concentration of IO_3^- after the mixing but before any reaction occurs (at time = 0).

b. At the same point in time (see Part **a**, time = 0), what is the molar concentration of H_2SO_3?

c. The rate of the reaction is measured by the disappearance of H_2SO_3. For the reaction mixture in this question, what is the reaction rate? Express the reaction rate with appropriate units.

6. The reactions in the Experimental Procedure, Parts C, E, and F are timed. Identify the visual "signal" to stop timing in each reaction.
   a. Part C.

b. Part E.

c. Part F.

Factors Affecting Reaction Rates

Date _____ Lab Sec. _____ Name _____ Desk No. _____

A. Nature of the Reactants

1. ①List the acids in order of decreasing reaction rate with magnesium: _____, _____, _____, _____

2. ②List the metals in order of decreasing reaction rate with 6 *M* HCl: _____, _____, _____

3. ③Identify the metals reacting in Figure 23.5 (from left to right). _____, _____, _____

B. Temperature of the Reaction: Hydrochloric Acid–Sodium Thiosulfate Reaction System

1. Time for Sulfur to Appear   Temperature of the Reaction

   ④_____ seconds   _____ °C

   ⑤_____ seconds   _____ °C

   ⑥_____ seconds   _____ °C

2. ⑦Plot temperature (*y* axis) versus time (*x* axis) for the three trials. Instructor's approval of graph: _____

3. From the plotted data, interpret the effect of temperature on reaction rate.

4. From your graph, estimate the temperature at which the appearance of sulfur should occur in 20 seconds. Assume no changes in concentration.

C. Temperature of the Reaction: Oxalic Acid–Potassium Permanganate Reaction System

1. Time for Permanganate Ion to Disappear   Temperature of the Reaction

   ⑧_____ seconds   _____ °C

   ⑨_____ seconds   _____ °C

   ⑩_____ seconds   _____ °C

2. ⑪Plot temperature (*y* axis) versus time (*x* axis) for the three trials. Instructor's approval of graph: _____

3. From your plotted data, interpret the effect of temperature on reaction rate.

4. From your graph, estimate the time for the disappearance of the purple permanganate ion at 15°C. Assume no changes in concentration.

D. Presence of a Catalyst

1. [12] What effect does the MnO_2 catalyst have on the rate of evolution of O_2 gas?

2. Write a balanced equation for the decomposition of H_2O_2.

E. Concentration of Reactants: Magnesium–Hydrochloric Acid System

Concentration of HCl	Volume of HCl	mol HCl	mass of Mg	[13] mol Mg	$\dfrac{\text{mol HCl}}{\text{mol Mg}}$	[14] Time (sec)
6 M						
4 M						
3 M						
1 M						

1. [15] Plot $\dfrac{\text{mol HCl}}{\text{mol Mg}}$ (y axis) versus time (x axis). Instructor's approval of graph: _____

2. From your graph, predict the time, in seconds, for 5 mg of Mg to react in 5 mL of 2.0 M HCl.

F. Concentration of Reactants: Iodic Acid–Sulfurous Acid System

Molar concentration of HIO_3 _____ Molar concentration of H_2SO_3 _____

[16] What is the volume (mL) per drop of the HIO_3 solution? _____ [17] What is the volume (mL) per drop of the water? _____

Test Tube	Drops of HIO_3	mL HIO_3	Drops of H_2O	mL H_2O	mL H_2SO_3	$[HIO_3]_0$[6] (diluted)	Time (sec)
[18] 1					1.0		
2					1.0		
3					1.0		
4					1.0		
5					1.0		

1. [19] Plot $[HIO_3]_0$ (y axis) versus time (x axis). Instructor's approval of graph: _____

2. How does a change in the molar concentration of HIO_3 affect the time required for the appearance of the deep-blue I_3^-•starch complex?

3. Estimate the time, in seconds, for the deep-blue I_3^-•starch complex to form when 10 drops of 0.01 M HIO_3 are used for the reaction. Assume all other conditions remain constant.

_____

[6]See footnote 5.

Experiment 24

A Rate Law and Activation Energy

Drops of blood catalyze the decomposition of hydrogen peroxide to water and oxygen gas.

OBJECTIVES

- To determine the rate law for a chemical reaction
- To utilize a graphical analysis of experimental data to
  —determine the order of each reactant in the reaction
  —determine the activation energy for the reaction

TECHNIQUES

The following techniques are used in the Experimental Procedure

INTRODUCTION

The rate of a chemical reaction is affected by a number of factors, most of which were observed in Experiment 23. The rate of a reaction can be expressed in a number of ways, depending on the nature of reactants being consumed or the products being formed. The rate may be followed as a change in concentration (mol/L) of one of the reactants or products per unit of time, the volume of gas produced per unit of time (Figure 24.1), or the change in color (measured as light absorbance) per unit of time, just to cite a few examples.

In this experiment, a quantitative statement as to *how* changes in reactant concentrations affect reaction rate is expressed in an experimentally derived rate law.

To assist in understanding the relationship between reactant concentration and reaction rate, consider the general reaction, $A_2 + 2 B_2 \rightarrow 2 AB_2$. The rate of this reaction is related, by some exponential power, to the initial concentration of each reactant. For this reaction, we can write the relationship as

$$\text{rate} = k [A_2]^p [B_2]^q \qquad (24.1)$$

This expression is called the **rate law** for the reaction. The value of k, the reaction **rate constant,** varies with temperature but is independent of reactant concentrations.

The superscripts p and q designate the **order** with respect to each reactant and are *always* determined experimentally. For example, if tripling the molar concentration of A_2 while holding the B_2 concentration constant increases the reaction rate by a factor of 9, then $p = 2$. In practice, when the B_2 concentration is in large excess relative to the A_2 concentration, the B_2 concentration remains essentially constant during the course of the reaction; therefore, the change in the reaction rate results from the more significant change in the smaller amount of A_2 in the reaction. An experimental study of the kinetics of any reaction involves determining the values of $k, p,$ and q.

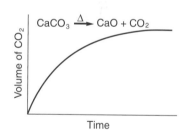

Figure 24.1 The rate of thermal decomposition of calcium carbonate is determined by measuring the volume of evolved carbon dioxide gas versus time.

Rate constant: a proportionality constant relating the rate of a reaction to the initial concentrations of the reactants

Order: the exponential factor by which the concentration of a substance affects reaction rate

In Parts A–D of this experiment the rate law for the reaction of hydrogen peroxide, H_2O_2, with potassium iodide, KI, is determined.[1] When these reactants are mixed, hydrogen peroxide slowly oxidizes iodide ion to elemental iodine, I_2. In the presence of excess iodide ion, molecular I_2 forms a water-soluble triiodide complex, I_3^- or $[I_2 \bullet I]^-$.

$$3\, I^-(aq) + H_2O_2(aq) + 2\, H_3O^+ (aq) \rightarrow I_3^-(aq) + 4\, H_2O(l) \qquad (24.2)$$

The rate of the reaction, governed by the molar concentrations of I^-, H_2O_2, and H_3O^+, is expressed by the rate law,

$$\text{rate} = k\, [I^-]^p[H_2O_2]^q[H_3O^+]^r \qquad (24.3)$$

Buffer: a solution that resists changes in acidity or basicity in the presence of added H^+ or OH^+ (Buffer solutions are studied in Experiment 16.)

When the $[H_3O^+]$ is greater than 1×10^{-3} mol/L (pH < 3), the reaction rate is too rapid to measure in the general chemistry laboratory; however, if the $[H_3O^+]$ is *less than* 1×10^{-3} mol/L (pH > 3), the reaction proceeds at a measurable rate. An acetic acid–sodium acetate **buffer** maintains a nearly constant $[H_3O^+]$ at about 1×10^{-5} mol/L (pH = ~5) during the experiment.[2] Since the molar concentration of H_3O^+ is held constant in the buffer solution and does not affect the reaction rate at the pH of the buffer, the rate law for the reaction becomes more simply

$$\text{rate} = k'\, [I^-]^p[H_2O_2]^q \qquad (24.4)$$

where $k' = k\, [H_3O^+]^r$.

In this experiment Parts B–D, the values of p, q, and k' are determined from the data analysis of Part A for the hydrogen peroxide–iodide ion system. Two sets of experiments are required: one set of experiments is designed to determine the value of p and the other to determine the value of q.

Determination of p, the Order of the Reaction with Respect to Iodide Ion

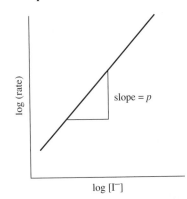

log (rate)

slope = p

log $[I^-]$

In the first set of experiments, the effect that iodide ion has on the reaction rate is observed in several kinetic trials. A "large" excess of hydrogen peroxide in a buffered system maintains the H_2O_2 and H_3O^+ concentrations essentially constant during each trial. Therefore, for this set of experiments, the rate law, Equation 24.4, reduces to the form

$$\text{rate} = k'\, [I^-]^p \bullet c \qquad (24.5)$$

c, a constant, equals $[H_2O_2]^q$.

In logarithmic form, Equation 24.5 becomes

$$\log (\text{rate}) = \log k' + p \log [I^-] + \log c \qquad (24.6)$$

Combining constants, we have the equation for a straight line:

$$\log (\text{rate}) = p \log [I^-] + C$$
$$y = \quad mx \qquad + b \qquad (24.7)$$

C equals $\log k' + \log c$ or $\log k' + \log [H_2O_2]^q$.

Therefore, a plot of log (rate) versus log $[I^-]$ produces a straight line with a slope equal to p, the order of the reaction with respect to the molar concentration of iodide ion. See margin figure.

Determination of q, the Order of the Reaction with Respect to Hydrogen Peroxide

In the second set of experiments, the effect that hydrogen peroxide has on the reaction rate is observed in several kinetic trials. A "large" excess of iodide ion in a buffered

---

[1]Your laboratory instructor may substitute $K_2S_2O_8$ for H_2O_2 for this experiment. The balanced equation for the reaction is
$$S_2O_8^{2-}(aq) + 3\, I^-(aq) \rightarrow 2\, SO_4^{2-}(aq) + I_3^-(aq)$$
[2]In general, a combined solution of H_2O_2 and I^- is only very slightly acidic and the acidity changes little during the reaction. Therefore, the buffer solution may not be absolutely necessary for the reaction. However, to ensure that change in H_3O^+ concentrations is *not* a factor in the reaction rate, the buffer is included as a part of the experiment.

system maintains the I^- and H_3O^+ concentrations essentially constant during each trial. Under these conditions the logarithmic form of the rate law (Equation 23.4) becomes

$$\log (\text{rate}) = q \log [H_2O_2] + C'$$
$$\phantom{\log (\text{rate})} y = mx + b \qquad\qquad (24.8)$$

C' equals $\log k' + \log [I^-]^p$.

A second plot, $\log (\text{rate})$ versus $\log [H_2O_2]$, produces a straight line with a slope equal to q, the order of the reaction with respect to the molar concentration of hydrogen peroxide.

Once the respective orders of I^- and H_2O_2 are determined (from the data plots) and the reaction rate for each trial has been determined, the values of p and q are substituted into Equation 24.4 to calculate a specific rate constant, k', for each trial.

Determination of the Specific Rate Constant, k'

Reaction rates are temperature dependent. Higher temperatures increase the kinetic energy of the (reactant) molecules, such that when two reacting molecules collide, they do so with a much greater force (more energy is dispersed within the collision system) causing bonds to rupture, atoms to rearrange, and new bonds (products) to form more rapidly. The energy required for a reaction to occur is called the **activation energy** for the reaction.

The relationship between the reaction rate constant, k', at a measured temperature, $T(K)$, and the activation energy, E_a, is expressed in the Arrhenius equation.

$$k' = Ae^{-E_a/RT} \qquad\qquad (24.9)$$

A is a collision parameter for the reaction and R is the gas constant ($=8.314$ J/mol•K). The logarithmic form of Equation 24.9 is

$$\ln k' = \ln A - \frac{E_a}{RT} \quad \text{or} \quad \ln k' = \ln A - \frac{E_a}{R}\left[\frac{1}{T}\right] \qquad (24.10)$$

Determination of Activation Energy, E_a

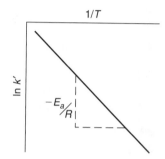

The latter equation of 24.10 conforms to the equation for a straight line, $y = b + mx$, where a plot of $\ln k'$ vs. $1/T$ yields a straight line with a slope of $-E_a/R$ and a y-intercept of $\ln A$.

As the temperature changes, the reaction rate also changes. A substitution of the "new" reaction rate at the "new" temperature into Equation 24.4 (with known orders of I^- and H_2O_2) calculates a "new" specific rate constant, k'. A data plot of these "new" specific rate constants ($\ln k'$) at these new temperatures ($1/T$) allows for the calculation of the activation energy, E_a, for the reaction. In Part E, the temperature of the solutions for kinetic trial 4 (Table 24.1) will be increased/decreased to determine rate constants at these new temperatures.

To follow the progress of the rate of the reaction, two solutions are prepared:

- Solution A: a diluted solution of iodide ion, starch, thiosulfate ion ($S_2O_3^{2-}$), and the acetic acid–sodium acetate buffer
- Solution B: the hydrogen peroxide solution

Observing the Rate of the Reaction

When Solutions A and B are mixed, the H_2O_2 reacts with the I^- (Equation 24.2).

$$3\,I^-(aq) + H_2O_2(aq) + 2\,H_3O^+(aq) \rightarrow I_3^-(aq) + 4\,H_2O(l) \quad \text{(repeat of Equation 24.2)}$$

To prevent an equilibrium (a back reaction) from occurring in Equation 24.2, the presence of thiosulfate ion removes I_3^- as it is formed:

$$2\,S_2O_3^{2-}(aq) + I_3^-(aq) \rightarrow 3\,I^-(aq) + S_4O_6^{2-}(aq) \qquad (24.11)$$

As a result, iodide ion is regenerated in the reaction system; this maintains a constant iodide ion concentration during the course of the reaction until the thiosulfate ion is consumed. When the thiosulfate ion has completely reacted in solution, the generated I_3^- combines with starch, forming a deep-blue I_3^-•starch complex. Its appearance signals a length of time for the reaction (Equation 24.2) to occur, and the length of time for the disappearance of the thiosulfate ion.

$$I_3^-(aq) + starch\ (aq) \rightarrow I_3^- \bullet starch\ (aq, deep\text{-}blue) \qquad (24.12)$$

For the reaction,

$$rate = \frac{\Delta\ mol\ I_3^-}{\Delta t}$$

The time required for a quantitative amount of thiosulfate ion to react is the time lapse for the appearance of the deep-blue solution. During that period a quantitative amount of I_3^- is generated; therefore, the rate of I_3^- production (mol I_3^- /time), and thus the rate of the reaction, is affected *only* by the initial concentrations of H_2O_2 and I^-.

Therefore, the rate of the reaction is followed by measuring the time required to generate a preset number of moles of I_3^-, *not* the time required to deplete the moles of reactants.

EXPERIMENTAL PROCEDURE

Procedure Overview: Measured volumes of several solutions having known concentrations of reactants are mixed in a series of trials. The time required for a visible color change to appear in the solution is recorded for the series of trials. The data are collected and plotted (two plots); from the plotted data, the order of the reaction with respect to each reactant is calculated and the rate law for the reaction is derived. After the rate law for the reaction is established, the reaction rate is observed at non-ambient temperatures. The plotted data produces a value for the activation energy of the reaction.

Read the entire procedure before beginning the experiment. Student pairs should gather the kinetic data.

A. Determination of Reaction Times

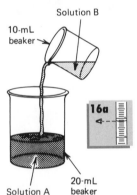

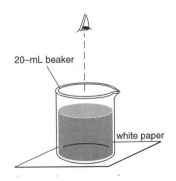

1. **Prepare Solutions A for the Kinetic Trials.** Table 24.1 summarizes the preparation of the solutions for the kinetic trials. Use previously-boiled deionized water. Measure the volumes of KI and $Na_2S_2O_3$ solutions with *clean*[3] pipets.[4] Burets or pipets can be used for the remaining solutions. Prepare, at the same time, all of the Solutions A for Kinetic Trials 1–8 in either clean and *labeled* 20-mL beakers or 150-mm test tubes. Trial 8 is to be of your design.

2. **Prepare Solutions for Kinetic Trial 4.**
   Solution A. Stir the solution in a small 20-mL beaker or 150-mm test tube.
   Solution B. Pipet 3.0 mL of 0.1 *M* H_2O_2 into a clean 10-mL beaker or 150-mm test tube.

3. **Prepare for the Reaction.** The reaction begins when the H_2O_2 (Solution B) is added to Solution A; be prepared to start timing the reaction *in seconds*. Place the beaker on a white sheet of paper so the deep-blue color change is more easily detected (Figure 24.2 or Figure 23.8). As one student mixes the solutions, the other notes the time. All of the solutions should be at ambient temperature before mixing. Record the temperature.

4. **Time the Reaction.** Rapidly add Solution B to Solution A—START TIME and swirl (once) the contents of the mixture. Continued swirling is unnecessary. The appearance of the deep-blue color is sudden. Be ready to STOP TIME. Record the time lapse to the nearest second on the Report Sheet. Repeat if necessary.

Notice! If the time for the color change of Trial 4 is less than 10 seconds, STOP; add an additional 10 mL of boiled, deionized water to each Solution A for each Kinetic Trial (total volume of the reaction mixtures will now 20 mL instead of 10 mL). A consequence of this dilution will result in a much longer time lapse for a color change in Trial 1 . . . be patient! Consult with your laboratory instructor before the addition of the 10 mL of boiled, deionized water.

---

[3]Cleanliness is important in preparing these solutions because H_2O_2 readily decomposes in the presence of foreign particles. Do *not* dry glassware with paper towels.

$$2\ H_2O_2 \xrightarrow{catalyst} 2\ H_2O + O_2$$

[4]5-mL *graduated* (±0.1 mL) pipets are suggested for measuring these volumes.

Figure 24.2 Viewing the appearance of the I_3^-•starch complex.

Table 24.1 Composition of Test Solutions

| Kinetic Trial | Solution A | | | | | Solution B* |
	Boiled, Deionized Water	Buffer**	0.3 M KI	0.02 M Na$_2$S$_2$O$_3$	Starch	0.1 M H$_2$O$_2$
1	4.0 mL	1.0 mL	1.0 mL	1.0 mL	5 drops	3.0 mL
2	3.0 mL	1.0 mL	2.0 mL	1.0 mL	5 drops	3.0 mL
3	2.0 mL	1.0 mL	3.0 mL	1.0 mL	5 drops	3.0 mL
4	1.0 mL	1.0 mL	4.0 mL	1.0 mL	5 drops	3.0 mL
5	2.0 mL	1.0 mL	1.0 mL	1.0 mL	5 drops	5.0 mL
6	0.0 mL	1.0 mL	1.0 mL	1.0 mL	5 drops	7.0 mL
7	5.0 mL	1.0 mL	1.0 mL	1.0 mL	5 drops	2.0 mL
8†	—	1.0 mL	—	1.0 mL	5 drops	—

*0.1 M K$_2$S$_2$O$_8$ may be substituted.
**0.5 M CH$_3$COOH and 0.5 M NaCH$_3$CO$_2$.
†You are to select the volumes of solutions for the trial.

5. **Repeat for the Remaining Kinetic Trials.** Mix and time the test solutions for the remaining seven kinetic trials. If the instructor approves, conduct additional kinetic trials, either by repeating those in Table 24.1 or by preparing other combinations of KI and H$_2$O$_2$. Make sure that the total diluted volume remains constant at 10 mL.

Disposal: Dispose of the solutions from the kinetic trials in the "Waste Iodide Salts" container.

CLEANUP: Rinse the beakers/test tubes twice with tap water and discard in the "Waste Iodide Salts" container. Dispose of two final rinses with deionized water in the sink.

B. Calculations for Determining the Rate Law

Perform the calculations, carefully *one step at a time*. Appropriate and correctly programmed software would be invaluable for completing this analysis. As you read through this section, complete the appropriate calculation and record it for each test solution on the Report Sheet.

1. **Moles of I$_3$$^-$ Produced.** Calculate the moles of S$_2$O$_3$$^{2-}$ consumed in each kinetic trial. From Equation 24.11 the moles of I$_3$$^-$ that form in the reaction equals one-half the moles of S$_2$O$_3$$^{2-}$ that react. This also equals the change in the moles of I$_3$$^-$, starting with none at time zero up until a final amount that was produced at the time of the color change. This is designated as "Δ(mol I$_3$$^-$)" produced.

2. **Reaction Rate.** The reaction rate for each kinetic trial is calculated as the ratio of the moles of I$_3$$^-$ produced, Δ(mol I$_3$$^-$), to the time lapse, Δt, for the appearance of the deep-blue color.[5] Compute these reaction rates, $\dfrac{\Delta(\text{mol I}_3{}^-)}{\Delta t}$, and the logarithms of the reaction rates (see Equations 24.7 and 24.8) for each kinetic trial and enter them on the Report Sheet. Because the total volume is a constant for all kinetic trials, we do *not* need to calculate the molar concentrations of the I$_3$$^-$ produced.

3. **Initial Iodide Concentrations.** Calculate the initial molar concentration, $[I^-]_0$, and the logarithm of the initial molar concentration, $\log[I^-]_0$, of iodide ion for each kinetic trial.[6]

4. **Initial Hydrogen Peroxide Concentrations.** Calculate the initial molar concentration, $[H_2O_2]_0$, and the logarithm of the initial molar concentration, $\log[H_2O_2]_0$, of hydrogen peroxide for each kinetic trial.[7]

[5]The moles of I$_3$$^-$ present initially, at time zero, is zero.
[6]Remember this is *not* 0.3 M I$^-$ because the total volume of the solution is 10 mL after mixing.
[7]Remember, too, this is *not* 0.1 M H$_2$O$_2$ because the total volume of the solution is 10 mL after mixing.

C. Determination of the Reaction Order, p and q, for Each Reactant

Appendix C

1. **Determination of p from Plot of Data.** Plot on the top half of a sheet of linear graph paper or preferably by using appropriate software log (Δmol I$_3^-$/Δt), which is log (rate) (y axis), versus log [I$^-$]$_0$ (x axis) at constant hydrogen peroxide concentration. Kinetic Trials 1, 2, 3, and 4 have the same H$_2$O$_2$ concentration. Draw the best straight line through the four points. Calculate the slope of the straight line. The slope is the order of the reaction, p, with respect to the iodide ion.

Appendix C

2. **Determination of q from Plot of Data.** Plot on the bottom half of the same sheet of linear graph paper or preferably by using appropriate software log (Δmol I$_3^-$/Δt) (y axis) versus log [H$_2$O$_2$]$_0$ (x axis) at constant iodide ion concentration using Kinetic Trials 1, 5, 6, and 7. Draw the best straight line through the four points and calculate its slope. The slope of the plot is the order of the reaction, q, with respect to the hydrogen peroxide.

3. **Approval of Graphs.** Have your instructor approve both graphs.

D. Determination of k′, the Specific Rate Constant for the Reaction

Appendix B

1. **Substitution of p and q into Rate Law.** Use the values of p and q (from Part C) and the rate law, rate $= \dfrac{\Delta(\text{mol I}_2)}{\Delta t} = k'$ [I$^-$]^p [H$_2$O$_2$]^q, to determine k' for the seven solutions. Calculate the average value of k' with proper units. Also determine the standard deviation and relative standard deviation (%RSD) of k' from your data.

2. **Class Data.** Obtain average k' values from other groups in the class. Calculate a standard deviation and relative standard deviation (%RSD) of k' for the class.

E. Determination of Activation Energy

1. **Prepare Test Solutions.** Refer to Table 24.1, Kinetic Trial 4. In separate, clean 150-mm test tubes prepare *two* additional sets of Solution A and Solution B. Place one (Solution A/Solution B) set in an ice bath. Place the other set in a warm water ($\sim$35°C) bath. Allow thermal equilibrium to be established for each set, about 5 minutes.

    Test solutions prepared at other temperatures are encouraged for additional data points.

2. **Mix Solutions A and B.** When thermal equilibrium has been established, quickly pour Solution B into Solution A, START TIME, and agitate the mixture. When the deep-blue color appears STOP TIME. Record the time lapse as before. Record the temperature of the water bath and use this time lapse for your calculations. Repeat to check reproducibility and for the other set(s) of solutions.

3. **The Reaction Rates and "New" Rate Constants.** The procedure for determining the reaction rates is described in Part B.2. Calculate and record the reaction rates for the (at least) two trials (two temperatures) from Part E.2 and re-record the reaction rate for the (room temperature) Kinetic Trial 4 in Part A.5. Carefully complete the calculations on the Report Sheet.

    Use the reaction rates at the three temperatures (ice, room, and $\sim$35°C temperatures) and the established rate law from Part C to calculate the rate constants, k' at these temperatures. Calculate the natural logarithm of these rate constants.

Appendix C

4. **Plot the Data.** Plot ln k' vs. $1/T(K)$ for the (at least) three trials at which the experiment was performed. Remember to express temperature in kelvins and $R = 8.314$ J/mol•K.

5. **Activation Energy.** From the data plot, determine the slope of the linear plot ($= -E_a/R$) and calculate the activation energy for the reaction. You may need to seek the advice of your instructor for completing the calculation on the Report Sheet.

The Next Step

The rate law for any number of chemical reactions can be studied in the same manner, e.g., see Experiment 23, Parts B, C, and F. Research the Internet for a kinetic study of interest (biochemical?) and design a systematic kinetic study of a chemical system.

A Rate Law and Activation Energy

Date _____ Lab Sec. _____ Name _____ Desk No. _____

1. Three data plots are required for analyzing the data in this experiment, two plots from the kinetic trials outlined in Table 24.1 and one plot from Part E. From each data plot a value is determined toward the completion of the analysis of the kinetic study for the reaction of I^- with H_2O_2. Complete the table in order to focus the analysis.

Source of Data	**y-axis label**	**x-axis label**	**datum to be obtained from the data plot**
Table 24.1, Trials 1–4	_____	_____	_____
Table 24.1, Trials 1, 5–7	_____	_____	_____
Part E	_____	_____	_____

2. a. In the collection of the rate data for the experiment, when does START TIME and when does STOP TIME occur for each kinetic trial in Table 24.1?

   b. What is the color of the solution at STOP TIME?

   c. Account for the color of the solution at STOP TIME.

3. In the kinetic analysis of this experiment for the reaction of iodide ion with hydrogen peroxide, state the purpose for each of the following solutions (see Table 24.1):
   a. deionized water

   b. buffer solution (acetic acid, sodium acetate mixture)

4. Refer to Experimental Procedure, Part A, Table 24.1, **Trial 1.** a. What is the function of the sodium thiosulfate in studying the kinetics of the hydrogen peroxide–iodide reaction?

   b. Calculate the moles of $S_2O_3^{2-}$ that are consumed during the course of the reaction.

c. Calculate the moles of I_3^- that are produced during the course of the reaction.

d. Calculate the initial molar concentration of I^- (at time = 0), $[I^-]_0$ (not 0.3 M, but after mixing Solutions A & B).

e. Calculate the initial molar concentration of H_2O_2 (at time = 0), $[H_2O_2]_0$ (not 0.1 M, but after mixing Solutions A & B).

5. How is the order of the reaction with respect to H_2O_2 determined in the experiment?

6. Explain how the rate constant, k', is determined for the rate law in the experiment.

7. From the following data plot, calculate the activation energy, E_a, for the reaction.

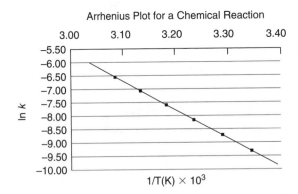

Arrhenius Plot for a Chemical Reaction

A Rate Law and Activation Energy

Date _____ Lab Sec. _____ Name _____ Desk No. _____

A. Determination of Reaction Times

Molar concentration of $Na_2S_2O_3$ _____; volume of $Na_2S_2O_3$ (L) _____; Ambient temperature _____ °C

Molar concentration of KI _____; Molar concentration of H_2O_2 _____; Total volume of kinetic trials (mL) _____

Kinetic Trial	1*	2	3	4	5	6	7	8
1. Time for color change, Δt (sec)								

B. Calculations for Determining the Rate Law

1. Moles of $S_2O_3^{2-}$ consumed (*mol*)								
2. Δ(mol I_3^-) produced								
3. $\dfrac{\Delta(\text{mol } I_3^-)}{\Delta t}$								
4. $\log \dfrac{\Delta(\text{mol } I_3^-)}{\Delta t}$								
5. Volume KI (*mL*)								
6. $[I^-]_0$ (*mol/L*)**								
7. $\log [I^-]_0$								
8. Volume H_2O_2 (*mL*)								
9. $[H_2O_2]_0$ (*mol/L*)**								
10. $\log [H_2O_2]_0$								

**Diluted* initial molar concentration.

*Calculations for Kinetic Trial 1.

C. Determination of the Reaction Order, p and q, for Each Reactant

Instructor's approval of graphs:

1. $\log (\Delta \text{mol } I_3{}^-/\Delta t)$ versus $\log [I^-]_0$ _____

2. $\log (\Delta \text{mol } I_3{}^-/\Delta t)$ versus $\log [H_2O_2]_0$ _____

3. value of p from graph _____;   value of q from graph _____

Write the rate law for the reaction.

D. Determination of k', the Specific Rate Constant for the Reaction

Kinetic Trial	1	2	3	4	5	6	7	8
1. Value of k'								

2. Average value of k' _____

3. Standard deviation of k' _____

4. Relative standard deviation of k' (%RSD) _____

Class Data/Group	1	2	3	4	5	6
Average value of k'						

Calculate the average value and the standard deviation of the reaction rate constant for the class. See Appendix B.

Calculate the relative standard deviation of k' (%RSD).

E. Determination of Activation Energy

	Time for color change	Reaction Rate	Calc. k'	ln k'	Temperature	$1/T(K)$
1. Trial 4	_____				_____	
2. Cold	_____				_____	
3. Warm	_____				_____	

4. Instructor's Approval of Data Plot _____

5. Value of $(-E_a/R)$ from ln k' vs 1/T graph _____

6. Activation Energy, E_a, from data plot. Show calculation. _____

Laboratory Questions

Circle the questions that have been assigned.

1. Part A.4. Describe the chemistry that was occurring in the experiment between the time when Solution A and Solution B were mixed and STOP TIME.

2. Part A.4. For Kinetic Trial 2, Alicia was distracted when the color change occurred, but decided to record the time lapse read from her watch. Will this distraction cause an increase or decrease in the slope of the log (rate) vs. log $[I^-]_o$? Explain.

3. Part A.4. How would the appearance of the solution change for each kinetic trial if
   a. the $Na_2S_2O_3$ solution were omitted from the experiment?
   b. the starch solution were omitted from the experiment?

4. Part C.2. Review the plotted data.
   a. What is the numerical value of the y-intercept?
   b. What is the kinetic interpretation of the value for the y-intercept?
   c. What does its value equal in Equation 24.8?

5. State the effect that each of the following changes has on the reaction rate in this experiment. (Assume no volume change for any of the concentration changes.)
   a. An increase in the H_2O_2 concentration. Explain.
   b. An increase in the volume of water in Solutions A. Explain.
   c. An increase in the $Na_2S_2O_3$ concentration. Explain.
   d. The substitution of a 0.5% starch solution for one at 0.2%. Explain.

*6. If 0.2 M KI replaced the 0.3 M KI in this experiment, how would this affect the following?
   a. The rate of the reaction.
   b. The slopes of the graphs used to determine p and q.
   c. The value of the reaction rate constant.

7. Part E.2. The temperature of the warm water bath is recorded too high. How will this technique error affect the reported activation energy for the reaction . . . too high or too low? Explain.

Calorimetry

A set of nested coffee cups is a good constant pressure calorimeter.

- To determine the specific heat of a metal
- To determine the enthalpy of neutralization for a strong acid–strong base reaction
- To determine the enthalpy of solution for the dissolution of a salt

The following techniques are used in the Experimental Procedure

Accompanying all chemical and physical changes is a transfer of heat (energy); heat may be either evolved (exothermic) or absorbed (endothermic). A **calorimeter** is the laboratory apparatus that is used to measure the quantity and direction of heat flow accompanying a chemical or physical change. The heat change in chemical reactions is quantitatively expressed as the **enthalpy (or heat) of reaction,** ΔH, at constant pressure. ΔH values are negative for exothermic reactions and positive for endothermic reactions.

ΔH values are often expressed as J/mol or kJ/mol

Three quantitative measurements of heat are detailed in this experiment: measurements of the specific heat of a metal, the heat change accompanying an acid–base reaction, and the heat change associated with the dissolution of a salt in water.

Specific Heat of a Metal

The energy (heat, expressed in joules, J) required to change the temperature of one gram of a substance by 1°C is the **specific heat**[1] of that substance:

$$\text{specific heat} \left(\frac{\text{J}}{\text{g} \cdot \text{°C}} \right) = \frac{\text{energy (J)}}{\text{mass (g)} \times \Delta T \, (\text{°C})} \qquad (25.1)$$

or, rearranging for energy,

$$\text{energy (J)} = \text{specific heat} \left(\frac{\text{J}}{\text{g} \cdot \text{°C}} \right) \times \text{mass (g)} \times \Delta T \, (\text{°C}) \qquad (25.2)$$

ΔT is the temperature change of the substance. Although the specific heat of a substance changes slightly with temperature, for our purposes, we assume it is constant over the temperature changes of this experiment.

The specific heat of a metal that does not react with water is determined by (1) heating a measured mass of the metal, M, to a known (higher) temperature, (2) placing it into a measured amount of water at a known (lower) temperature, and (3) measuring the final equilibrium temperature of the system after the two are combined.

---

[1]The specific list of a substance is an intensive property (independent of sample size), as are its melting point, boiling point, density, and so on.

The following equations, based on the law of conservation of energy, show the calculations for determining the specific heat of a metal. Considering the direction of energy flow by the conventional sign notation of energy loss being "negative" and energy gain being "positive," then

$$-\text{energy (J) lost by metal}_M = \text{energy (J) gained by water}_{H_2O} \qquad (25.3)$$

Substituting from Equation 25.2,

$$-\text{specific heat}_M \times \text{mass}_M \times \Delta T_M = \text{specific heat}_{H_2O} \times \text{mass}_{H_2O} \times \Delta T_{H_2O} \qquad (25.4)$$

Rearranging Equation 25.4 to solve for the specific heat of the metal$_M$ gives

Equation 25.4 is often written as $-c_{p,M} \times m_M \times \Delta T_M = c_{p,H_2O} \times m_{H_2O} \times \Delta T_{H_2O}$

$$\text{specific heat}_M = -\frac{\text{specific heat}_{H_2O} \times \text{mass}_{H_2O} \times \Delta T_{H_2O}}{\text{mass}_M \times \Delta T_M} \qquad (25.5)$$

In the equation, the temperature change for either substance is defined as the difference between the final temperature, T_f, and the initial temperature, T_i, of the substance:

$$\Delta T = T_f - T_i \qquad (25.6)$$

These equations assume no heat loss to the calorimeter when the metal and the water are combined. The specific heat of water is 4.18 J/g • °C.

Enthalpy (Heat) of Neutralization of an Acid–Base Reaction

Enthalpy of neutralization: energy released per mole of water formed in an acid–base reaction—an exothermic quantity

The reaction of a strong acid with a strong base is an exothermic reaction that produces water and heat as products.

$$H_3O^+ (aq) + OH^- (aq) \rightarrow 2\,H_2O(l) + \text{heat} \qquad (25.7)$$

The **enthalpy (heat) of neutralization,** ΔH_n, is determined by (1) assuming the density and the specific heat for the acid and base solutions are equal to that of water and (2) measuring the temperature change, ΔT (Equation 25.6), when the two are mixed.

$$\text{energy change}_n = -\text{specific heat}_{H_2O} \times \textit{combined}\ \text{masses}_{\text{acid + base}} \times \Delta T \qquad (25.8)$$

ΔH_n is generally expressed in units of kJ/mol of water that forms from the reaction. The mass (grams) of the solution equals the *combined* masses of the acid and base solutions.

Enthalpy (Heat) of Solution for the Dissolution of a Salt

Lattice energy: energy required to vaporize one mole of salt into its gaseous ions—an endothermic quantity

Hydration energy: energy released when one mole of a gaseous ion is attracted to and surrounded by water molecules forming one mole of hydrated ion in aqueous solution—an exothermic quantity

When a salt dissolves in water, energy is either absorbed or evolved depending upon the magnitude of the salt's lattice energy and the hydration energy of its ions. For the dissolution of KI:

$$KI(s) \xrightarrow{\ H_2O\ } K^+(aq) + I^-(aq) \qquad \Delta H_s = +\,13\ \text{kJ/mol} \qquad (25.9)$$

The **lattice energy** (an endothermic quantity) of a salt, ΔH_{LE}, and the **hydration energy** (an exothermic quantity), ΔH_{hyd}, of its composite ions account for the amount of heat evolved or absorbed when one mole of the salt dissolves in water. The **enthalpy (heat) of solution,** ΔH_s, is the sum of these two terms (for KI, see Figure 25.1).

$$\Delta H_s = \Delta H_{LE} + \Delta H_{hyd} \qquad (25.10)$$

Whereas ΔH_{LE} and ΔH_{hyd} are difficult to measure in the laboratory, ΔH_s is easily measured.

The enthalpy of solution for the dissolution of a salt, ΔH_s, is determined experimentally by adding the heat changes of the salt and the water when the two are mixed. ΔH_s is expressed in units of kilojoules per mole of salt.

$$\text{total enthalpy change} = (-\text{energy change}_{H_2O}) + (-\text{energy change}_{\text{salt}}) \qquad (25.11)$$

$$\Delta H_s = \left(-\frac{\text{specific heat}_{H_2O} \times \text{mass}_{H_2O} \times \Delta T_{H_2O}}{\text{mole}_{\text{salt}}} \right)$$

$$+ \left(-\frac{\text{specific heat}_{\text{salt}} \times \text{mass}_{\text{salt}} \times \Delta T_{\text{salt}}}{\text{mole}_{\text{salt}}} \right) \qquad (25.12)$$

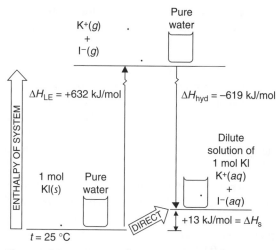

Figure 25.1 Energy changes in the dissolving of solid KI in water.

Refer to Equation 25.6 for an interpretation of ΔT. A temperature rise for the dissolution of a salt, indicating an exothermic process, means that the ΔH_{hyd} is greater than the ΔH_{LE} for the salt; conversely, a temperature decrease in the dissolution of the salt indicates that ΔH_{LE} is greater than ΔH_{hyd} and ΔH_s is positive.

The specific heats of some salts are listed in Table 25.1.

Table 25.1 Specific Heat of Some Salts

Salt	Formula	Specific Heat (J/g·°C)
ammonium chloride	NH_4Cl	1.57
ammonium nitrate	NH_4NO_3	1.74
ammonium sulfate	$(NH_4)_2SO_4$	1.409
calcium chloride	$CaCl_2$	0.657
lithium chloride	$LiCl$	1.13
sodium carbonate	Na_2CO_3	1.06
sodium hydroxide	$NaOH$	1.49
sodium sulfate	Na_2SO_4	0.903
sodium thiosulfate pentahydrate	$Na_2S_2O_3{\cdot}5H_2O$	1.45
potassium bromide	KBr	0.439
potassium nitrate	KNO_3	0.95

Procedure Overview: Three different experiments are completed in a "double" coffee cup calorimeter. Each experiment requires careful mass/volume measurements and temperature measurements before and after the mixing of the respective components. Calculations are based on an interpretation of plotted data.

Ask your instructor which parts of this experiment you are to complete.

You and a partner are to complete at least two trials for each part assigned.

- Part A: The same metal sample is used. A boiling water bath is required.
- Part B: New solutions are required for each trial; therefore initially obtain about 110 mL of 1.1 M HCl, 110 mL of 1.1 M HNO$_3$, and 210 mL of 1.0 M NaOH from the stock reagents and plan your work schedule carefully.
- Part C: Two trials for a single salt are required; therefore determine the masses of two samples of the same salt while using the balance.

The temperature vs. time curves to be plotted in Parts A.5, B.4, and C.4 can be established by using a thermal probe that is connected directly to either a calculator or computer with the appropriate software. If this thermal sensing/recording apparatus is available in the laboratory, consult with your instructor for its use and adaptation to the experiment. The probe merely replaces the glass or digital thermometer in Figure 25.4.

A. Specific Heat of a Metal

The temperature is to be recorded with the correct number of significant figures

Use a stirring rod to assist in the gentle transfer of the metal into the water of the calorimeter

1. **Prepare the Metal.** Obtain 10–30 g of an unknown metal[2] from your instructor. Record the number of the unknown metal on the Report Sheet. Use weighing paper to measure its mass on your assigned balance. Transfer the metal to a dry, 200-mm test tube. Place the 200-mm test tube in a 400-mL beaker filled with water, well above the level of the metal sample in the test tube (Figure 25.2). Heat the water to boiling and maintain this temperature for at least 5 minutes so that the metal reaches thermal equilibrium with the boiling water. Proceed to Part A.2 while the water is heating.

2. **Prepare the Water in the Calorimeter.** The apparatus for the calorimetry experiment appears in Figure 25.4. Obtain two 6- or 8-oz Styrofoam coffee cups, a plastic lid, stirrer and a 110° glass or a digital thermometer. Thoroughly clean the styrofoam cups with several rinses of deionized water. Measure and record the combined mass (± 0.01 g) of the calorimeter (the two Styrofoam cups, the plastic lid, and stirrer).

   Using a graduated cylinder, add ~20.0 mL of water and measure the mass of the calorimeter *plus* water. Secure the glass or digital (Figure 25.3) thermometer with a clamp and position the bulb or thermal sensor below the water surface. (**Caution:** *Carefully handle a glass thermometer. If the thermometer is accidentally broken, notify your instructor immediately.*)

3. **Measure and Record the Temperatures of the Metal and Water.** Once thermal equilibrium has been reached in Parts A.1 and A.2, measure and record the temperatures of the *boiling* water from Part A.1 and of the water in the calorimeter from Part A.2. Record the temperatures "using all certain digits *plus* one uncertain digit."

4. **Transfer the Hot Metal to the Cool Water and Record the Data.** Remove the test tube from the boiling water and *quickly* transfer *only* the metal to the water in the calorimeter.[3] Replace the lid and swirl the contents gently. Record the water

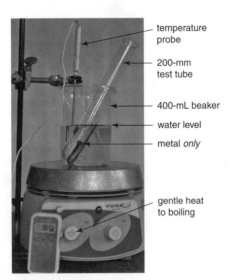

Figure 25.2 Placement of the metal in the dry test tube below the water surface in the beaker. A Bunsen flame may replace the hot plate.

temperature
probe

200-mm
test tube

400-mL beaker

water level

metal *only*

gentle heat
to boiling

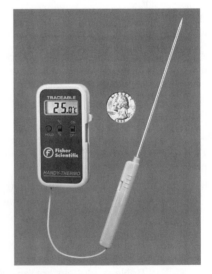

Figure 25.3 A modern digital thermometer can be substituted for a glass thermometer.

[2]Ask your instructor to determine the approximate mass of metal to use for the experiment.
[3]Be careful *not* to splash out any of the water in the calorimeter. If you do, you will need to repeat the entire procedure. Also, be sure that the metal is fully submerged in the calorimeter.

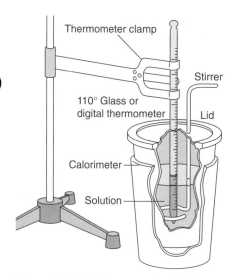

Figure 25.4 Schematic of a "coffee cup" calorimeter (see opening photo).

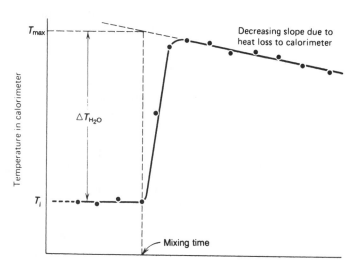

Figure 25.5 Extrapolation of temperature vs. time data (not to scale) for an exothermic process.

temperature as a function of time (about 5-second intervals for 1 minute and then 15-second intervals for ~5 minutes) on the table at the end of the Report Sheet.

5. **Plot the Data.** Plot the temperature (y axis) versus time (x axis) on the top half of a sheet of linear graph paper or by using appropriate software. The maximum temperature is the intersection point of two lines: (1) the best line drawn through the data points on the cooling portion of the curve and (2) a line drawn perpendicular to the time axis at the mixing time [when the metal is added to the water (Figure 25.5)].[4] Have your instructor approve your graph.

Appendix C

6. **Do It Again.** Repeat Parts A.1 through A.5 for the same metal sample. Plot the data on the bottom half of the same sheet of linear graph paper.

Disposal: Return the metal to the appropriately labeled container, as advised by your instructor.

1. **Measure the Volume and Temperature of the HCl.** Measure 50.0 mL of 1.1 *M* HCl in a *clean*, graduated cylinder. Measure and record its temperature.

2. **Measure the Volume and Temperature of the NaOH.** Using a second *clean*, graduated cylinder transfer 50.0 mL of a **standard** 1.0 *M* NaOH **solution** to the *dry* calorimeter (see Figure 25.4). Record the temperature and exact molar concentration of the NaOH solution.

3. **Collect the Data.** Carefully, but quickly, add the acid to the base, replace the calorimeter lid, and swirl gently. Read and record the temperature and time every 5 seconds for 1 minute and thereafter every 15 seconds for ~5 minutes.

4. **Plot the Data.** Plot the temperature (y axis) versus time (x axis) on the top half of a sheet of linear graph paper or by using appropriate software. Determine the maximum temperature as was done in Part A.5. Have your instructor approve your graph.

5. **Do It Again.** Repeat the acid-base experiment, Parts B.1 through B.4. Plot the data on the bottom half of the same sheet of graph paper.

B. Enthalpy (Heat) of Neutralization for an Acid–Base Reaction

Standard solution: a solution with a very accurately measured concentration of a solute

Appendix C

---

[4]The maximum temperature is never recorded because of some, albeit very small, heat loss to the calorimeter wall.

6. **Change the Acid; Repeat the Neutralization Reaction.** Repeat Parts B.1 through B.5, substituting 1.1 M HNO_3 for 1.1 M HCl. On the Report Sheet, compare the ΔH_n values for the two strong acid-strong base reactions.

Disposal: Discard the neutralized solutions contained in the calorimeter into the "Waste Acids" container. Rinse the calorimeter twice with deionized water.

C. Enthalpy (Heat) of Solution for the Dissolution of a Salt

1. **Prepare the Salt.** On weighing paper, measure about 5.0 g (± 0.001 g) of the assigned salt. Record the name of the salt and its mass on the Report Sheet.

2. **Prepare the Calorimeter.** Measure the mass of the *dry* calorimeter. Using your clean, graduated cylinder, add ~20.0 mL of deionized water to the calorimeter. Measure the combined mass of the calorimeter and water. Secure the thermometer with a clamp and position the bulb or thermal sensor below the water surface (see Figure 25.4) and record its temperature.

3. **Collect the Temperature Data.** Carefully add (do not spill) the salt to the calorimeter, replace the lid, and swirl gently. Read and record the temperature and time at 5-second intervals for 1 minute and thereafter every 15-seconds for ~5 minutes.

Appendix C

4. **Plot the Data.** Plot the temperature (y axis) versus time (x axis) on the top half of a sheet of linear graph paper or by using appropriate software. Determine the maximum (for an exothermic process) or minimum (for an endothermic process) temperature as was done in Part A.5. Have your instructor approve your graph.

5. **Do It Again.** With a fresh sample, repeat the dissolution of your assigned salt, Parts C.1 through C.4. Plot the data on the bottom half of the same sheet of linear graph paper.

Disposal: Discard the salt solution into the "Waste Salts" container, followed by additional tap water. Consult with your instructor.

CLEANUP: Rinse the coffee cups twice with tap water and twice with deionized water, insert the thermometer into its carrying case, and return them.

The Next Step

Heat is evolved or absorbed in all chemical reactions. (1) Since heat is transferred to/from the calorimeter, design an experiment to determine the calorimeter constant (called its heat capacity) for a calorimeter. (2) An analysis of the combustion of different fuels is an interesting, yet challenging, project. Design an apparatus and develop a procedure for the thermal analysis (kilojoules/gram) of various combustible materials, e.g, alcohol, gasoline, coal, wood.

Date _____ Lab Sec. _____ Name _____ Desk No. _____

1. A 34.44-g sample of a metal is heated to 98.6°C in a hot water bath until thermal equilibrium is reached. The metal sample is quickly transferred to 50.0 mL of water at 22.2°C contained in a calorimeter. The thermal equilibrium temperature of the metal sample plus water mixture is 28.3°C. What is the specific heat of the metal?

2. Will the recorded temperature change for an exothermic reaction performed in a metal calorimeter be greater or less than that in a Styrofoam "coffee cup" calorimeter? Explain. Assume metal to be a better conductor of heat than Styrofoam.

3. a. Experimental Procedure, Part A.1. What is the procedure for heating a metal to an exact, but measured, temperature?

   b. Experimental Procedure, Part A.1. How can "bumping" be avoided when heating water in a beaker?

4. Three student chemists measured 50.0 mL of 1.00 M NaOH in separate Styrofoam "coffee cup" calorimeters (Part B). Brett added 50.0 mL of 1.10 M HCl to his solution of NaOH; Dale added 45.5 mL of 1.10 M HCl (equal moles) to his NaOH solution. Lyndsay added 50.0 mL of 1.00 M HCl to her NaOH solution. Each student recorded the temperature change and calculated the enthalpy of neutralization.

   Two of the chemists will report, within experimental error, the same temperature change for the HCl/NaOH reaction. Identify the two students and explain.

5. Angelina observes a temperature increase when her salt dissolves in water (Part C).
   a. Is the lattice energy for the salt greater or less than the hydration energy for the salt? Explain.

   b. Will the solubility of the salt increase or decrease with temperature increases? Explain.

6. A 4.50-g sample of LiCl at 25.0°C dissolves in 25.0 mL of water also at 25.0°C. The final equilibrium temperature of the resulting solution is 60.8°C. What is the enthalpy of solution, ΔH_s, of LiCl expressed in kilojoules per mole?

Calorimetry

Date _____ Lab Sec. _____ Name _____ Desk No. _____

A. Specific Heat of a Metal

Unknown No. _____

	Trial 1	*Trial 2*
1. Mass of metal (*g*)	_____	_____
2. Temperature of metal (boiling water) (°*C*)	_____	_____
3. Mass of calorimeter (*g*)	_____	_____
4. Mass of calorimeter + water (*g*)	_____	_____
5. Mass of water (*g*)	███████████	███████████
6. Temperature of water in calorimeter (°*C*)	_____	_____
7. Maximum temperature of metal and water from graph (°*C*)	_____	_____
8. Instructor's approval of graph	_____	_____

Calculations for Specific Heat and the Molar Mass of a Metal

	Trial 1	*Trial 2*
1. Temperature change of water, ΔT (°*C*)	███████████	███████████
2. Heat *gained* by water (*J*)	███████████	███████████
3. Temperature change of metal, ΔT (°*C*)	███████████	███████████
4. Specific heat of metal, Equation 25.5 ($J/g\bullet°C$)	███████████ *	███████████
5. Average specific heat of metal ($J/g\bullet°C$)	███████████	

*Show calculations for Trial 1.

B. Enthalpy (Heat) of Neutralization for an Acid–Base Reaction

	HCl + NaOH		HNO₃ + NaOH	
	Trial 1	*Trial 2*	*Trial 1*	*Trial 2*
1. Volume of acid (*mL*)	_____	_____	_____	_____
2. Temperature of acid (°C)	_____	_____	_____	_____
3. Volume of NaOH (*mL*)	_____	_____	_____	_____
4. Temperature of NaOH (°C)	_____	_____	_____	_____
5. Exact molar concentration of NaOH (*mol/L*)	_____		_____	
6. Maximum temperature from graph (°C)	_____	_____	_____	_____
7. Instructor's approval of graph	_____	_____	_____	_____

Calculations for Enthalpy (Heat) of Neutralization for an Acid–Base Reaction

	HCl + NaOH		HNO₃ + NaOH	
1. *Average* initial temperature of acid and NaOH (°C)	▭	▭	▭	▭
2. Temperature change, ΔT (°C)	▭	▭	▭	▭
3. Volume of final mixture (*mL*)	▭	▭	▭	▭
4. Mass of final mixture (*g*) (Assume the density of the solution is 1.0 g/mL.)	▭	▭	▭	▭
5. Specific heat of mixture	4.18 J/g•°C		4.18 J/g•°C	
6. Heat evolved (*J*)	▭	▭	▭	▭
7. Moles of OH⁻ reacted, the limiting reactant (*mol*)	▭	▭	▭	▭
8. Moles of H₂O formed (*mol*)	▭	▭	▭	▭
9. Heat evolved per mole of H₂O, ΔH_n (*kJ/mol H₂O*)	▭	▭ *	▭	▭
10. Average ΔH_n (*kJ/mol H₂O*)	▭		▭	

*Show calculations for Trial 1.

Comment on your two values of ΔH_n.

C. Enthalpy (Heat) of Solution for the Dissolution of a Salt

Name of salt _____

	Trial 1	Trial 2
1. Mass of salt (g)		
2. Moles of salt (mol)		
3. Mass of calorimeter (g)		
4. Mass of calorimeter + water (g)		
5. Mass of water (g)		
6. Initial temperature of water ($°C$)		
7. Final temperature of mixture from graph ($°C$)		
8. Instructor's approval of graph		

Calculations for Enthalpy (Heat) of Solution for the Dissolution of a Salt

	Trial 1	Trial 2
1. Change in temperature of solution, ΔT ($°C$)		
2. Heat change of water (J)		
3. Heat change of salt (J) (Obtain its specific heat from Table 25.1.)		
4. *Total* enthalpy change, Equation 25.11 (J)		
5. ΔH_s, Equation 25.12 (J/mol salt)	*	
6. Average ΔH_s (J/mol salt)		

*Show calculations for Trial 1.

Specific Heat of a Metal				Enthalpy (Heat) of Neutralization for an Acid–Base Reaction								Enthalpy (Heat) of Solution for the Dissolution of a Salt			
Trial 1		Trial 2		Trial 1		Trial 2		Trial 1		Trial 2		Trial 1		Trial 2	
Time	Temp	Time	Temp	Time	Temp	Time	Temp	Time	Temp	Time	Temp	Time	Temp	Time	Temp

Laboratory Questions

Circle the questions that have been assigned.

1. Part A.1. The 200-mm test tube also contained some water (besides the metal) which was subsequently added to the calorimeter (in Part A.4). Considering a higher specific heat for water, will the temperature change in the calorimeter be higher, lower, or unaffected by this technique error? Explain.

2. Part A.4. When a student chemist transferred the metal to the calorimeter, some water splashed out of the calorimeter. Will this technique error result in the specific heat of the metal being reported too high or too low? Explain.

3. Part A.5. In measuring the specific heat of a metal, a student chemist used the highest *recorded* temperature for calculating the metal's specific heat rather than the extrapolated temperature. Will this error result in a larger or smaller specific heat value reported in the experiment? Explain.

4. Part B. The enthalpy of neutralization for *all* strong acid–strong base reactions should be the same within experimental error. Explain. Will that also be the case for all weak acid–strong base reactions? Explain.

5. Part B. Heat is lost to the Styrofoam calorimeter. Assuming a 6.22°C temperature change for the reaction of HCl(*aq*) with NaOH(*aq*), calculate the heat loss to the inner 2.35-g Styrofoam cup. The specific heat of Styrofoam is 1.34 J/g•°C.

6. Part B.3. A student chemist carelessly added only 40.0 mL (instead of the recommended 50.0 mL) of 1.1 *M* HCl to the 50.0 mL of 1.0 *M* NaOH. Explain the consequence of the error.

7. Part C.3. If some of the salt remains adhered to the weighing paper (and therefore is *not* transferred to the calorimeter), will the enthalpy of solution for the salt be reported too high or too low? Explain.

8. Part C. Because heat is absorbed from the calorimeter from an endothermic dissolving process, is the calculated enthalpy of solution, ΔH_s, more positive or more negative than what it should be? Explain.

Experiment 26

Thermodynamics of the Dissolution of Borax

Borax is a commonly added to (clothes) wash water to increase the pH for more effective cleansing.

OBJECTIVES

- To standardize a hydrochloric acid solution
- To determine the solubility product of borax as a function of temperature
- To determine the standard free energy, standard enthalpy, and standard entropy changes for the dissolution of borax in an aqueous solution

TECHNIQUES

The following techniques are used in the Experimental Procedure

INTRODUCTION

Large deposits of borax are found in the arid regions of the southwestern United States, most notably in the Mojave Desert (east central) region of California. Borax is obtained as tincal, $Na_2B_4O_5(OH)_4 \cdot 8H_2O$, and kernite, $Na_2B_4O_7 \cdot 4H_2O$, from an open pit mine near Boron, California, and as tincal from brines from Searles Lake near Trona, California. Borax, used as a washing powder for laundry formulations, is commonly sold as 20-Mule Team Borax®. Historically, borax was mined in Death Valley, California, in the late nineteenth century. To transport the borax from this harsh environment, teams of 20 mules were used to pull a heavy wagon loaded with borax (and a water wagon) across the desert and over the mountains to railroad depots for shipment to other parts of the world. Borax is used as a cleansing agent, in the manufacture of glazing paper and varnishes, and as a flux in soldering and brazing; however, its largest current use is in the manufacture of borosilicate glass.

The free energy change of a chemical process is proportional to its equilibrium constant according to the equation

$$\Delta G° = -RT \ln K \qquad (26.1)$$

where R, the gas constant, is 8.314×10^{-3} kJ/mol·K and T is the temperature in kelvins. The equilibrium constant, K, is expressed for the equilibrium system when the reactants and products are in their **standard states.** For a slightly soluble salt in an

Standard state: the state of a substance at one atmosphere (and generally 25°C)

aqueous system, the precipitate and the ions in solution correspond to the standard states of the reactants and products, respectively.

The "standard state" equilibrium for the slightly soluble silver chromate salt is

$$Ag_2CrO_4(s) \rightleftharpoons 2\ Ag^+(aq) + CrO_4^{2-}(aq) \tag{26.2}$$

The solubility product, K_{sp}, is set equal to the product of the molar concentrations of the ions, each raised to the power of their respective coefficients in the balanced equation—this is the mass action expression for the system:

$$K_{sp} = [Ag^+]^2\ [CrO_4^{2-}] \tag{26.3}$$

Free energy change, $\Delta G°$: negative value, spontaneous process; positive value, nonspontaneous process

and the **free energy change** for the equilibrium is

$$\Delta G° = -RT \ln K_{sp} = -RT \ln [Ag^+]^2\ [CrO_4^{2-}] \tag{26.4}$$

Enthalpy change, $\Delta H°$: negative value, exothermic process; positive value, endothermic process

Additionally, the free energy change of a chemical process is also a function of the **enthalpy change** and the **entropy change** of the process:

$$\Delta G° = \Delta H° - T\Delta S° \tag{26.5}$$

Entropy change, $\Delta S°$: negative value, decrease in randomness of the process; positive value, increase in randomness of the process

When the two free energy expressions are set equal for a slightly soluble salt, such as silver chromate, then

$$-RT \ln K_{sp} = \Delta H° - T\Delta S° \tag{26.6}$$

Rearranging and solving for $\ln K_{sp}$,

$$\ln K_{sp} = -\frac{\Delta H°}{R}\left(\frac{1}{T}\right) + \frac{\Delta S°}{R} \text{ (analogous to the equation } y = mx + b) \tag{26.7}$$

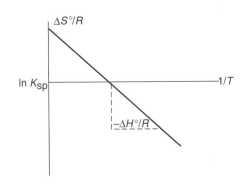

This equation can prove valuable in determining the thermodynamic properties of a chemical system, such as that of a slightly soluble salt. A linear relationship exists when the values of $\ln K_{sp}$ obtained at various temperatures are plotted as a function of the reciprocal temperature. The (negative) slope of the line equals $-\Delta H°/R$, and the y-intercept (where $x = 0$) equals $\Delta S°/R$. Since R is a constant, the $\Delta H°$ and the $\Delta S°$ for the equilibrium system can easily be calculated.

Since the values of $\ln K_{sp}$ may be positive or negative for slightly soluble salts and T^{-1} values are always positive, the data plot of $\ln K_{sp}$ versus $1/T$ appears in the first and fourth quadrants of the Cartesian coordinate system.

The Borax System

Borax is often given the name sodium tetraborate decahydrate and the formula $Na_2B_4O_7 \cdot 10H_2O$. However, according to its chemical behavior, a more defining formula for borax is $Na_2B_4O_5(OH)_4 \cdot 8H_2O$, also called tincal.

In this experiment the thermodynamic properties, $\Delta G°$, $\Delta H°$, and $\Delta S°$, are determined for the aqueous solubility of borax (tincal), $Na_2B_4O_5(OH)_4 \cdot 8H_2O$. Borax dissolves and dissociates in water according to the equation

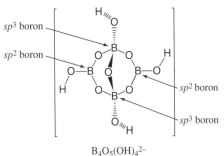

$B_4O_5(OH)_4^{2-}$

$$Na_2B_4O_5(OH)_4 \cdot 8H_2O(s) \rightleftharpoons 2\ Na^+(aq) + B_4O_5(OH)_4^{2-}(aq) + 8\ H_2O(l) \tag{26.8}$$

The mass action expression, set equal to the solubility product at equilibrium, for the solubility of borax is

$$K_{sp} = [Na^+]^2\ [B_4O_5(OH)_4^{2-}] \tag{26.9}$$

The $B_4O_5(OH)_4^{2-}$ anion, because it is the conjugate base of the weak acid boric acid, is capable of accepting two protons from a strong acid in an aqueous solution:

$$B_4O_5(OH)_4^{2-}(aq) + 2\,H^+(aq) + 3\,H_2O(l) \rightleftharpoons 4\,H_3BO_3(aq) \qquad (26.10)$$

Therefore, the molar concentration of the $B_4O_5(OH)_4^{2-}$ anion in a saturated borax solution can be measured with a titrimetric analysis of the saturated borax solution using a standardized hydrochloric acid solution as the titrant.

$$\text{mol }B_4O_5(OH)_4^{2-} = \text{volume (L) HCl} \times \frac{\text{mol HCl}}{\text{volume (L) HCl}} \times \frac{1\,\text{mol }B_4O_5(OH)_4^{2-}}{2\,\text{mol HCl}} \qquad (26.11)$$

$$[B_4O_5(OH)_4^{2-}] = \frac{\text{mol }B_4O_5(OH)_4^{2-}}{\text{volume (L) sample}} \qquad (26.12)$$

This analysis is also a measure of the molar solubility of borax in water at a given temperature—according to the stoichiometry, one mole of the $B_4O_5(OH)_4^{2-}$ anion forms for every mole of borax that dissolves.

$$[B_4O_5(OH)_4^{2-}] = \text{molar solubility of borax} \qquad (26.13)$$

Temperature changes do affect the molar solubility of most salts, and borax is no exception. For example, the solubility of borax is 2.01 g/100 mL at 0°C and is 170 g/100 mL at 100°C.[1]

As a consequence of the titration and according to the stoichiometry of the dissolution of the borax (Equation 26.8), the molar concentration of the sodium ion in the saturated solution is twice that of the experimentally determined $B_4O_5(OH)_4^{2-}$ anion concentration.

$$[Na^+] = 2 \times [B_4O_5(OH)_4^{2-}] \qquad (26.14)$$

The solubility product for borax at a measured temperature is, therefore,

$$K_{sp} = [Na^+]^2[B_4O_5(OH)_4^{2-}] = [2 \times [B_4O_5(OH_4^{2-}]]^2[B_4O_5(OH)_4^{2-}]$$
$$= 4[B_4O_5(OH)_4^{2-}]^3 = 4[\text{molar solubility of borax}]^3 \qquad (26.15)$$

To obtain the thermodynamic properties for the dissolution of borax, values for the molar solubility and the solubility product for borax are determined over a range of temperatures. An interpretation of a data plot for $\ln K_{sp}$ vs. $1/T$ enables the chemist to determine the standard enthalpy change, $\Delta H°$, the standard entropy change, $\Delta S°$, and ultimately to calculate the standard free energy change, $\Delta G°$, for the dissolution of borax.

Standardized HCl Solution

A standardized HCl solution is prepared using anhydrous sodium carbonate as the primary standard. Sodium carbonate samples of known mass are transferred to Erlenmeyer flasks, dissolved in deionized water, and titrated to a methyl orange endpoint (pH range 3.1 to 4.4) with the prepared hydrochloric acid solution.

$$CO_3^{2-}(aq) + 2\,H_3O^+(aq) \rightarrow 3\,H_2O(l) + CO_2(g) \qquad (26.16)$$

The flask is heated to near boiling close to the stoichiometric point of the analysis to remove the carbon dioxide gas produced in the reaction.

Analysis of Data

The objectives of this experiment are fourfold: (1) determine the molar concentration of the $B_4O_5(OH)_4^{2-}$ anion and the molar solubility of borax for *at least* five different

[1]The solubility of borax at room temperature is about 6.3 g/100 mL.

temperatures using the titration technique with a standardized hydrochloric acid solution; (2) calculate the solubility product of borax at each temperature; (3) plot the natural logarithm of the solubility product versus the reciprocal temperature of the measurements; (4) extract from the plotted data the thermodynamic properties of $\Delta H°$ and $\Delta S°$ for the dissolution of borax, from which its $\Delta G°$ is calculated.

EXPERIMENTAL PROCEDURE

Procedure Overview: This experiment is to be completed in cooperation with other chemists/chemist groups in the laboratory. In Part A, a standardized solution of hydrochloric acid is to be prepared. In Part B, four warm water baths are to be set up (see Part B.3 and Figure 26.3) at the beginning of the laboratory, each at a different temperature, but at a maximum of 60°C. *The water baths then are to be shared.* Consult with your colleagues (and your laboratory instructor) in coordinating the setup of the apparatus. In conjunction with the four warm water baths for Part B, about 150 mL of warm (~55°C) deionized water is to be prepared for Part C.1.

A. Standardized HCl Solution

Keep the dry primary standard sample in a desiccator.

1. **Dry the Primary Standard.** Dry 2–3 g of anhydrous sodium carbonate, Na_2CO_3, for several hours in a drying oven set at about 110°C. Cool the sample in a desiccator. Stockroom personnel may have previously dried the sodium carbonate. Check with your laboratory instructor.

2. **Prepare the HCl Solution.** Prepare 200 mL of ~0.20 M HCl starting with concentrated (12.1 M) HCl. (**Caution:** *Concentrated HCl causes severe skin burns.*)

3. **Prepare the Primary Standard.** Calculate the mass of sodium carbonate that neutralizes 15–20 mL of 0.20 M HCl at the stoichiometric point. Show the calculation on the Report Sheet. Measure this mass (± 0.001 g) on a tared piece of weighing paper or dish and transfer to a 125-mL Erlenmeyer flask. Prepare *at least* three samples of sodium carbonate for the analysis of the HCl solution.

4. **Prepare the Buret.** Clean a buret and rinse with several 3- to 5-mL portions of the ~0.20 M HCl solution. Use a clean funnel to fill the buret with the ~0.20 M HCl solution. After 10–15 seconds use the proper technique to read and record the volume of HCl solution in the buret, "using all certain digits *plus* one uncertain digit."

5. **Titrate the Primary Standard.** To each solid sodium carbonate sample add ~50 mL of deionized water and several drops of methyl orange indicator.[2] Place a sheet of white paper beneath the Erlenmeyer flask. Dispense the HCl solution from the buret, swirling the Erlenmeyer flask during the addition.

    Very near or at the apparent endpoint of the indicator, heat the flask to near boiling (see Figures 17.2 or 20.3) to drive off the carbon dioxide gas. Carefully (dropwise) add additional HCl titrant until the endpoint is reached and the color persists for 30 seconds (a color change caused by the addition of one "additional" drop of the HCl solution from the buret). Stop the addition of the HCl titrant. After 10–15 seconds read and record the volume of HCl in the buret.

6. **Repeat the Analysis.** Refill the buret and complete the standardization procedure with the remaining sodium carbonate samples.

7. **Do the Calculations.** Calculate the molar concentration of the prepared HCl solution from the sodium carbonate samples. The molar concentrations of the HCl solution from the trials should agree within $\pm 1\%$; if not, complete additional trials as necessary.

Disposal: Dispose of the analyzed samples in the "Waste Salts" container.

---

[2]Methyl orange indicator changes color over the pH range of 3.1–4.4; its color appears yellow at a pH greater than 4.4, but red-orange at a pH less than 3.1.

Five or six borax samples can be prepared for the experiment. The sixth sample (a saturated solution in an ice bath) will have a *very* low solubility of borax. Consult with your instructor.

1. **Calibrate Test Tubes.** Pipet 5 mL of deionized water into five (or six) *clean* medium-sized (~150 mm) test tubes (Figure 26.1). Mark the bottom of the meniscus (use a marking pen or tape). Discard the water and allow the test tubes to air- or oven-dry. Label the test tubes.

2. **Prepare Stock Solution of Borax.** Using a 125- or 250-mL Erlenmeyer flask, add 30–35 g of borax to 100 mL of deionized water. Stopper the flask and agitate the mixture for several minutes to prepare the saturated solution.

3. **Prepare the Test Solutions of Borax.** a. *Second set of test tubes.* Label a second set of clean, medium-sized test tubes for use in the setup shown in Figure 26.3.

   b. *Half-fill the test tubes with stock solution.* Again, thoroughly agitate the borax stock solution and then, with the stock solution, half-fill this second set of medium-sized test tubes. Place the test tubes in the respective baths shown Figure 26.3. The temperatures of the baths need not be exactly those indicated, but the measured temperature is important (Figure 26.2). The "first" bath should *not* exceed 60°C.[3] Share the water baths.

4. **Prepare Saturated Solutions of Borax.** a. *Saturate the solutions.* Occasionally agitate the test tubes in the baths (for 10–15 minutes), assuring the formation of a saturated solution—solid borax should always be present—add more solid borax if necessary.

   b. *Allow sample to settle (not change temperature).* Allow the borax to settle until the solution is clear (this will require several minutes, be patient!) and has reached thermal equilibrium. Allow 10–15 minutes for thermal equilibrium to be established.

5. **Prepare Borax Samples for Analysis.** Quickly, but carefully, transfer 5 mL of the clear solutions to the correspondingly labeled, calibrated test tubes from Part B.1 and Figure 26.1 (do *not* transfer any of the solid borax!).
   Record the *exact* temperature of the respective water baths.
   Remove the heat from the water baths.

B. Preparation of Borax Solutions

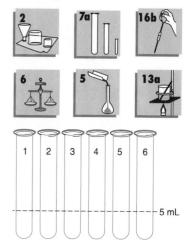

Figure 26.1 Calibrate six, clean 150-mm test tubes at the 5-mL mark. (Part B.1)

Figure 26.2 A stirring hot plate with a thermal sensor can be used to set water bath temperatures.

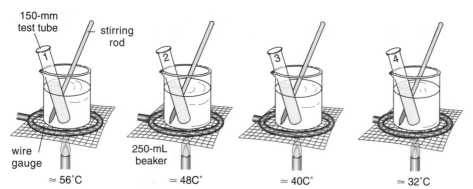

ambient   ≈ 5°C

Figure 26.3 Six (different temperature) baths for preparing saturated borax solutions. (Part B.3). A hot plate may be substituted for the Bunsen flame.

Labels in figure: 150-mm test tube, stirring rod, wire gauge, 250-mL beaker, ≈ 56°C, ≈ 48°C, ≈ 40°C, ≈ 32°C

---

[3]Borax (tincal) is stable in aqueous solutions at temperatures less than 61°C; at higher temperatures, some dehydration of the $B_4O_5(OH)_4{}^{2-}$ anion occurs.

C. Analysis of Borax Test Solutions

1. **Transfer the Samples.** a. *Prepare for sample transfer.* Set up and label a set of five (or six) clean labeled (1–6), 125- or 250-mL Erlenmeyer flasks. As the samples (Part B.5) cool to room temperature in the test tubes, some borax may crystallize. Return those samples to a *warm* (~55°C) water bath until all of the solid dissolves.

   b. *Transfer the samples.* After the sample is clear, transfer it to the correspondingly labeled Erlenmeyer flask, rinse the test tube with two or three ~5-mL portions of *warm* deionized water, and combine the washings with the sample.

2. **Titrate the Samples.** Dilute each sample to about 25 mL with warm, deionized water. Add 2–3 drops of bromocresol green.[4] Titrate each of the five samples to a yellow endpoint with the standardized HCl solution prepared in Part A. Remember to record the buret readings before and after each analysis of a sample.

> *Disposal:* Dispose of the analyte in the "Waste Salts" container and the titrant in the "Waste Acids" container.

D. Data Analysis

Appendix C

Five (or six) repeated calculations are required in order to establish the data plot for the determination of $\Delta H°$ and $\Delta S°$ for the dissolution of borax. The lengthy task of completing the calculations and for minimizing errors in the calculations can be reduced with the use of an Excel (or similar) spreadsheet. The data from the calculations can then be plotted using the embedded graphing capabilities of Excel.

The calculations and the analyses that are required are:

1. Calculate the molar solubility of borax at each of the measured temperatures.

2. Calculate the solubility product of borax at each of the measured temperatures.

3. Plot the natural logarithm of the solubility product versus the reciprocal temperature (K^{-1}) for each sample and draw the "best straight line" through the data points.

4. Determine the slope of the linear plot and calculate the standard enthalpy of solution for borax.

5. Determine the y-intercept (at $x = 0$) of the linear plot and calculate the standard entropy of solution for borax.

The Next Step

An equilibrium constant is determined spectrophotometrically in Experiment 34. What modifications would need to be made to study the thermodynamics of that equilibrium? Design the experiment.

---

[4]The pH range for the color change of bromocresol green is 4.0 to 5.6. The indicator appears yellow at a pH less than 4.0 and blue at a pH greater than 5.6.

Thermodynamics of the Dissolution of Borax

Date _____ Lab Sec. _____ Name _____ Desk No. _____

1. The standard free energy change for the decomposition of two moles of hydrogen peroxide at 25°C is −224 kJ.

$$2 \, H_2O_2(l) \rightarrow 2 \, H_2O(l) + O_2(g) \quad \Delta G° = -234 \text{ kJ}$$

a. Calculate the equilibrium constant for the reaction.

b. The standard enthalpy change, $\Delta H°$, for the decomposition of two moles of hydrogen peroxide is −196.1 kJ. Determine the standard entropy change, $\Delta S°$, for the reaction at 25°C.

c. i) What is the meaning of the sign and magnitude of the standard free energy change?

 ii) What is the meaning of the sign and magnitude of the standard enthalpy change?

 iii) What is the meaning of the sign and magnitude of the standard entropy change?

2. a. Experimental Procedure, Part A.2. Describe the preparation of 200 mL of 0.20 *M* HCl, starting with *conc* HCl (12.1 *M*).

 b. Experimental Procedure, Part A.3. How many grams of the primary standard sodium carbonate, Na_2CO_3 (molar mass = 105.99 g/mol) is needed to react with 15 mL of 0.20 *M* HCl?

3. The $B_4O_5(OH)_4^{2-}$ ion, present in 5.0 mL of a saturated $Na_2B_4O_5(OH)_4$ solution at a measured temperature, is titrated to the bromocresol green endpoint with 6.51 mL of 0.146 M HCl.
   a. How many moles of $B_4O_5(OH)_4^{2-}$ are present in the sample?

   b. What is the molar concentration of $B_4O_5(OH)_4^{2-}$ in the sample?

   c. Calculate the K_{sp} for $Na_2B_4O_5(OH)_4$ from this data.

   d. What is the free energy change for the dissolution of $Na_2B_4O_5(OH)_4$ at 25°C?

4. Experimental Procedure, Part C. a. How many saturated solutions of borax are to be titrated?

   b. Each saturated solution of borax is prepared at a different temperature. How many trials per saturated solution are to be completed?

   c. What values are to be calculated from the data of the titrations?

5. Experimental Procedure, Part D. A data plot is to be constructed to determine the thermodynamic values of $\Delta G°$, $\Delta H°$, and $\Delta S°$ for the dissolution of borax.
   a. What are the coordinates of the data plot?

   b. How many data points are on the plot?

   c. How is $\Delta H°$ determined from the data plot?

   d. How is $\Delta S°$ determined from the data plot?

Experiment 26 *Report Sheet*

Thermodynamics of the Dissolution of Borax

Date __________ Lab Sec. ______ Name ____ Desk No. __________

A. Standardized HCl Solution

Calculate the mass of sodium carbonate sample required in Part A.3.

	Trial 1	*Trial 2*	*Trial 3*	*Trial 4*
1. Tared mass of Na_2CO_3 (*g*)				
2. Moles of Na_2CO_3 (*mol*)				
3. Buret reading, *initial* (*mL*)				
4. Buret reading, *final* (*mL*)				
5. Volume of HCl added (*mL*)				
6. Moles of HCl added (*mol*)				
7. Molar concentration of HCl (*mol/L*)				
8. Average molar concentration of HCl (*mol/L*)				

B. Preparation of Borax Solution

Sample number	*1*	*2*	*3*	*4*	*5*	*6*
1. Volume of sample (*mL*)	5.00	5.00	5.00	5.00	5.00	5.00
2. Temperature of sample (°*C*)						

C. Analysis of Borax Test Solutions

1. Buret reading, *initial* (*mL*)						
2. Buret reading, *final* (*mL*)						
3. Volume of HCl added (*mL*)						

D. Data Analysis

Sample number	1	2	3	4	5	6
1. Temperature (K)						
2. $1/T$ (K^{-1})						
3. Moles of HCl used (mol)						
4. Moles of $B_4O_5(OH)_4^{2-}$ (mol)						
5. $[B_4O_5(OH)_4^{2-}]$ (mol/L)						
6. Molar solubility of borax (mol/L)						
7. Solubility product, K_{sp}						
8. $\ln K_{sp}$						

9. Instructor's approval of plotted data   _____

10. $-\Delta H°/R$ (from data plot)   _____

11. $\Delta S°/R$ (from data plot)   _____

12. $\Delta H°$ (kJ/mol)

13. $\Delta S°$ (J/mol•K)

14. $\Delta G°$ (kJ), at 298 K

Laboratory Questions

Circle the questions that have been assigned.

1. Part A.1. No desiccator is available. The sodium carbonate is cooled to room temperature, but the humidity in the room is high. How will this affect the reported molar concentration of the hydrochloric acid solution in Part A.7 . . . too high, too low, or unaffected? Explain.

2. Part A.5. The endpoint in the titration is "overshot."
   a. Is the reported molar concentration of the hydrochloric acid solution too high or too low? Explain.
   b. As a result of this poor titration technique, is the reported number of moles of $B_4O_5(OH)_4^{2-}$ in each of the analyses (Part C.2) too high or too low? Explain.

3. Part B.2. The solid borax reagent is contaminated with a water-soluble substance that does not react with hydrochloric acid. How will this contamination affect the reported solubility product of borax? Explain.

4. Part B.4. For the borax solution at 48°C, no solid borax is present in the test tube. Five milliliters is transferred to the corresponding calibrated test tube and subsequently titrated with the standardized hydrochloric acid solution. How will this oversight in technique affect the reported K_{sp} of borax at 48°C . . . too high, too low, or unaffected? Explain.

5. Part B.5. A "little more" than 5 mL of a saturated solution is transferred to the corresponding calibrated test tube and subsequently titrated with the standardized hydrochloric acid solution. How will this "generosity" affect the reported molar solubility of borax for that sample . . . too high, too low, or unaffected? Explain.

6. Part C.2. The saturated solution of borax is diluted with "more than" 25 mL of deionized water. How does this dilution affect the reported number of moles of $B_4O_5(OH)_4^{2-}$ in the saturated solution . . . too high, too low, or unaffected? Explain.

7. Explain why the slope of a "hand-drawn" straight line may be more representative of the data than a slope calculated from a least squares program (e.g., a trendline on Excel).

Experiment 27

Oxidation–Reduction Reactions

Zinc metal is placed into a blue copper(II) ion solution (before, left). Copper metal collects on the zinc strip and the colorless zinc(II) ion goes into solution (after, right). Zinc has a greater activity than does copper.

OBJECTIVES

- To observe and predict products of oxidation–reduction reactions
- To determine the relative reactivity of a series of metallic elements

TECHNIQUES

The following techniques are used in the Experimental Procedure

INTRODUCTION

Most chemical reactions are classified as being either acid–base (Experiment 6) or oxidation–reduction (**redox**). Reactions may also be classified as synthesis, single displacement, double displacement, or decomposition.[1]

Redox reactions are often accompanied by spectacular color changes, generally more so than what is observed in acid–base reactions. The changing of the color of the leaves each fall season, the rusting of our automobiles, the detonation of fireworks, the formation of "brown" smog, and the commercial production of copper, aluminum, and iron are all examples of redox reactions.

In an acid–base reaction, protons (H^+) are transferred; in a redox reaction, electrons (e^-) are transferred from one substance to another, resulting in changes in oxidation numbers (see Dry Lab 2A) of two or more elements in the chemical reaction.

In a redox reaction the substance that experiences an increase in oxidation number is said to be **oxidized**—to do so it must have donated or lost electrons, a process of **oxidation.** Conversely, the substance that experiences a decrease in oxidation number is said to be **reduced**—as a result it must have accepted or gained electrons, a process of **reduction.**

In all redox reactions the substance that is oxidized must lose its electrons to the substance that is reduced (or gains the electrons). Therefore the substance being oxidized is causing a reduction and therefore called a **reducing agent.** Conversely, the substance that is reduced in the reaction must gain electrons from the substance that is oxidized, thereby causing oxidation and called an **oxidizing agent** (Figure 27.1).

Electrons are never considered a reactant or product in a chemical reaction but are merely transferred—the total number of electrons donated must equal the total number of electrons gained in a redox reaction.

For example, zinc metal reacts with copper(II) ion (see opening photo):

$$Zn(s) + Cu^{2+}(aq) \rightarrow Zn^{2+}(aq) + Cu(s) \tag{27.1}$$

Oxidation: a process whereby a substance loses electrons, but increases in oxidation number

Reduction: a process whereby a substance gains electrons, but decreases in oxidation number

Reducing agent: a substance that donates electrons causing reduction of another substance to occur (the reducing agent is therefore oxidized)

Oxidizing agent: a substance that accepts electrons causing oxidation of another substance to occur (the oxidizing agent is therefore reduced)

---

[1]Synthesis and decomposition reactions were observed in Experiment 7 and double displacement reactions in Experiment 6.

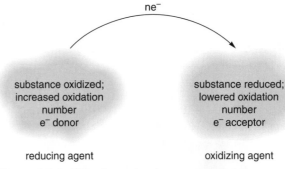

ne⁻

| substance oxidized; increased oxidation number e⁻ donor | substance reduced; lowered oxidation number e⁻ acceptor |

reducing agent   oxidizing agent

Figure 27.1 The relationship between oxidation, reduction, oxidizing agent, and reducing agent.

In the reaction, zinc has increased its oxidation number from 0 to +2, by releasing 2 moles of electrons per mole of zinc; zinc has therefore been oxidized:

$$Zn(s) \rightarrow Zn^{2+}(aq) + 2\,e^- \quad \text{an } \textit{oxidation} \text{ half-reaction} \qquad (27.2)$$

In addition, copper(II) ion has decreased in oxidation number from +2 to 0 by accepting 2 moles of electrons per mole of copper(II) ion; copper(II) ion has therefore been reduced:

$$Cu^{2+}(aq) + 2\,e^- \rightarrow Cu(s) \quad \text{a } \textit{reduction} \text{ half-reaction} \qquad (27.3)$$

An overview of the reaction suggests that the presence of zinc caused the reduction of copper(II) ion—*zinc is the reducing agent;* the presence of copper(II) ion caused the oxidation of zinc—*copper(II) ion is the oxidizing agent.* The "total" balanced redox reaction (Equation 27.1) is a sum of the two half-reactions, the oxidation half-reaction and the reduction half-reaction.

Reducing Agents and Oxidizing Agents

A common laboratory reducing agent (in addition to zinc metal) is the thiosulfate ion, $S_2O_3^{2-}$. Its half-reaction for oxidation in an aqueous solution is

$$2\,S_2O_3^{2-}(aq) \rightarrow S_4O_6^{2-}(aq) + 2\,e^- \qquad (27.4)$$

Reducing agents with their corresponding half-reactions used in this experiment are:

Cu(s, copper-colored) → Cu^{2+}(aq, blue) + 2 e⁻
3 I⁻(aq, colorless) → I_3^-(aq, purple) + 2 e⁻
Fe^{2+}(aq, light green to colorless) → Fe^{3+}(aq, red-brown) + e⁻
$C_2O_4^{2-}$(aq, colorless) → 2 CO_2(g, colorless) + 2 e⁻
2 $S_2O_3^{2-}$(aq, colorless) → $S_4O_6^{2-}$(aq, colorless) + 2 e⁻

A common laboratory oxidizing agent is the permanganate ion, MnO_4^-. Its half-reaction for reduction in an acidic solution is

$$MnO_4^-(aq) + 8\,H^+(aq) + 5\,e^- \rightarrow Mn^{2+}(aq) + 4\,H_2O(l) \qquad (27.5)$$

Oxidizing agents with their corresponding half-reactions used in this experiment are:

NO_3^-(aq, colorless) + 2 H^+(aq) + e⁻ → NO_2(g, red-brown) + H_2O(l)
NO_3^-(aq, colorless) + 4 H^+(aq) + 3 e⁻ → NO(g, colorless) + 2 H_2O(l)
MnO_4^-(aq, purple) + 8 H^+(aq) + 5 e⁻ → Mn^{2+}(aq, light pink to colorless) + 4 H_2O(l)
O_2(g, *colorless*) + 4 H^+(aq) + 4 e⁻ → 2 H_2O(l)
H_2O_2(aq, *colorless*) + 2 H^+(aq) + 2 e⁻ → 2 H_2O(l)
ClO^-(aq, colorless) + 2 H^+(aq) + 2 e⁻ → Cl^-(aq, colorless) + H_2O(l)

In Part A, several redox reactions involving these reducing and oxidizing agents are observed and studied, reactants and products are recorded. You may have to consult with your instructor when writing the balanced redox equation for the reactions that are studied.

Displacement Reactions (Activity Series)

Activity series: a listing of (generally) metals in order of decreasing chemical reactivity

Parts B and C of the experiment seek to establish the relative chemical reactivity of several metals and hydrogen. The metals and hydrogen, listed in order of decreasing activity (decreasing tendency to react), constitute an abbreviated **activity series.**

The result of one metal being placed into a solution containing the cation of another metal establishes the relative reactivity of the two metals. A displacement reaction occurs if there is evidence of a chemical change. For example, if metal A is more reactive (has a greater tendency to lose electrons to form its cation) than metal B, then metal A displaces B^{n+} from an aqueous solution. Metal A is oxidized to A^{m+} in solution and B^{n+} is reduced to metal B.

$$A(s) + B^{n+}(aq) \rightarrow A^{m+}(aq) + B(s) \qquad (27.6)$$

A specific example is the relative reactivity of iron versus lead: experimentally, when iron metal is placed in a solution containing lead(II) ion, iron is oxidized (loses two electrons) to form the iron(II) ion and the lead(II) ion is reduced (gains two electrons) to form lead metal (Figure 27.2, right). The equation for the reaction is

$$Fe(s) + Pb^{2+}(aq) \rightarrow Fe^{2+}(aq) + Pb(s) \qquad (27.7)$$

On the other hand, when lead metal is placed into a solution containing iron(II) ion, *no reaction* occurs (Figure 27.2, left):

$$Pb(s) + Fe^{2+}(aq) \rightarrow \text{no reaction} \qquad (27.8)$$

Consequently, iron prefers to be ionic, whereas lead prefers the metallic state—iron loses electrons more easily than lead. Iron, therefore, is more easily oxidized than lead and is said to have a greater **activity** (chemical reactivity) than lead.

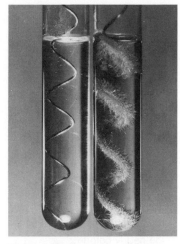

Figure 27.2 Lead metal does *not* react in a solution containing iron(II) ion (left). Iron metal does react in a solution containing lead(II) ion (right).

EXPERIMENTAL PROCEDURE

Procedure Overview: Observations of a number of redox reactions are analyzed and equations are written. The relative reactivity of several metals is determined from a series of oxidation–reduction/displacement reactions.

Perform the experiment with a partner. At each circled superscript [1-9] in the procedure, *stop,* and record your observation on the Report Sheet. Discuss your observations with your lab partner and your instructor.

A. Oxidation–Reduction Reactions

1. **Oxidation of Magnesium.** Half-fill a 20-mL beaker with deionized water and *carefully* heat to near boiling—test the water with litmus or pH paper. Grip a 2-cm piece of magnesium ribbon with crucible tongs. Heat it directly in a Bunsen burner flame until it ignites. (**Caution:** *Do not look directly at the flame.*) The white-ash product is magnesium oxide. Allow the ash to drop into the 20-mL beaker. Swirl and again test the solution with litmus or pH paper at the end of the laboratory period. Compare your observations for the two litmus or pH paper tests. Record your observations on the Report Sheet.[1]

2. **Oxidation of Copper.** In three separate small test tubes add approximately 1 mL of 6 M HCl, 1 mL of 6 M HNO$_3$, and 1 mL of conc HNO$_3$. (**Caution:** *Avoid skin contact. Flush affected areas with large amounts of water.*) Place the three test tubes in a fume hood and add a 1-cm wire strip of copper metal to each. Record your observations on the Report Sheet.[2]

3. **A Series of Redox Reactions.** Refer to Table 27.1. Clean and label eight small test tubes. Place about 1 mL of each solution listed as Solution A in Table 27.1 into the test tube. (**Caution:** *Avoid skin contact with all solutions!*) Slowly add up to 1 mL of Solution B until a permanent change is observed. For test tubes 5 and 6, after the addition of Solution B, stopper the test tube and shake vigorously. Some reactions may be slow to develop. Record your observations.[3]

Disposal: Dispose of the test solutions in the "Waste Salts" container.

CLEANUP: Rinse the test tubes twice with tap water and twice with deionized water. Discard the rinses in the "Waste Salts" container.

B. Reactions with Hydronium Ion

1. **Reactivity of Ni, Cu, Zn, Fe, Al, and Mg with H_3O^+.** Obtain 1-cm strips of Ni, Cu, Zn, Fe, Al, and Mg. Place about 1.5 mL of 6 M HCl into each of six small

Figure 27.3 The deep purple permanganate ion, added to an acidic iron(II) solution, is reduced to the nearly colorless manganese(II) ion.

Table 27.1. Preparation of Several Oxidation–Reduction Reactions

Test Tube No.	Solution A	Addition of Solution B
1	Chlorine bleach and 2 drops of 6 M HCl (the active ingredient in chlorine bleach is ClO^-)	Drops of 0.1 M Fe(NH$_4$)$_2$SO$_4$
2	Chlorine bleach and 2 drops of 6 M HCl	Add 1 drop of starch solution followed by drops of 0.1 M KI
3	0.01 M KMnO$_4$ and 2 drops of 6 M H$_2$SO$_4$	Drops of 0.1 M Fe(NH$_4$)$_2$SO$_4$ (Figure 27.3)
4	0.01 M KMnO$_4$ and 2 drops of 6 M H$_2$SO$_4$	Drops of 1 M K$_2$C$_2$O$_4$
5	Deionized water and 2 drops of 6 M H$_2$SO$_4$	Drops of 0.1 M Fe(NH$_4$)$_2$SO$_4$
6	Deionized water and 2 drops of 6 M H$_2$SO$_4$	Add 1 drop of starch solution followed by drops of 0.1 M KI
7	0.1 M H$_2$O$_2$ and 2 drops of 6 M H$_2$SO$_4$	Drops of 0.1 M Fe(NH$_4$)$_2$SO$_4$
8	0.1 M H$_2$O$_2$ and 2 drops of 6 M H$_2$SO$_4$	Add 1 drop of starch solution followed by drops of 0.1 M KI

test tubes (Figure 27.4). Polish each metal (with steel wool or sandpaper) to remove any oxide coating and immediately place it in a test tube of hydrochloric acid.[2] Allow 10–15 minutes for any reactions to occur. Record what you see—*look closely!*[4]

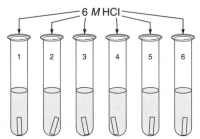

Figure 27.4 Setup for observing the reactivity of six metals in hydrochloric acid.

C. Displacement Reactions between Metals and Metal Cations

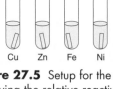

Cu   Zn   Fe   Ni

Figure 27.5 Setup for the observing the relative reactivity of metals and metal ions.

1. **Reactivity of Cu, Zn, and Fe with Ni^{2+}.** Obtain 1-cm strips of Cu, Zn, Fe, and Ni. Place about 1 mL of 0.1 M NiSO$_4$ in each of three test tubes. Place a short strip of freshly polished Cu, Zn, and Fe in successive test tubes (Figure 27.5). Observation of a tarnishing or dulling of the metal or color change of the solution indicates that a reaction has occurred. Allow 5–10 minutes to observe the reaction. Record.[5]

2. **Reactivity of Cu, Zn, Fe, and Ni with Other Cations.** Follow the procedure in Part C.2 (and correspondingly on the Report Sheet) to test the reactivity of the metals in the following 0.1 M test solutions: Cu(NO$_3$)$_2$,[6] Zn(NO$_3$)$_2$,[7] and Fe(NH$_4$)$_2$(SO$_4$)$_2$.[8] This series of tests will fill the table on the Report Sheet. The metal strip may be reused if it is unreacted after the previous test and is rinsed with deionized water or if it is again freshly polished. Record your observations.

3. **Relative Activity of the Metals.** List the four metals in order of *decreasing activity.*[9]

Disposal: Dispose of the waste test solutions and one rinse of the test tubes in the "Waste Salts" container. Place the unreacted metals in the "Waste Solids" container.

CLEANUP: Rinse the test tubes twice with tap water and twice with deionized water. Discard the rinses in the sink.

The Next Step

(1) Oxidation-reduction reactions are involved in respiration and photosynthesis, corrosion, bleaching (Experiment 29), water purification, electroplating, and so on. Research an area of interest where oxidation-reduction reactions are an integral part of the study. (2) What metals are used to inhibit the corrosion of underground (iron/steel) storage tanks, of metal structures below water in harbors?

[2]The polishing of the metal and quick placement in HCl are especially critical for aluminum and magnesium as they form quick, tough protective oxide coatings.

Oxidation–Reduction Reactions

Date _____ Lab Sec. _____ Name _____ Desk No. _____

1. a. Oxygen is a common oxidizing agent in nature. What change (increase or decrease) in the oxidation number of oxygen must occur if it is to be an oxidizing agent? Explain.

   b. If oxygen gas were to oxidize copper metal, what change (increase or decrease) in oxidation number must occur of the copper metal?

2. Zinc metal is a common reducing agent in analytical chemistry. What does it mean for a substance to be a reducing agent?

3. Each of the following processes is a likely change in a redox reaction. Label the chemical change as an oxidation process, a reduction process, or neither:

   a. $ClO_2^- \rightarrow ClO_3^-$ _____

   b. $Co^{3+} \rightarrow Co^{2+}$ _____

   c. $Cr_2O_7^{2-} \rightarrow Cr^{3+}$ _____

   d. $SO_3 \rightarrow SO_4^{2-}$ _____

   e. $H^+ \rightarrow H_2$ _____

4. Cite the part of the Experimental Procedure for three **Cautions**! in this experiment. Identify the reason or response for each caution.

5. The following equation is not balanced for both mass and charge! Explain.

$$MnO_4^-(aq) + H_2C_2O_4(aq) + 6 H^+(aq) \rightarrow Mn^{2+}(aq) + 2 CO_2(g) + 4 H_2O(l)$$

What is the correct balanced equation?

6. Experimental Procedure, Part C. List the generic chemicals M, Q, Y, and Z in order of deceasing activity on the basis of the following reactions:

$Z + Q^+ \rightarrow$ no reaction

$Y + Q^+ \rightarrow Y^+ + Q$

$M + Z^+ \rightarrow$ no reaction

$Q + M^+ \rightarrow Q^+ + M$

most active _____, _____, _____, _____, least active

*7. Along the 800-mile Alyeska (Alaska) pipeline transporting oil from Prudhoe Bay (north) to the Valdez Marine Terminal (south), zinc ribbon is buried to inhibit the corrosion of the below-ground sections of the (iron/steel) pipeline. Explain how zinc serves in this function.

Corrosion of the Alyeska (Alaskan) pipeline is inhibited by the presence of zinc metal.

Oxidation–Reduction Reactions

Date _____ Lab Sec. _____ Name _____ Desk No. _____

A. Oxidation–Reduction Reactions

1. ①**Oxidation of Magnesium.** Write a description of the reaction. What did the litmus tests reveal?

   Write a balanced equation for the reaction of magnesium in air. $\boxed{\text{Box}}$ the oxidizing agent in the equation.

2. ②**Oxidation of Copper.** Describe your observations of each test tube.

   Comment on the relative oxidizing strengths of the three acids in the test tubes.

3. ③**A Series of Redox Reactions.** On a separate sheet of paper, organize your data to record the test tube number and your observations for each mixture that shows a reaction. Write a balanced redox equation for each observed reaction: $\boxed{\text{Box}}$ the oxidizing agent in each written equation. Submit this with the completed Report Sheet.

B. Reactions with Hydronium Ion

1. ④**Reactivity of Ni, Cu, Zn, Fe, Al, and Mg with H_3O^+**

   Which metals show a definite reaction with HCl?

   Record this information on the table in Part C of the Report Sheet.

   Arrange the metals that *do react* in order of decreasing activity.

   Write a balanced equation for the reaction that occurs between Mg and H_3O^+.

C. Displacement Reactions between Metals and Metal Cations

Complete this table with NR (no reaction) or R (reaction) where appropriate. For all observed reactions, write a balanced *net ionic* equation. Use additional paper if necessary. Box the oxidizing agent in each written equation.

	Ni	Cu	Zn	Fe	Al	Mg
④HCl	_____	_____	_____	_____	_____	_____
⑤NiSO₄	NR	_____	_____	_____		
⑥Cu(NO₃)₂	_____	NR	_____	_____		
⑦Zn(NO₃)₂	_____	_____	NR	_____		
⑧Fe(NH₄)₂(SO₄)₂	_____	_____	_____	NR		

⑨List the four metals along with Al, Mg, and hydrogen in order of *decreasing activity*.

_____, _____, _____, _____, _____, _____, _____.

Balanced net ionic equations.

Laboratory Questions

Circle the questions that have been assigned.

1. Part A.1. Sodium metal is also readily oxidized by oxygen. If the product of the reaction were dissolved in water, what would be the color of the litmus for a litmus test? Explain. What is the product?

2. Part A.2. Oxygen gas has an oxidizing strength comparable to that of nitric acid. Patina is a green or greenish-blue coating that forms on copper metal in the environment. Account for its formation.

3. Part A.3, Test Tube #4. a. Does the $KMnO_4$ solution function as an oxidizing agent or a reducing agent? Explain.
   b. What was the color change of the $KMnO_4$ in the reaction?

4. Part A.3. Test Tube #7. Does the ferrous ion in the $Fe(NH_4)_2SO_4$ solutions function as an oxidizing agent or a reducing agent? Explain.

5. Part B.1. Eliseo couldn't find the 6 *M* HCl and so used 6 *M* HNO_3 for testing the metals instead. His logic? Both are strong acids. Explain how the results of the experiment would have been different.

6. Part C. Single displacement, double displacement, and decomposition reactions may all be redox reactions. Identify the type of redox reactions in Part C. Explain.

7. Part C. a. On the basis of your intuitive understanding of the chemical properties of sodium and gold, *where* in your activity series would you place sodium and gold?
   b. Will hydrochloric acid react with gold metal to produce gold(III) ions and hydrogen gas? Explain.

Experiment 28

Chemistry of Copper

The reaction of copper with nitric acid is spontaneous, producing nitrogen dioxide gas and copper(II) ion.

OBJECTIVES

- To observe the chemical properties of copper through a series of chemical reactions
- To use several separation and recovery techniques to isolate the copper compounds from solution
- To determine percent recovery of copper through a cycle of reactions

TECHNIQUES

The following techniques are used in the Experimental Procedure

INTRODUCTION

Copper is an element that is chemically combined into a variety of compounds in nature, but most commonly in the form of a sulfide, as in chalcocite, Cu_2S, and chalcopyrite, $FeCuS_2$. Copper metal is an excellent conductor of heat and electricity and is an **alloying element** in bronze and brass. Copper is a soft metal with a characteristic bright orange-brown color, which we often call "copper color" (Figure 28.1). Copper is relatively inert chemically; it does not readily air oxidize (react with oxygen in air) and is not attacked by simple inorganic acids such as sulfuric and hydrochloric acids. Copper metal that does oxidize in air is called **patina.**

Alloying element: an element of low percent composition in a mixture of metals, the result of which produces an alloy with unique, desirable properties

Copper(II) ion forms a number of very colorful compounds; most often, these compounds are blue or blue-green, although other colors are found depending on the copper(II) compound.

We will observe several chemical and physical properties of copper through a sequence of redox, precipitation, decomposition, and acid-base reactions that produce a number of colorful compounds.

Starting with metallic copper at the top, the sequence of products formed is shown on the diagram:

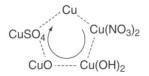

Figure 28.1 The penny is made of zinc (bottom) with a thin copper coating (top).

Dissolution of Copper Metal

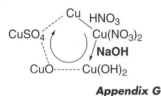

Copper reacts readily with oxidizing agents (substances that remove electrons from other substances i.e., Cu → Cu²⁺ + 2e⁻). In this experiment aqueous nitric acid, HNO_3, oxidizes copper metal to the copper(II) ion (opening photo).

$$Cu(s) + 4\,HNO_3(aq) \rightarrow Cu(NO_3)_2(aq) + 2\,NO_2(g) + 2\,H_2O(l) \qquad (28.1)$$

The products of this reaction are copper(II) nitrate, $Cu(NO_3)_2$ (a water-soluble salt that produces a blue solution), and nitrogen dioxide, NO_2 (a dense, toxic, red-brown gas). The solution remains acidic because an excess of nitric acid is used for the reaction.

Precipitation of Copper(II) Hydroxide from Solution

Appendix G

For Part B of the Experimental Procedure, the solution containing the soluble $Cu(NO_3)_2$ is treated with sodium hydroxide, NaOH, a base. Copper(II) hydroxide, $Cu(OH)_2$, a light blue solid, precipitates from the solution.

$$Cu(NO_3)_2(aq) + 2\,NaOH(aq) \rightarrow Cu(OH)_2(s) + 2\,NaNO_3(aq) \qquad (28.2)$$

Sodium nitrate, $NaNO_3$, is a colorless salt that remains dissolved in solution as $Na^+(aq)$ and $NO_3^-(aq)$. . . . The spectator ions.

The net ionic equation is

$$Cu^{2+}(aq) + 2\,OH^-(aq) \rightarrow Cu(OH)_2(s) \qquad (28.3)$$

The writing of net ionic equations and the recognition of spectator ions were presented in Experiment 6.

Conversion of Copper(II) Hydroxide to a Second Insoluble Salt

Heat applied to solid copper(II) hydroxide causes black, insoluble copper(II) oxide, CuO, to form and H_2O to vaporize.

$$Cu(OH)_2(s) \xrightarrow{\Delta} CuO(s) + H_2O(g) \qquad (28.4)$$

Dissolution of Copper(II) Oxide

Copper(II) oxide dissolves readily with the addition of aqueous sulfuric acid, H_2SO_4, forming a sky-blue solution as a result of the formation of the water-soluble salt, copper(II) sulfate, $CuSO_4$.

$$CuO(s) + H_2SO_4(aq) \rightarrow CuSO_4(aq) + H_2O(l) \qquad (28.5)$$

The net ionic equation is

$$CuO(s) + 2\,H^+(aq) \rightarrow Cu^{2+}(aq) + H_2O(l) \qquad (28.6)$$

Reformation of Copper Metal

Finally, the addition of magnesium metal, Mg, to the copper(II) sulfate solution completes the copper cycle.

In this reaction, magnesium serves as a **reducing agent** (a substance that provides electrons to another substance i.e., $Cu^{2+} + 2e^- \rightarrow Cu$). Magnesium, being a more active metal than copper, reduces copper(II) ion from the copper(II) sulfate solution to copper metal and water-soluble magnesium sulfate, $MgSO_4$.

$$CuSO_4(aq) + Mg(s) \rightarrow Cu(s) + MgSO_4(aq) \qquad (28.7)$$

The net ionic equation is

$$Cu^{2+}(aq) + Mg(s) \rightarrow Cu(s) + Mg^{2+}(aq) \qquad (28.8)$$

Magnesium metal also reacts with sulfuric acid. Therefore when magnesium metal is added to the acidic copper(II) sulfate solution, a second reaction occurs that produces hydrogen gas, H_2, and additional magnesium sulfate.

$$Mg(s) + H_2SO_4(aq) \rightarrow H_2(g) + MgSO_4(aq) \qquad (28.9)$$

Therefore hydrogen gas bubbles are observed during this reaction step of the cycle. This reaction also removes any excess magnesium metal that remains after the copper metal has been recovered.

Procedure Overview: Copper metal is successively treated with nitric acid, sodium hydroxide, heat, sulfuric acid, and magnesium in a cycle of chemical reactions to regenerate the copper metal. The chemical properties of copper are observed in the cycle and the percent recovery is determined. Three trials are recommended.

You will need to obtain an instructor approval after each step in the Experimental Procedure. Perform the experiment with a partner. At each circled superscript^{①–③} in the procedure, *stop,* and record your observation on the Report Sheet. Discuss your observations with your lab partner and your instructor.

A. Copper Metal to Copper(II) Nitrate

Perform the series of reactions in a test tube that is compatible with your laboratory centrifuge. Consult with your laboratory instructor.

1. **Prepare the Copper Metal Sample.** Obtain a *less than* 0.02-g sample of Cu wool. Measure the mass (± 0.001 g) of the selected test tube.[1] Roll and place the Cu wool into the test tube, measure and record the mass of the test tube and copper sample.

2. **Reaction of the Copper Metal.** Hold the test tube with a test tube clamp for the remainder of the experiment, do *not* use your fingers.

Perform this step in the fume hood because of the evolution of toxic $NO_2(g)$. (**Caution:** *Do not inhale the evolved nitrogen dioxide gas.*) Using a dropper bottle or a dropper pipet, add drops (≤ 10 drops) of *conc* HNO_3 to the copper sample until no further evidence of a chemical reaction is observed. Do not add an excess! (**Caution:** *Concentrated* HNO_3 *is very corrosive. Do not allow it to touch the skin. If it does, wash immediately with excess water. Nitric acid will turn your skin yellow, a way to check your laboratory technique!*)

At this point the Cu metal has completely reacted. What is the color of the gas? Add 10 drops of deionized water. Show the resulting solution to your laboratory instructor for approval^① and save the solution for Part B.

B. Copper(II) Nitrate to Copper(II) Hydroxide

1. **Preparation of Copper(II) Hydroxide.** Agitate or continuously stir with a stirring rod the solution from Part A.2 while slowly adding 10 drops of 6 *M* NaOH. (**Caution:** *Wash with water immediately if the* NaOH *comes into contact with the skin.*) This forms the $Cu(OH)_2$ precipitate. After the first 10 drops are added, add 10 more drops of 6 *M* NaOH. Using a wash bottle and deionized water, rinse the stirring rod, allowing the rinse water to go into the test tube. Centrifuge the solution for 30 seconds (ask your instructor for instructions in operating the centrifuge).

2. **A Complete Precipitation.** Test for a complete precipitation of $Cu(OH)_2$ by adding 2–3 more drops of 6 *M* NaOH to the **supernatant.** If additional precipitate forms, add 4–5 more drops and again centrifuge. Repeat the test until no further formation of the $Cu(OH)_2$ occurs. The solution should appear colorless and the precipitate should be light blue. Obtain your laboratory instructor's approval^② and save for Part C.

Supernatant: the clear solution in the test tube

---

[1] Consult with your instructor.

C. Copper(II) Hydroxide to Copper(II) Oxide

Cool flame: an adjusted Bunsen flame having a low supply of fuel

1. **Heat the Sample.** Decant (pour off) and discard the supernatant from the test tube. *Carefully* (***very carefully!***) and slowly heat the test tube with a **cool flame**[2] until the $Cu(OH)_2$ precipitate changes color. **Read footnote 2!** You need *not* heat the contents to dryness. Avoid ejection (and projection) of your copper compound by *not* holding the test tube over the direct flame for a prolonged period of time. If the contents of the test tube are ejected, you will need to restart the Experimental Procedure at Part A. Obtain your instructor's approval[3] and save for Part D.

D. Copper(II) Oxide to Copper(II) Sulfate

1. **Dissolution of Copper(II) Oxide.** To the solid CuO in the test tube from Part C, add drops ($\leq$20 drops, 1 mL) of 6 M H_2SO_4 with agitation until the CuO dissolves. (**Caution:** *Do not let sulfuric acid touch the skin!*) (Slight heating *may* be necessary, but be careful not to eject the contents!) The solution's sky-blue appearance is evidence of the presence of soluble $CuSO_4$. Obtain your instructor's approval[4] and save for Part E.

E. Copper(II) Sulfate to Copper Metal

1. **Formation of Copper Metal.** Using fine steel wool (or sandpaper), polish about 5–7 cm of Mg ribbon. Cut or tear the ribbon into 1-cm lengths. Dilute the solution from Part D with deionized water until the test tube is half-full. Add a 1-cm Mg strip to the solution. When the Mg strip has reacted (disappeared), add a second Mg strip and so on until the "blue" has disappeared from the solution. Describe what is happening. What is the coating on the magnesium ribbon? What is the gas?[5]

    If a white, milky precipitate forms [from the formation of magnesium hydroxide, $Mg(OH)_2$], add several drops of 6 M H_2SO_4. (**Caution:** *Avoid skin contact.*) Break up the red-brown Cu coating on the Mg ribbon with a stirring rod. *After* breaking up the Cu metal *and* after adding several pieces of Mg ribbon, centrifuge the mixture.

2. **Washing.** Add drops of 6 M H_2SO_4 to dissolve any excess Mg ribbon. (**Caution:** *Avoid skin contact.*) Do this by breaking up the Cu metal with a stirring rod to expose the Mg ribbon, coated with Cu metal, to the H_2SO_4 solution. Centrifuge for 30 seconds, decant, and discard the supernatant. Be careful to keep the Cu metal in the test tube. Wash the red-brown Cu metal with three 1-mL portions of deionized water.[3] Rinse the stirring rod in the test tube. Centrifuge, decant, and discard each washing.

  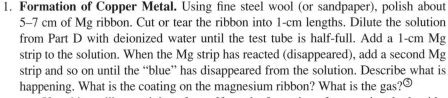

3. **Determination of the Mass of Recovered Copper.** *Very gently* dry the Cu in the test tube over a *cool* flame (***see footnote 2***). Allow the tube and contents to cool and determine the mass ($\pm$0.001 g). Repeat the heating procedure until a reproducibility in mass of $\pm$1% is obtained. Record the mass of Cu recovered in the experiment.

> *Disposal:* All solutions used in the procedure can be disposed of in the "Waste Salts" container. Dispose of the copper metal in the "Waste Solids" container. Check with your instructor.

CLEANUP: Rinse all glassware twice with tap water and twice with deionized water. Discard all rinses in the sink.

The Next Step

Many alloys other than coinage alloys have varying amounts of copper, e.g., brass. What are some other methods for determining the amount of copper in a sample—gravimetrically, spectrophotometrically (see Experiment 35), or volumetrically (see Experiments 23, 24, or 29—also copper(II) ion reacts with iodide ion to produce I_3^-)?

---

[2]From Technique 13C, "If you can feel the heat of the flame with the hand holding the test tube clamp, the flame is too hot!"

[3]Wash the Cu metal by adding water, stirring the mixture with a stirring rod, and allowing the mixture to settle.

Date _____ Lab Sec. _____ Name _____ Desk No. _____

1. Match each of the following equations appearing in the Introduction with the reaction types (more than one reaction type may be appropriate for each equation):

Equation matches with	Reaction Types
28.1 _____	1. acid-base
28.2 _____	2. single displacement
28.4 _____	3. double displacement
28.5 _____	4. oxidation-reduction (redox)
28.7 _____	5. decomposition
28.9 _____	6. precipitation

2. Copper forms many different compounds in this experiment.
   a. Identify the *oxidizing agent* in the conversion of copper metal to copper(II) ion.

   b. Identify the *oxidizing agent* in the conversion of copper(II) ion to copper metal.

   c. Classify the type of reaction for the conversion of copper(II) oxide to copper(II) sulfate.

3. A number of **"Caution"** chemicals and solutions are used in this experiment. Refer to the Experimental Procedure and the corresponding sections to identify the specific chemical or solution that must be handled with care.

Experimental Procedure	Precautionary Chemical/Solution
Part A.2	
Part A.2	
Part B.1	
Part D.1	
Part E.1	

4. Experimental Procedure, Parts C.1 and E.3. Extreme caution *must* be observed when heating a solution in a test tube.
    a. What criterion indicates that you are heating the solution in the test tube with a "cool flame"?

    b. At what angle should the test tube be held while moving the test tube circularly in and out of the cool flame?

    c. What is the consequence of not using this technique properly?

5. a. What function does a centrifuge perform?

    b. Describe the technique for "balancing a centrifuge" when centrifuging a sample.

6. A 0.0226 g sample of copper metal is recycled through the series of reactions. If 0.0214 g of copper is then recovered after the series of reactions in this experiment, what is the percent recovery of the copper metal?

*7. What volume, in drops, of 16 M (*conc*) HNO_3 is required to react with 0.0214 g of Cu metal? See Equation 28.1. Assume 20 drops per milliliter.

Chemistry of Copper

Date _____ Lab Sec. _____ Name _____ Desk No. _____

Data for Copper Cycle	*Trial 1*	*Trial 2*	*Trial 3*
1. Mass of test tube (g)	_____	_____	_____
2. Mass of test tube (g) + copper (g)	_____	_____	_____
3. Mass of copper sample (g)	_____	_____	_____

Synthesis of	**Observation**	**Instructor Approval**	**Balanced Equation**
①A. $Cu(NO_3)_2$ (aq)	_____	_____	_____
②B. $Cu(OH)_2$ (s)	_____	_____	_____
③C. CuO (s)	_____	_____	_____
④D. $CuSO_4$ (aq)	_____	_____	_____

⑤E. Copper(II) sulfate to copper metal. Write a full description of the reactions that occurred. Include a balanced equation in your discussion.

4. Mass of test tube and recovered copper (g)   1st mass (g)   _____   _____   _____

    2nd mass (g)   _____   _____   _____

    3rd mass (g)   _____   _____   _____

5. Final mass of Cu recovered (g)   _____   _____   _____

6. Percent recovery, $\dfrac{\text{mass Cu (recovered)}}{\text{mass Cu (original)}} \times 100$

7. Average percent recovery

8. Account for the percent recovery being equal to or less than 100%.

Laboratory Questions

Circle the questions that have been assigned.

1. Part A.2 What is the formula *and* the color of the gas that is evolved?

2. Part B.1. When the NaOH solution is added, $Cu(OH)_2$ does not precipitate immediately. What else present in the reaction mixture from Part A reacts with the NaOH before the copper (II) ion? Explain.

3. Part C.1. The sample in Part B was *not* centrifuged. Why? Perhaps the student chemist had to be across campus for another appointment. Because of the student's "other priorities" how will this
   a. affect the experimental procedure for Part C? Explain.
   b. affect the percent recovery of copper in the experiment? Explain.

4. Part D.1. All of the CuO does *not* react with the sulfuric acid. Will the reported percent recovery of copper in the experiment be too high or too low? Explain.

5. Part E. Sulfuric acid has a dual role in the chemistry. What are its two roles in the recovery of the copper metal?

6. Part E.2. Jacob couldn't find the 6 *M* H_2SO_4, so instead substituted the 6 *M* HNO_3 that was available. What change was most likely observed as a result of this decision? Explain.

7. Part E.3. The percent recovery of copper metal is calculated to be greater than 100%! Identify two reasons why this might be so.

Experiment 29

Bleach Analysis

Sodium hypochlorite, the bleaching agent of bleach, oxidizes the dye and "whitens" the fabric.

OBJECTIVES

- To determine the "available chlorine" in commercial bleaching agents
- To prepare a standardized sodium thiosulfate solution

TECHNIQUES

The following techniques are used in the Experimental Procedure

INTRODUCTION

Natural fibers, such as cotton and wool, and paper products tend to retain some of their original color during the manufacturing process. For example, brown paper is more "natural" than white paper, and original cotton fibers tend *not* to be "white."

Natural dyes and, in general, all of nature's colors exist because of the presence of molecules with multiple bonds. Similarly, stains and synthetic colors are due to molecules with one or more double and/or triple bonds. The electrons of the multiple bond readily absorb visible radiation, reflecting (or transmitting) the colors of the visible radiation that are *not* absorbed.[1] These electrons of the double/triple bonds are readily removed by oxidizing agents. Because the oxidized form[2] of the molecules causing the colors and stains no longer absorb visible radiation, the fabric or paper appears whiter.

All of the halogens and solutions of the oxyhalides are excellent oxidizing agents. Chlorine, as an **oxidant,** is an excellent bactericide for use in swimming pools, drinking water, and sewage treatment plants.

Most commercial **bleaching agents** contain the hypochlorite ion, ClO^-. The ClO^- ion removes light-absorbing electrons from the multiple bonds of molecules. Therefore, the absorption of visible radiation does not occur; all visible light is reflected (or transmitted), giving the object a "white" appearance.

$$ClO^-(aq) + H_2O(l) + 2\,e^- \rightarrow Cl^-(aq) + 2\,OH^-(aq) \qquad (29.1)$$

The textile and paper industries, as well as home and commercial laundries, use large quantities of hypochlorite solutions to oxidize the color- and stain-causing molecules. The hypochlorite ion is generally in the form of a sodium or a calcium salt in the bleaching agent.

Oxidant: oxidizing agent, a substance that causes the oxidation of another substance

Bleaching agent: an oxidizing agent

---

[1] In Dry Lab 3, the explanation of light absorption in the visible region is discussed in terms of an absorption of energy (from the visible region) by electrons.
[2] The stains or impurities may be oxidized to form several products.

Strengths of Bleaching Agents

Mass ratio: mass of Cl_2/mass of sample (solution or solid)

Electrolysis and electrolytic cells are studied in Experiment 33

Industry expresses the strength of the bleaching agents as the mass of molecular *chlorine*, Cl_2, per unit mass of solution (or mass of powder) regardless of the actual chemical form of the oxidizing agent. This **mass ratio,** expressed as a percent, is called the percent "available chlorine" in the bleach solution.

As an example, "ultra" liquid laundry bleach is generally a 6.00% NaClO aqueous solution by mass. Liquid laundry bleach is manufactured by the **electrolysis** of a cold, stirred, concentrated sodium chloride solution. The chlorine gas and the hydroxide ion that are generated at the electrodes in the electrolysis cell combine to form the hypochlorite ion, ClO^-, and in the presence of Na^+, a sodium hypochlorite solution.

$$Cl_2(aq) + 2\,OH^-(aq) \rightarrow ClO^-(aq) + Cl^-(aq) + H_2O(l) \qquad (29.2)$$

The percent "available chlorine" in a 6.00% NaClO solution is calculated as

$$\frac{\text{grams } Cl_2}{\text{grams soln}} \times 100 = \frac{6.00 \text{ g NaClO}}{100 \text{ g soln}} \times \frac{1 \text{ mol NaClO}}{74.44 \text{ g NaClO}} \times \frac{1 \text{ mol } Cl_2}{1 \text{ mol NaClO}}$$
$$\times \frac{70.90 \text{ g } Cl_2}{\text{mol } Cl_2} \times 100 = 5.71\% \ Cl_2$$

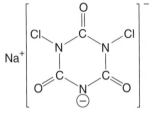

sodium dichloro-*s*-triazinetrione

Solid bleaching powders and water disinfectants (e.g., for home swimming pools or hot tubs) contain a variety of compounds that serve as oxidizing agents. Solids such as $Ca(ClO)Cl/Ca(ClO)_2$ mixtures and sodium dichloro-*s*-triazinetrione, commonly found in household cleansers and water disinfectants, are substances that release ClO^- in solution. Solid household bleaches often contain sodium perborate, $NaBO_2 \cdot H_2O_2 \cdot 3H_2O$, in which the bleaching agent is hydrogen peroxide, a compound that is also found in hair bleaches.

Analysis of Bleach with Potassium Iodide and Sodium Thiosulfate Solutions

$3\,I^-(aq) \rightarrow I_3^-(aq) + 2\,e^-$

thiosulfate ion
Stoichiometric point: point at which the mole ratio of reactants in a titration is the same as the mole ratio in the balanced equation

The oxidizing property of the hypochlorite ion is used to perform an analysis for the amount of "available chlorine" in a bleach sample for this experiment. In the analysis, the hypochlorite ion oxidizes iodide ion to molecular iodine—the hypochlorite ion is the limiting reactant in the analysis. In the presence of an excess amount of iodide ion, the molecular iodine actually exists as a water-soluble triiodide complex, I_3^- (or $[I_2 \cdot I]^-$).

$$ClO^-(aq) + 3\,I^-(aq) + H_2O(l) \rightarrow I_3^-(aq) + Cl^-(aq) + 2\,OH^-(aq) \qquad (29.3)$$

The amount of I_3^- produced in the reaction is determined by its subsequent reaction with a standard sodium thiosulfate, $Na_2S_2O_3$, solution.

$$I_3^-(aq) + 2\,S_2O_3^{2-}(aq) \rightarrow 3\,I^-(aq) + S_4O_6^{2-}(aq) \qquad (29.4)$$

The **stoichiometric point** for this redox titration is detected by adding a small amount of starch. Starch forms a rather strong bond to I_3^-; whereas I_3^- has a yellow-brown color, the $I_3^- \cdot$ starch complex is a deep, and very intense, blue. Just prior to the disappearance of the yellow-brown color of the I_3^- ion in the titration, the starch is added to intensify the presence of the I_3^- still to be titrated (Figure 29.1).

$$I_3^-(aq) + \text{starch}(aq) \rightarrow I_3^- \cdot \text{starch}(aq, \text{deep-blue}) \qquad (29.5)$$

Titrant: the solution being added from the buret

The addition of the sodium thiosulfate **titrant** is then resumed until the deep-blue color of the $I_3^- \cdot$ starch complex disappears.

$$I_3^- \cdot \text{starch}(aq) + 2\,S_2O_3^{2-}(aq) \rightarrow 3\,I^-(aq) + S_4O_6^{2-}(aq) + \text{starch}(aq, \text{colorless}) \ (29.6)$$

 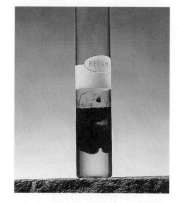

Figure 29.1 I_3^- without starch added (left); I_3^- with starch added (right).

From the stoichiometries of Equations 29.4 (2 mol $S_2O_3^{2-}$ to 1 mol I_3^-), 29.3 (1 mol I_3^- from 1 mol ClO^-), and 29.2 (1 mol ClO^- from 1 mol Cl_2) in that sequence, the moles of "available chlorine" in the sample can be calculated. From the molar mass of chlorine and the mass of the sample used for the analysis, the percent "available chlorine" is calculated.

Note that the chemical analysis of a bleach solution follows the same chemistry as was applied to the chemical systems in Experiments 23 (Part F) and 24 where the oxidizing agents were $HIO_3(aq)$ and $H_2O_2(aq)$, respectively and again in Experiment 31 when the oxidizing agent is oxygen.

$S_4O_6^{2-}$, tetrathionate ion

A Standard Solution of Sodium Thiosulfate, $Na_2S_2O_3$

The concentration of a standard $Na_2S_2O_3$ solution (used in Equation 29.4)[3] is determined by its reaction with solid potassium iodate, KIO_3, a **primary standard** (Figure 29.2). A measured mass of potassium iodate is dissolved in an acidic solution containing an excess of potassium iodide, KI. Potassium iodate, a strong oxidizing agent, oxidizes I^- to the water-soluble, yellow-brown $I_3^-(aq)$ in solution:

$$IO_3^-(aq) + 8\,I^-(aq) + 6\,H^+(aq) \rightarrow 3\,I_3^-(aq) + 3\,H_2O(l) \qquad (29.7)$$

The I_3^- is then titrated with a prepared sodium thiosulfate solution, where the I_3^- is reduced to I^- and the thiosulfate ion, $S_2O_3^{2-}$, is oxidized to the tetrathionate ion, $S_4O_6^{2-}$ (Equation 29.4).

The stoichiometric point is detected using a starch solution. Just prior to the disappearance of the yellow-brown iodine in solution, starch is added to form a deep-blue complex with iodine (Equation 29.5).

The addition of the thiosulfate solution is then continued until the remaining I_3^- is converted to I^- and the deep-blue color disappears (Equation 29.6).

Note that the reaction of IO_3^- with I^- (Equation 29.7) for the standardization of the sodium thiosulfate solution substitutes for the reaction of ClO^- with I^- (Equation 29.3) in the analysis of the bleaching agent.

Primary standard: a substance that has a known high degree of purity, a relatively large molar mass, is nonhygroscopic, and reacts in a predictable way

Figure 29.2 Potassium iodate has the necessary properties of a primary standard.

Procedure Overview: A solution of $Na_2S_2O_3$ is prepared and standardized using solid KIO_3 as a primary standard. A bleach sample is mixed with an excess of iodide ion, generating I_3^-. The amount of I_3^- generated is analyzed by titration with the standard $Na_2S_2O_3$ solution.

Quantitative, reproducible data are the objective of this experiment. Practice good laboratory techniques. Three trials are necessary. Ask your instructor about the completion of both Parts C and D in this experiment. Nearly 800 mL of boiled, deionized

EXPERIMENTAL PROCEDURE

---

[3]The $Na_2S_2O_3$ solution is a secondary standard solution in that its concentration can only be determined from an analysis; it cannot be prepared directly from solid $Na_2S_2O_3$.

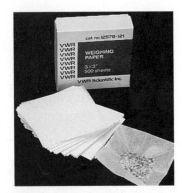

Figure 29.3 Weighing paper for measuring the KIO₃ primary standard.

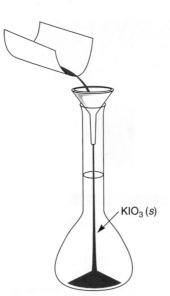

KIO₃ (s)

Figure 29.4 Transferring a solid reagent to a volumetric flask (left) and diluting to volume (right).

water cooled to room temperature is required for this experiment. Begin preparing this at the beginning of the laboratory period.

A. Preparation of the KIO₃ Primary Standard

1. **Prepare a Primary Standard Solution.** Use weighing paper (Figure 29.3) to measure the mass (± 0.001 g) of KIO₃ (previously dried[4] at 110°C) that is required to prepare 100 mL of 0.01 M KIO₃. *Show this calculation on the Report Sheet.* Dissolve the solid KIO₃ and dilute to volume with the freshly boiled, deionized water in a 100-mL volumetric flask (Figure 29.4). Calculate and record its exact molar concentration.

B. Standard 0.1 M Na₂S₂O₃ Solution

1. **Prepare a Na₂S₂O₃ Solution.**[5] This solution should be prepared no more than 1 week in advance because unavoidable decomposition occurs. Measure the mass (± 0.01 g) of Na₂S₂O₃·5H₂O needed to prepare 250 mL of a 0.1 M Na₂S₂O₃ solution. *Show this calculation on the Report Sheet.* Dissolve the solid Na₂S₂O₃·5H₂O with the freshly boiled, deionized water and dilute to 250 mL; an Erlenmeyer flask can be used to prepare and store the solution. Agitate until the salt dissolves.

2. **Prepare a Buret.** Prepare a *clean,* 50-mL buret for titration. Rinse the clean buret with two or three 5-mL portions of the Na₂S₂O₃ solution, draining each rinse through the buret tip. Fill the buret with the Na₂S₂O₃ solution, drain the tip of air bubbles, and after 10–15 seconds, record the volume "using all certain digits (from the labeled calibration marks on the buret) *plus* one uncertain digit (the last digit which is the best estimate between the calibration marks)."

3. **Titrate the KIO₃ Solution.** Pipet 25 mL of the standard KIO₃ solution (from Part A) into a 125-mL Erlenmeyer flask, add ~2 g (± 0.01 g) of solid KI, and swirl to dissolve. Add ~10 mL of 0.5 M H₂SO₄. Begin titrating the KIO₃ solution immediately. When the yellow-brown I₃⁻ nearly disappears (pale yellow color appears),[6]

---

[4]The KIO₃ should have been dried prior to your coming to the laboratory.
[5]Ask your instructor about this solution; it may have already been prepared.
[6]Consult your instructor at this point.

add 2 mL of starch solution.[7] Stirring constantly, continue titrating *slowly*[8] until the deep-blue color completely (and suddenly) disappears.

4. **Repeat the Titration with Another KIO₃ Sample.** Repeat twice the procedure in Part B.3 by rapidly adding the $Na_2S_2O_3$ titrant until ~1 mL before the yellow-brown color of I_3^- disappears. Add the starch solution and continue titrating until the solution is colorless. Three trials for the standardization of $Na_2S_2O_3$ should yield molar concentrations that agree to within ±1%.

5. **Calculate Concentration of Na₂S₂O₃.** Calculate the molar concentration of the (now) standardized $Na_2S_2O_3$ solution.

Disposal: Dispose of the test solutions as directed by your instructor.

6. **Prepare the Buret for Bleach Analysis.** Prepare the 50-mL buret with the $Na_2S_2O_3$ solution for titration as described in Part B.2. This is to be ready for immediate use in Part E. Ten to fifteen seconds after filling the buret, read and record (Report Sheet, E.1) the volume of $Na_2S_2O_3$ solution.

C. Preparation of Liquid Bleach for Analysis

1. **Obtain Bleach Samples.** Two trials for each bleach are suggested. Obtain about 12 mL of each liquid bleach from your instructor.

2. **Prepare the Bleach Solution for Analysis.** Pipet 10 mL of bleach into a 100-mL volumetric flask and dilute to the 100-mL mark with the boiled, deionized water. Mix the solution thoroughly. Pipet 25 mL of this diluted bleach solution into a 250-mL Erlenmeyer flask; add 20 mL of deionized water, 2 g of KI, and 10 mL of 0.5 M H_2SO_4. A yellow-brown color indicates the presence of I_3^- that has been generated by the ClO^- ion. Proceed *immediately* to Part E.

D. Preparation of Solid Bleach for Analysis

1. **Obtain Bleach Samples.** Two trials for each bleach are suggested. Obtain about 5 g of each powdered bleach from your instructor.

2. **Prepare a Solution from the Powdered Bleach.** Place about 5 g of a powdered bleach sample into a mortar and grind (Figure 29.5). Measure the mass (±0.001 g) of the finely pulverized sample on weighing paper and transfer it to a 100-mL volumetric flask fitted with a funnel. Dilute the sample to 100 mL with the boiled, deionized water. Mix thoroughly. If an abrasive is present in the powder, the sample may *not* dissolve completely. Allow the insolubles to settle.

Figure 29.5 A mortar and pestle is used to grind the powder.

3. **Prepare the Bleach Solution for Analysis.** Pipet 25 mL of this bleach solution (be careful to *not* draw any solids into the pipet!) into a 250-mL Erlenmeyer flask and add 20 mL of deionized water, 2 g of KI, and 10 mL of 0.5 M H_2SO_4. A yellow-brown color indicates the presence of I_3^- that has been generated by the ClO^- ion. Proceed *immediately* to Part E.

---

[7] If the deep-blue color of I_3^-•starch does *not* appear, you have passed the stoichiometric point in the titration and you will need to discard the sample and prepare another.
[8] The I_3^-•starch complex is slow to dissociate.

E. Analysis of Bleach Samples

1. **Titrate the Bleach Sample.** *Immediately* titrate the liberated I_3^- with the standardized $Na_2S_2O_3$, prepared in Part B.6, until the yellow-brown I_3^- color is almost discharged.

2. **Perform the Final Steps in the Analysis.** Add 2 mL of starch solution to form the deep-blue I_3^-•starch complex.[9] While swirling the flask, continue titrating *slowly* until the deep-blue color disappears. Refer to the titration procedure in Part B.3. Record the final volume of the titrant in the buret.

3. **Repeat with Another Sample.** Repeat the experiment with a second sample of the same bleach.

4. **Repeat the Analysis for Another Bleach.** Analyze another bleaching agent (liquid, Part C, or solid, Part D) to determine the amount of "available chlorine." Compare the available chlorine content of the two bleaches.

Disposal: Dispose of the KIO_3 solution and the remaining bleach solutions in the "Waste Oxidants" container. Dispose of the $Na_2S_2O_3$ solution in the buret and the Erlenmeyer flask in the "Waste Thiosulfate" container. Dispose of the sulfuric acid in the "Waste Acids" container. Dispose of all test solutions as directed by your instructor.

CLEANUP: Flush the buret twice with tap water and twice with deionized water. Repeat the cleaning operation with the pipets. Discard all washings as directed by your instructor. Clean up the balance area, discarding any solids in the "Waste Solids" container.

The Next Step

What other oxidants are used as bleaches or as bactericides in the treatment of water? Are they more or less effective than hypochlorite bleach? For example, is this procedure applicable for the analysis for sodium perborate in powdered bleaches or the disinfectants used in home and public swimming pools? Devise a procedure for the analysis of a series of commercial or industrial oxidants.

---

[9]If you are analyzing a powdered bleach, the 2 mL of starch may be added immediately after the $0.5\ M\ H_2SO_4$ in Part D.3.

Bleach Analysis

Date _____ Lab Sec. _____ Name _____ Desk No. _____

A. Preparation of the KIO$_3$ Primary Standard

Calculate the mass of KIO$_3$ (molar mass = 214.02 g/mol) required for Part A.1.

1. Tared mass of KIO$_3$ (g) _____

2. Moles of KIO$_3$ (*mol*)

3. Molar concentration of KIO$_3$ solution (*mol/L*)

B. Standard 0.1 M Na$_2$S$_2$O$_3$ Solution

Calculate the mass of Na$_2$S$_2$O$_3$•5H$_2$O (molar mass = 248.19 g/mol) required for Part B.1.

	Trial 1	Trial 2	Trial 3
1. Volume of KIO$_3$ solution (*mL*)	25.0	25.0	25.0
2. Moles of KIO$_3$ titrated (*mol*)			
3. Buret reading, *initial* (*mL*)			
4. Buret reading, *final* (*mL*)			
5. Volume of Na$_2$S$_2$O$_3$ added (*mL*)			
6. Moles of Na$_2$S$_2$O$_3$ added (*mol*)			
7. Molar concentration of Na$_2$S$_2$O$_3$ (*mol/L*)			
8. Average molar concentration of Na$_2$S$_2$O$_3$ (*mol/L*)			
9. Standard deviation			
10. Relative standard deviation (%RSD)			

C. Preparation of Liquid Bleach for Analysis

	Trial 1	Trial 2	Trial 1	Trial 2
Sample Name or Description				
1. Volume of original bleach titrated (*mL*)	2.5	2.5	2.5	2.5
2. Mass of bleach titrated (g) (assume the density of liquid bleach is 1.084 g/mL)				

D.   Preparation of Solid Bleach for Analysis

	Trial 1	Trial 2	Trial 1	Trial 2
Sample Name or Description	_____	_____	_____	_____
1. Tared mass of bleach (g)	_____	_____	_____	_____

E.   Analysis of Bleach Samples

	Trial 1	Trial 2	Trial 1	Trial 2
1. Buret reading, *initial* (mL)	_____	_____	_____	_____
2. Buret reading, *final* (mL)	_____	_____	_____	_____
3. Volume of $Na_2S_2O_3$ added (mL)				

Data Analysis

	Trial 1	Trial 2	Trial 1	Trial 2
1. Moles of $Na_2S_2O_3$ added (mol)				
2. Moles of ClO^- reacted (mol)				
3. Moles of "available chlorine" (mol)				
4. Mass of "available chlorine" (g)				
5. Mass percent "available chlorine" (**liquid**) (assume the density of liquid bleach is 1.084 g/mL) (%)				
6. Average mass percent "available chlorine" (**liquid**) (%)				
7. Mass percent "available chlorine" (**solid**) (%)				
8. Average mass percent "available chlorine" (**solid**) (%)				

Laboratory Questions

Circle the questions that have been assigned.

1. Part A.1. The solid KIO_3 is diluted shy of the 100-mL mark in the volumetric flask.
   a. How does this technique error affect the molar concentration of the KIO_3 solution?
   b. Will the reported molar concentration of the standardized $Na_2S_2O_3$ solution be too high, too low, or unaffected? Explain.

2. Part B.1. In preparing the sodium thiosulfate solution, the solid $Na_2S_2O_3 \cdot 5H_2O$ is not dried (as was the KIO_3 in Part A.1). Will this oversight cause the reported molar concentration of the standard $Na_2S_2O_3$ solution to be too high, too low, or unaffected? Explain.

3. Part B.3. Deionized water from a wash bottle was added to wash down the wall of the receiving flask during the titration. How does this affect the reported molar concentration of the sodium thiosulfate solution? Explain.

4. Part B.3. The KIO_3 primary standard solution can be titrated without the presence of the starch solution. Explain.

5. Part C.1. The bleach is diluted but beyond the 100-mL mark in the volumetric flask. Will the percent "available chlorine" calculated after the analysis in Part E be reported as being too high or too low? Explain.

6. Part C.2 (or D.3). The starch solution is added immediately after the sulfuric acid. What is observed? Because the $I_3^- \cdot$starch complex is slow to dissociate, what difficulties might be encountered in Part E.2 if the starch were added too early in the analysis? Explain.

7. Part E.2. The endpoint of the titration is surpassed. How will this technique error affect the reported percent "available chlorine" in the bleach sample?

8. Part E.2. An air bubble initially trapped in the tip of the buret is released during the titration. Will the reported percent "available chlorine" in the bleach sample be reported too high or too low? Explain.

Experiment 30

Vitamin C Analysis

Vitamin C is avilable in tablets or from natural sources such as citrus fruits.

- To determine the amount of vitamin C in a vitamin tablet, a fresh fruit, or a fresh vegetable sample **OBJECTIVE**

The following techniques are used in the Experimental Procedure **TECHNIQUES**

PRINCIPLES

The human body does not synthesize vitamins; therefore the vitamins that we need for catalyzing specific biochemical reactions are gained only from the food we eat. Vitamin C can be obtained from a variety of fresh fruits (most notably, citrus fruits) and vegetables.

Vitamin C plays a vital role for the proper growth and development of teeth, bones, gums, cartilage, skin, and blood vessels. It is an important anti-oxidant (reducing agent) that helps protect against cancers, heart disease, and stress. Vitamin C is also critical in enabling the body to absorb iron and folic acid.

The recommended daily allowance (RDA) of vitamin C for an adult is 75–90 mg, however levels as high as 200 mg are proven to be beneficial. Table 30.1 lists the concentration ranges of vitamin C for various fruits and vegetables (Figure 30.1).

Vitamin C, also called **ascorbic acid**, is one of the more abundant and easily obtained vitamins in nature. It is a colorless, water-soluble acid that, in addition to its acidic properties, is a powerful biochemical reducing agent, meaning it readily undergoes oxidation, even from the oxygen of the air.

Even though ascorbic acid is an acid, its *reducing* properties are used in this experiment to analyze its concentration in various samples. There are many other acids present in foods (e.g., citric acid) that could interfere with an acid analysis and

Figure 30.1 Vegetables are a good supply of vitamin C

Table 30.1 Vitamin C in Foods

<10 mg/100 g	beets, carrots, eggs, milk
10–25 mg/100 g	asparagus, cranberries, cucumbers, green peas, lettuce, pineapple
25–100 mg/100 g	Brussels sprouts, citrus fruits, tomatoes, spinach
100–350 mg/100 g	chili peppers, sweet peppers, turnip, greens, kiwi

not permit a selective determination of the ascorbic acid content. The equation for the oxidation of ascorbic acid is

$$\text{or } C_6H_8O_6(aq) + H_2O(l) \rightarrow C_6H_8O_7(aq) + 2\,H^+(aq) + 2\,e^-$$

Analysis of Vitamin C

A sample containing ascorbic acid is prepared as the analyte, dissolved in an acidic solution to which is added solid KI. The titrant is a standard potassium iodate, KIO_3, solution.

The analysis of Vitamin C in a sample is completed in three steps:

First, as the KIO_3 solution is added to the analyte, the iodide ion, I^-, is oxidized to the water-soluble, triiodide complex, I_3^- (or $[I_2 \cdot I]^-$).

$$IO_3^-(aq) + 8\,I^-(aq) + 6\,H^+(aq) \rightarrow 3\,I_3^-(aq) + 3\,H_2O(l) \qquad (30.2)$$

Secondly, ascorbic acid, $C_6H_8O_6$, being a reducing agent, immediately reduces the tri-iodide ion, I_3^-, as it forms the colorless iodide ion, I^-, according to Equation 30.3.

$$C_6H_8O_6(aq) + I_3^-(aq) + H_2O(l) \rightarrow C_6H_8O_7(aq) + 3\,I^-(aq) + 2\,H^+(aq) \quad (30.3)$$

And finally, when the ascorbic acid in the analyte is consumed (at the stoichiometric point) by the I_3^-, the KIO_3 titrant generates an excess of I_3^- (again according to Equation 30.2) producing the yellow-brown color of I_3^- in solution.

The sudden appearance of the excess I_3^- is enhanced by the presence of an added starch solution to the analyte. Starch forms a deep-blue $I_3^- \cdot$starch complex that is easier to view than the yellow-brown color of I_3^-.

$$I_3^-(aq, yellow\text{-}brown) + starch(aq) \rightarrow I_3^- \cdot starch(aq, deep\text{-}blue) \qquad (30.4)$$

To summarize, (1) IO_3^-, the titrant, generates I_3^- (Equation 30.2) in the analyte; (2) $C_6H_8O_6$ immediately reacts with the I_3^- (Equation 30.3) until it is depleted from the analyte; and (3) the excess I_3^- forms a deep-blue complex to signal the depletion of the ascorbic acid in the sample (Equation 30.4).

Standard Solution of Potassium Iodate, KIO₃

Primary standard: a substance that has a known high degree of purity, a relatively large molar mass, is nonhygroscopic, and reacts in a predictable way

Potassium iodate, KIO_3, is a **primary standard** (Figure 30.2). A dried, accurately measured mass of previously dried KIO_3 is diluted to volume in a volumetric flask. This solution becomes the standard solution (as the titrant) in the analysis of ascorbic acid in a sample.

Figure 30.2 Potassium iodate has the necessary properties of a primary standard

EXPERIMENTAL PROCEDURE

A standard solution of potassium iodate is prepared from solid KIO_3, a primary standard. The standard KIO_3 solution in conjunction with KI is used to analyze an ascorbic acid sample. The excess I_3^- (not having reacted with ascorbic acid) is titrated with the standardized KIO_3 solution, using starch as an indicator.

In Part B, which source of Vitamin C are you to analyze? Consult with your instructor. Quantitative, reproducible data are objectives of this experiment; practice good laboratory techniques. Nearly 300 mL of boiled and cooled, deionized water is required for this experiment. Begin preparing this at the beginning of the laboratory period.

A. Preparation of the KIO₃ Primary Standard

1. **Prepare the Primary Standard Solution.** Use weighing paper to measure the tared mass (± 0.001 g) of KIO_3 (previously dried[1] at 110°C) that is required to prepare 250 mL of 0.01 M KIO_3. *Show this calculation on the Report Sheet.* Dissolve the solid KIO_3 and dilute to volume with the freshly boiled, deionized water in a 250-mL volumetric flask. Calculate and record its *exact* molar concentration.

2. **Prepare the Buret for the Ascorbic Acid Analysis.** Prepare a *clean*, 50-mL buret for the standard KIO_3 solution. Add 2–3 mL of the KIO_3 solution to the buret, wet the wall of the buret and drain through the buret tip. Repeat once. Fill the buret and eliminate all air bubbles in the buret tip. After 10–15 seconds, read and record the volume of titrant.

Read and record the volume in the buret to the correct number of significant figures.

B. Sample Preparation

1. **Vitamin C tablet.** Read the label on the bottle to determine the approximate mass of vitamin C in each tablet. Measure (± 0.001 g) the fraction of the total mass of a tablet that corresponds to ~100 mg of ascorbic acid. Dissolve the sample in a 250-mL Erlenmeyer flask with 40 mL of 0.5 M H_2SO_4[2] and then about 0.5 g $NaHCO_3$.[3] Kool-Aid™, Tang™, or Gatorade™ may be substituted as dry samples, even though their ascorbic acid concentrations are much lower. Fresh Fruit™ has a very high concentration of ascorbic acid. Proceed immediately to Part C.

2. **Fresh fruit sample.** Filter 125–130 mL of freshly squeezed juice through several layers of cheesecloth (or vacuum-filter). Measure the mass (± 0.01 g) of a clean, dry 250-mL Erlenmeyer flask. Add about 100 mL of filtered juice and again determine the mass. Add 40 mL of 0.5 M H_2SO_4 and 0.5 g $NaHCO_3$. Concentrated fruit juices may also be used as samples. Proceed immediately to Part C.

3. **Fresh vegetable sample.** Measure about 100 g (± 0.01 g) of a fresh vegetable. Transfer the sample to a mortar[4] (Figure 30.3) and grind. Add 5 mL of 0.5 M H_2SO_4 and continue to pulverize the sample. Add another 15 mL of 0.5 M H_2SO_4, stir, and filter through several layers of cheesecloth (or vacuum-filter). Add 20 mL of 0.5 M H_2SO_4 to the mortar, stir, and pour through the same filter. Repeat the washing of the mortar with 20 mL of the freshly boiled, deionized water. Combine all of the washings in a 250-mL Erlenmeyer flask and add 0.5 g $NaHCO_3$. Proceed immediately to Part C.

Figure 30.3 A mortar and pestle is used to grind the sample.

C. Vitamin C Analysis

1. **Prepare the Vitamin C Sample for Analysis.** To the sample from Part B add ~1 g of KI and ~2 mL of the starch solution.

2. **Immediately Titrate the Sample.** While swirling the flask, *slowly* dispense the standard KIO_3 solution—you will note that the deep-blue color appears but then disappears near the endpoint. Continue swirling the flask while adding the KIO_3 titrant until the deep-blue color persists for at least 20 seconds. Read and record the final volume in the buret.

---

[1] The KIO_3 should have been previously dried.

[2] Remember that vitamin tablets contain binders and other material that may be insoluble in water—*do not heat* in an attempt to dissolve the tablet!

[3] The $NaHCO_3$ reacts in the acidic solution to produce $CO_2(g)$, providing an inert atmosphere above the solution, minimizing the possibility of the air oxidation of the ascorbic acid.

[4] A blender may be substituted for the mortar and pestle.

3. **Repeat the Analysis.** Repeat the sample preparation (Part B) and the analysis twice in order to complete the three trials. Repeated analyses of the ascorbic acid content of the sample should be $\pm 1\%$.

Disposal: Dispose of the remaining KIO_3 solution in the "Waste Oxidants" container. Dispose of the excess $Na_2S_2O_3$ solution in the "Waste Reducing Agent" container.

CLEANUP: Flush the buret twice with tap water and twice with deionized water. Repeat the cleaning operation with the pipets. Discard all washings as advised by your instructor. Clean up the balance area, discarding any solids in the "Waste Solids" container.

The Next Step

Vitamin C is present in many foods (Table 30.1) at various concentrations and in vitamin supplements. What are those concentrations? What effects do weather conditions, soils, geographic regions, and seasonal variations have on vitamin C levels in fruits and vegetables? Is the concentration in vitamin supplements what the manufacturer claims? What other experimental procedures are used for vitamin C analysis? Design a study to further investigate vitamin C levels.

Vitamin C Analysis

Date _____ Lab Sec. _____ Name _____ Desk No. _____

1. Vitamin C is a reducing agent, an anti-oxidant. What ion does it reduce in this experiment? Explain.

2. a. Experimental Procedure, Part C. What is the oxidizing agent in the titration?

   b. What is the color change of the starch indicator that signals the stoichiometric point in Part C?

3. One calculation is to be completed before experimental work can begin. Complete that calculation here.

4. Experimental Procedure, Part B.1. What is the purpose of adding $NaHCO_3$ to the reaction mixture? Explain. Write a balanced equation for its reaction.

5. Eight ounces (1 fl. oz = 29.57 mL) of cranberry juice contains 130% of the recommended daily allowance of vitamin C (RDA = 75–90 mg for adults). How many milliliters of the cranberry juice will provide 100% of the recommended daily allowance?

6. A 53.06-g sample of a citrus juice was prepared for the analysis procedure described in this experiment. A 8.73-mL volume of 0.0107 M KIO_3 titrated the sample to the deep-blue I_3^-•starch endpoint.

   a. How many moles of I_3^- were generated for its reaction with the ascorbic acid?

   b. How many moles of ascorbic acid are in the sample?

   c. How many milligrams of ascorbic acid (molar mass = 176.1 g/mol) are in the sample?

   d. Calculate the number of milligrams of ascorbic acid per 100 g of citrus juice. Compare the result to the values listed in Table 30.1.

Vitamin C Analysis

Date _____ Lab Sec. _____ Name _____ Desk No. _____

A. Preparation of the KIO₃ Primary Standard

Calculate the mass of KIO_3 (molar mass $= 214.02$ g/mol) required for Part A.1.

1. Tared mass of KIO_3 (*g*)   _____
2. Moles of KIO_3 (*mol*)   _____
3. Molar concentration of KIO_3 solution (*mol/L*)   _____

B. Sample Preparation

Sample Name: _____

1. Mass of sample (*g*)   _____

C. Vitamin C Analysis

	Trial 1	Trial 2	Trial 3
1. Buret reading, *initial* (*mL*)	_____	_____	_____
2. Buret reading, *final* (*mL*)	_____	_____	_____
3. Volume of KIO_3 added (*mL*)			
4. Moles of IO_3^- dispensed (*mol*)			
5. Moles of I_3^- produced (*mol*)			
6. Moles of $C_6H_8O_6$ in sample (*mol*)			
7. Mass of $C_6H_8O_6$ in sample (*g*)	*		
8. Mass ratio (*mg $C_6H_8O_6$/100 g sample*)			

9. Average mass ratio of $C_6H_8O_6$ in sample (mg/100 g). See Table 30.1.   _____

*Show calculation on next page.

Calculation for Trial 1.

Laboratory Questions

Circle the questions that have been assigned.

1. Part A.1. The KIO_3 sample was not sufficiently dried. Will the reported molar concentration of the KIO_3 solution (Part B.3) be too high, too low, or unaffected? Explain.

2. Part B.1. The $NaHCO_3$ solution is omitted in an analysis of the sample. Will the reported amount of ascorbic acid in the sample be too high, too low, or unaffected? Explain.

3. Part C.1. Explain why the mass of KI ($\sim$1g), and the volume of starch ($\sim$2 mL) are only approximate even though the analysis is quantitative.

4. Part C.2. After adding the standard solution of KIO_3 to the prepared analyte, the sample solution *never* forms the deep-blue color. What modification of the Experimental Procedure or error corrected can be made in order to complete the analysis?

5. Part C.2. The deep-blue color of the I_3^-•starch complex does not appear but a yellow-brown does appear! What next? Should you continue titrating with the standard KIO_3 solution or discard the sample? Explain.

6. Part C.2. The final buret reading is read and recorded as 27.43 mL instead of the correct 28.43 mL. Will the reported amount of ascorbic acid in the sample be too high or too low? Explain.

Experiment 31

Dissolved Oxygen Levels in Natural Waters

The dissolved oxygen levels in natural waters are dependent on the temperature and water movement and flow.

- To develop a proper technique for obtaining a "natural" water sample
- To determine the dissolved oxygen concentration of a "natural" water sample
- To learn the chemical reactions involved in "fixing" and analyzing a water sample for dissolved oxygen using the Winkler method

The following techniques are used in the Experimental Procedure

Streams, rivers, lakes, and oceans play vital roles in our quality of life. Not only are they a source of food supplies with the likes of shrimp and salmon but they also provide recreational opportunities in the forms of boating and swimming. Additionally the larger bodies of water such as lakes and oceans affect the seasonal weather patterns, producing changes in rainfall and snowfall and generating conditions for hurricanes and typhoons.

The aesthetic appearance of smaller bodies of water such as rivers and lakes indicates an immediate perception of the quality of the water. Color, surface growth, and odor are early indicators of the quality of the water and the nature of its marine life. As the public water supplies of most larger cities rely on the presence of surface water, water chemists must be keenly aware of the makeup of that water. "How must the water be treated to provide safe and clean water to the consumers?"

A number of water quality parameters are of primary interest in analyzing a "natural" water sample: pH, dissolved oxygen, alkalinity, and hardness are but a few. pH, a quick test, is generally determined with a previously calibrated pH meter; dissolved oxygen concentrations can be completed with a dissolved oxygen meter (Figure 31.1) although its availability is less likely than that of a pH meter. Alkalinity and hardness levels are determined using the titrimetric technique (see Experiments 20 and 21).

The concentration of dissolved oxygen in a water sample is an important indicator of water quality. Waters with high oxygen concentrations indicate aerobic conditions . . . clean, clear, and unpolluted. Low oxygen concentrations indicate anaerobic conditions . . . high turbidity, foul odors, extensive plant growth on the surface. Dissolved oxygen levels that drop to less than 5 ppm can stress the existing aquatic life.

The solubilities of oxygen in fresh water (saturated solution) at various temperatures are listed in Table 31.1.

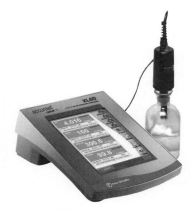

Figure 31.1 Dissolved oxygen meters can be used for determining $O_2(aq)$ levels in water samples.

Table 31.1 Solubility of oxygen in fresh water at various temperatures

Temperature (°C)	ppm O_2	Temperature (°C)	ppm O_2
0	14.6	24	8.5
5	12.8	27	8.1
10	11.3	30	7.6
15	10.2	35	7.1
18	9.5	40	6.5
21	9.0	45	6.0

Winkler Method of Analysis

The Winkler method of analysis for dissolved oxygen, developed by Lajos Winkler in 1888, is the standard experimental procedure for determining the dissolved oxygen concentration in water.

The Winker test is performed in two parts: (1) the water sample is gathered "in the field" where the dissolved oxygen is "fixed" with two reagents and (2) the sample is titrated for final analysis in the laboratory within a 48-hour period.

The "natural" water sample is carefully collected onsite such that no air bubbles remain trapped in the flask after collection. The oxygen is "fixed" by an immediate reaction with manganese(II) sulfate in a basic solution:

$$4\,MnSO_4(aq) + O_2(aq) + 8\,NaOH(aq) + 2\,H_2O(l)$$
$$\rightarrow 4\,Mn(OH)_3(s) + 4\,Na_2SO_4(aq) \quad (31.1)$$

The oxygen is "fixed" as the manganese(III) hydroxide,[1] an orange-brown color precipitate—the more precipitate, the greater is the dissolved oxygen concentration.

While "onsite" a basic solution of KI-NaN$_3$ is also added to the sample.[2] The manganese(III) hydroxide oxidizes the iodide ion to the triiodide ion, I_3^-, while the manganese(III) reduces to the manganese(II) ion.

$$2\,Mn(OH)_3(s) + 3\,I^-(aq) + 6\,H^+(aq) \rightarrow I_3^-(aq) + 6\,H_2O(l) + 2\,Mn^{2+}(aq) \quad (31.2)$$

The resulting solution now has a slight yellow-brown color due to the presence of I_3^- ($[I_2 \cdot I]^-$).

The remainder of the dissolved oxygen analysis is completed in the laboratory (but within 48 hours). The sample is acidified with sulfuric acid to dissolve any precipitate. A titration of the sample with a standardized sodium thiosulfate solution in the presence of a starch indicator determines the amount of I_3^- generated in the reactions and provides a direct determination of the dissolved oxygen concentration in the water sample.

$$I_3^-(aq) + 2\,S_2O_3^{2-}(aq) \rightarrow 3\,I^-(aq) + S_4O_6^{2-}(aq) \quad (31.3)$$

The starch indicator forms a deep-blue complex with I_3^-, but is colorless in the presence of I^-.

$$I_3^- \cdot starch\ (deep\text{-}blue) \rightarrow 3\,I^- + starch\ (colorless) \quad (31.4)$$

From Equations 31.1–31.3, 1 mole O_2 reacts to produce 4 moles of $Mn(OH)_3$, of which 2 moles of $Mn(OH)_3$ reacts to produce 1 mole of I_3^-. The I_3^-, which is the result of the "fixing" of the dissolved oxygen, reacts with 2 moles of $S_2O_3^{2-}$ in the titration.

From the data collected and analyzed, the moles of O_2 converted to milligrams divided by the volume of the water sample (in liters) that is titrated results in the dissolved oxygen concentration expressed in mg/L or ppm (parts per million) O_2.

$$\frac{mg\ O_2}{L\ sample} = ppm\ O_2 \quad (31.5)$$

---

[1] There is uncertainty among chemists as to the oxidation number of manganese in the precipitate—$MnO(OH)_2$, the hydrated form of MnO_2, often represents the form of the precipitate.

[2] Sodium azide, NaN_3, is added to eliminate interference in the dissolved oxygen analysis caused by the presence of nitrite ion, NO_2^-, common in wastewater samples.

A sodium thiosulfate solution is standardized with potassium iodate, KIO_3, a primary standard. In the presence of iodide ion, KIO_3 generates a quantified concentration of triiodide ion, I_3^-.

Standard Solution of Sodium Thiosulfate[3]

$$IO_3^-(aq) + 8\,I^-(aq) + 6\,H^+(aq) \rightarrow 3\,I_3^-(aq) + 3\,H_2O(l) \qquad (31.6)$$

This solution is then titrated to the starch endpoint with the prepared sodium thiosulfate solution.

$$I_3^-(aq) + 2\,S_2O_3^{2-}(aq) \rightarrow 3\,I^-(aq) + S_4O_6^{2-}(aq) \qquad (31.7)$$

For the analysis of the dissolved oxygen concentration in a water sample, the standard $Na_2S_2O_3$ solution should have a molar concentration of $0.025\ M$ or less.

Procedure Overview. A water sample is collected from a source that is selected either by the student chemist or the instructor. At *least* three samples are carefully collected in 250-mL Erlenmeyer flasks while avoiding the presence of air bubbles. The dissolved oxygen in the sample is immediately "fixed" with the addition of a basic solution of manganese(II) sulfate and a basic solution of KI-NaN_3. In the laboratory the sample is acidified and titrated with a standard sodium thiosulfate solution. The dissolved oxygen concentrations are reported in units of parts per million (ppm) O_2.

Ask your instructor if a standard solution of $Na_2S_2O_3$ is available. If so, proceed to Part B of the Experimental Procedure.

EXPERIMENTAL PROCEDURE

Create and design your own Report Sheet for this part of the experiment.

A. A Standard 0.025 M $Na_2S_2O_3$ Solution

1. **Preparation and Standardization of 0.1 M $Na_2S_2O_3$ Solution.** Refer to Experiment 29, Parts A and B of the Experimental Procedure for the preparation and standardization of a 0.1 M $Na_2S_2O_3$ solution. Prepare only 100 mL of the $Na_2S_2O_3$ of the solution described in Experiment 29, Part B.1 and standardize the solution using KIO_3 as the primary standard solution (Part B.3-4). Calculate the average concentration of the $Na_2S_2O_3$ solution.

2. **Preparation of a Standard 0.025 M $Na_2S_2O_3$ Solution.** Using a pipet and 100-mL volumetric flask, prepare a 0.025 M $Na_2S_2O_3$ solution from the standardized 0.1 M $Na_2S_2O_3$.

Disposal: Dispose of the test solutions as directed by your instructor.

1. **Prepare the Flask for Sampling.** Thoroughly clean and rinse at *least* three 250-mL Erlenmeyer flasks and rubber stoppers to fit. Allow to air dry.

B. Collection of Water Sample

2. **Collect the Water Sample.** Gently lay the flask along the horizontal surface of the water. See Figure 31.2. Slowly and gradually turn the flask upright as the flask fills being careful not to allow any air bubbles to form in the flask. Fill the flask to overflowing.

3. **"Fix" the Dissolved Oxygen.** *Below the surface* of the water sample, pipet $\sim$1 mL of the basic 2.1 M $MnSO_4$ solution into the sample (some overflowing will occur). Similarly pipet $\sim$1 mL of the basic KI-NaN_3 solution.

4. **Secure the Sample.** a. Carefully stopper the sample to ensure that no air bubbles become entrapped beneath the stopper in the water sample. Again, some overflowing will occur.

---

[3]See Experiment 29 for further explanation and Experimental Procedure

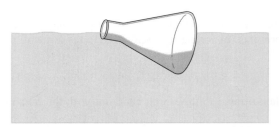

Figure 31.2 Allow a gentle flow of water into the flask. Slowly turn the flask upright as it fills to overflowing.

    b. Invert and roll the flask to thoroughly mix the reagents. Once the precipitate settles, repeat the mixing process.

    c. Label the sample number for each of the flasks. Store the sample in the dark and, preferably, in a cool/cold location.

5. **Temperature.** Read and record the temperature at the sample site. Also, write a brief description of the sample site.

C. Sample Analysis

Read and record the buret to the correct number of significant figures.

1. **Prepare the Titrant.** Prepare a clean buret. Add 3 to 5 mL of the standard $Na_2S_2O_3$ solution to the buret, roll the solution to wet the wall of the buret and dispense through the buret tip and discard. Use a clean funnel to fill the buret . . . dispense a small portion through the buret tip. Read and record the volume of $Na_2S_2O_3$ solution in the buret (Technique 16A.2), "using all certain digits plus one uncertain digit."
    Place a white sheet of paper beneath the receiving flask.

2. **Prepare Sample #1.** a. Remove the stopper from the 250-mL Erlenmeyer flask. To the collected water sample, add ~1 mL of "conc" H_2SO_4 (**Caution!**) and stir or swirl to dissolve any precipitate. The sample can now be handled in open vessels.

    b. Transfer a known, measured volume (~200 mL, ±0.1 mL) to a receiving flask (either a beaker or Erlenmeyer flask) for the titrimetric analysis (Part C.3).

3. **Titrate Water Sample #1.** Slowly dispense the $Na_2S_2O_3$ titrant into the water sample. Swirl the flask as titrant is added (Technique 16C.4). When the color of the analyte fades to a light yellow-brown, add ~1 mL of the starch solution. Continue slowly adding titrant—when one drop (ideally half-drop) results in the disappearance of the deep-blue color of the I_3^-•starch complex, stop the titration and again (after ~15 seconds) read and record the volume of titrant in the buret.

4. **Additional Trials.** Repeat the analysis for the two remaining samples.

5. **Calculations.** Calculate the dissolved oxygen concentration for each sample expressed in ppm O_2 (mg O_2/L sample).

> *Disposal:* Dispose of the test solutions as directed by your instructor.

The Next Step

The Biological Oxygen Demand (BOD) of a water sample is a measure of the organic material in a water sample that is consumable by aerobic bacteria. The $O_2(aq)$ concentration is measured when a sample is taken and then again five (5) days later, five days being the incubation period for the aerobic bacteria to consume a portion of the $O_2(aq)$ to biodegrade the organic material. Research the importance and significance of BOD levels in natural waters and develop an experiment to determine the BOD for a water analysis.

Dissolved Oxygen Levels in Natural Waters

Date _____ Lab Sec. _____ Name _____ Desk No. _____

1. For a natural water sample, what range of dissolved concentrations may you expect? Explain your reasoning.

2. How does the dissolved oxygen concentration in a water sample change (if at all) with
   a. ambient temperature changes?

   b. atmospheric pressure changes?

   c. the volume of the flask collecting the water sample?

   d. amount of organic matter in the water sample?

   e. the depth of the body of water (e.g., lake, river, or ocean)?

3. A solution of $MnSO_4$ is added to "fix" the dissolved oxygen in the collected sample.
   a. What is the meaning of the expression, "fix the dissolved oxygen" and why is it so important for the analysis of dissolved oxygen in a water sample?

   b. Describe the preparation of 25.0 mL of a 2.1 M $MnSO_4$ solution starting with $MnSO_4 \cdot H_2O$ (molar mass = 169.01 g/mol).

   c. The "exact" volume of $MnSO_4$ solution added to the collected sample is not critical. Explain.

4. A water chemist obtained a 250-mL sample from a nearby lake and "fixed" the oxygen "on site" with alkaline solutions of $MnSO_4$ and $KI-NaN_3$. Returning to the laboratory a 200-mL sample was analyzed by acidifying the sample with "conc" H_2SO_4 and then titrating with 14.4 mL of 0.0213 M $Na_2S_2O_3$ solution to the starch endpoint. (This is a calculation similar to the one for this experiment.)

   a. Calculate the number of moles of I_3^- that reacted with the $Na_2S_2O_3$ solution.

   b. Calculate the number of moles of $Mn(OH)_3$ that were produced from the reduction of the dissolved oxygen.

   c. Calculate the number of moles and milligrams of O_2 present in the titrated sample.

   d. What is the dissolved oxygen concentration in the sample, expressed in ppm O_2?

5. a. What is the procedure for preparing 250 mL of 0.0210 M $Na_2S_2O_3$ for this experiment from a 100-mL volume of standard 0.106 M $Na_2S_2O_3$.

   b. Unfortunately, for the preparation of the 0.0210 M solution, only a 25.0-mL calibrated volumetric pipet is available. Explain how you would prepare the diluted $Na_2S_2O_3$ solution using the 25.0-mL pipet and what would be its *exact* molar concentration?

6. A 100-mL volume of a primary standard 0.0110 M KIO_3 solution is prepared. A 25.0-mL aliquot of this solution is used to standardize a prepared $Na_2S_2O_3$ solution. A 15.6-mL volume of the $Na_2S_2O_3$ solution titrated the KIO_3 solution to the starch endpoint. What is the molar concentration of the $Na_2S_2O_3$ solution?

$$IO_3^-(aq) + 8\,I^-(aq) + 6\,H^+(aq) \rightarrow 3\,I_3^-(aq) + 3\,H_2O(l)$$
$$I_3^-(aq) + 2\,S_2O_3^{2-}(aq) \rightarrow 3\,I^-(aq) + S_4O_6^{2-}(aq)$$

Dissolved Oxygen Levels in Natural Waters

Date _____ Lab Sec. _____ Name _____ Desk No. _____

A. A Standard 0.025 M $Na_2S_2O_3$ Solution

Prepare a self-designed Report Sheet for this part of the experiment. Review the Report Sheet of Experiment 29 for guidance. Submit this with the completed Report Sheet.

B. Collection of Water Sample

Sampling Site: Temperature: _____°C

Characterize/describe the sampling site.

C. Sample Analysis

	Sample 1	Sample 2	Sample 3
1. Sample volume (mL)	_____	_____	_____
2. Buret reading, initial (mL)	_____	_____	_____
3. Buret reading, final (mL)	_____	_____	_____
4. Volume $Na_2S_2O_3$ dispensed (mL)			
5. Molar concentration of $Na_2S_2O_3$ (M), Part A			
6. Moles of $Na_2S_2O_3$ dispensed (mol)			
7. Moles of I_3^- reduced by $S_2O_3^{2-}$ (mol)			
8. Moles of O_2 (mol)			
9. Mass of O_2 (mg)			
10. Dissolved oxygen, ppm O_2			
11. Average dissolved oxygen, ppm O_2			
12. Standard deviation			
13. Relative standard deviation (%RSD)			

Write a short summary based upon an interpretation of your analytical data.

Laboratory Questions

Circle the questions that have been assigned.

1. Part B. The water chemist waits until returning to the laboratory to "fix" the water sample for the dissolved oxygen analysis. Will the reported dissolved oxygen concentration be reported too high, too low, or remain unchanged? Explain.

2. Part B.4. No precipitate forms! Assuming the reagents were properly prepared and dispensed into the sample, what might be predicted about its dissolved oxygen concentration? Explain.

3. Part C.3. The color of the analyte did not fade to form the light yellow-brown color but remained intense even after the addition of a full buret of the $S_2O_3^{2-}$ titrant, even though a precipitate formed in Part B.4. What can be stated about the dissolved oxygen concentration of the sample? Explain.

4. Assuming a dissolved oxygen concentration of 7.0 ppm (mg/L) in a 300-mL water sample,
   a. how many moles of $Mn(OH)_3$ will be produced with the addition of the $MnSO_4$ solution?
   b. how many moles of I_3^- will be produced when the KI-NaN$_3$ solution is added to the solution above (Part 4.a)?
   c. how many moles of $S_2O_3^{2-}$ will be needed to react with the I_3^- that is generated?
   d. assuming the concentration of the $S_2O_3^{2-}$ titrant to be 0.025 M, how many milliliters of titrant will be predictably used?

5. A nonscientist brings a water sample to your laboratory and asks you to determine why there was a fish kill in the nearby lake. Having recently finished this experiment, what might you tell that person about the legitimacy of a test for dissolved oxygen? What reasoning would you use to maintain the integrity of your laboratory?

6. a. Fish kills are often found near the discharge point of water from cooling waters at electrical generating power plants. Explain why this occurrence may occur.
   b. Fish kills are often found in streams following heavy rainfall in a watershed dominated by farmland or denuded forestland. Explain why this occurrence may occur.

7. Explain how the dissolved oxygen concentrations may change starting at the headwaters of a river and ending at the ocean. Account for the changes.

8. Salt (ocean) water generally has a lower dissolved oxygen concentration than fresh water at a given temperature. Explain why this is generally observed.

Galvanic Cells, the Nernst Equation

Copper metal spontaneously oxidizes to copper(II) ion in a solution containing silver ion. Silver metal crystals form on the surface of the copper metal.

OBJECTIVES

- To measure the relative reduction potentials for a number of redox couples
- To develop an understanding of the movement of electrons, anions, and cations in a galvanic cell
- To study factors affecting cell potentials
- To estimate the concentration of ions in solution using the Nernst equation

TECHNIQUES

The following techniques are used in the Experimental Procedure

INTRODUCTION

Electrolyic cells are of two types, galvanic and electrolysis, both employing the principle of oxidation-reduction (redox) reactions. In galvanic (or voltaic) cells (this experiment), redox reactions occur spontaneously as is common with all portable batteries of which we are very familiar. Electric cars, flashlights, watches, and power tools operate because of a specific spontaneous redox reaction. Electrolysis cells (Experiment 33) are driven by nonspontaneous redox reactions, reactions that require energy to occur. The recharging of batteries, electroplating and refining of metals, and generation of various gases all require the use of energy to cause the redox reaction to proceed.

Experimentally, when copper wire is placed into a silver ion solution (see opening photo), copper atoms *spontaneously* lose electrons (copper atoms are oxidized) to the silver ions (which are reduced). Silver ions migrate to the copper atoms to pick up electrons and form silver atoms at the copper metal/solution **interface;** the copper ions that form then move into the solution away from the interface. The overall reaction that occurs at the interface is (see opening photo).

Interface: the boundary between two phases; in this case, the boundary that separates the solid metal from the aqueous solution

$$Cu(s) + 2\,Ag^+(aq) \rightarrow 2\,Ag(s) + Cu^{2+}(aq) \tag{32.1}$$

This redox reaction can be divided into an oxidation and a reduction half-reaction. Each half-reaction, called a **redox couple,** consists of the reduced state and the oxidized state of the substance.

Redox couple: an oxidized and reduced form of an ion/substance appearing in a reduction or oxidation half-reaction, generally associated with galvanic cells

$$Cu(s) \rightarrow Cu^{2+}(aq) + 2\,e^- \qquad \text{oxidation half-reaction (redox couple)} \tag{32.2}$$

$$2\,Ag^+(aq) + 2\,e^- \rightarrow 2\,Ag(s) \qquad \text{reduction half-reaction (redox couple)} \tag{32.3}$$

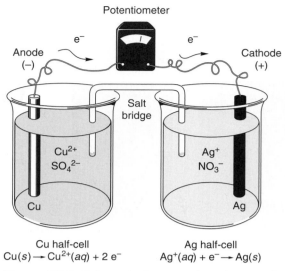

Potentiometer

Anode (−) e⁻   e⁻ Cathode (+)

Salt bridge

Cu^{2+}
SO_4^{2-}

Ag^+
NO_3^-

Cu   Ag

Cu half-cell   Ag half-cell
$Cu(s) \rightarrow Cu^{2+}(aq) + 2\ e^-$ $Ag^+(aq) + e^- \rightarrow Ag(s)$

Figure 32.1 Schematic diagram of a galvanic cell.

A **galvanic cell** is designed to take advantage of this *spontaneous transfer of electrons.* Instead of electrons being transferred at the interface between the copper metal and the silver ions in solution, a galvanic cell separates the copper metal from the silver ions to force the electrons to pass externally through a wire, an external circuit. Figure 32.1 is a schematic diagram of a galvanic cell setup for these two redox couples.

The two redox couples are placed in separate compartments, called **half-cells.** Each half-cell consists of an electrode, usually the metal (reduced state) of the redox couple, and a solution containing the corresponding cation (oxidized state) of the redox couple. The electrodes of the half-cells are connected by a wire through which the electrons flow, providing current for the **external circuit.**

A **salt bridge** that connects the two half-cells completes the construction of the galvanic cell (and the circuit). The salt bridge permits limited movement of ions from one half-cell to the other, the **internal circuit,** so that when the cell operates, electrical neutrality is maintained in each half-cell. For example, when copper metal oxidizes to copper(II) ions in the Cu^{2+}/Cu half-cell, either anions must enter or copper(II) ions must leave the half-cell to maintain neutrality. Similarly, when silver ions reduce to form silver metal in its half-cell, either anions must leave or cations must enter its half-cell to maintain neutrality.

The electrode at which reduction occurs is called the **cathode;** the electrode at which oxidation occurs is called the **anode.** Because oxidation releases electrons to the electrode to provide a current in the external circuit, the anode is designated the *negative* electrode in a galvanic cell. The reduction process draws electrons from the circuit and supplies them to the ions in solution; the cathode is the *positive* electrode. This sign designation allows us to distinguish the anode from the cathode in a galvanic cell.

Half-cell: a part of the galvanic cell that hosts a redox couple

External circuit: the movement of charge as electrons through a wire connecting the two half-cells, forming one half of the electrical circuit in a galvanic cell

Salt bridge: paper moistened with a salt solution, or an inverted tube containing a salt solution, that bridges two half-cells to complete the solution part of an electrical circuit

Internal circuit: the movement of charge as ions through solution from one half-cell to the other, forming one half of the electrical circuit in a galvanic cell

Cell Potentials

Different metals, such as copper and silver, have different tendencies to oxidize; similarly, their ions have different tendencies to undergo reduction. The **cell potential** of a galvanic cell is due to the difference in tendencies of the two metals to oxidize (lose electrons) or of their ions to reduce (gain electrons). Commonly, a measured **reduction potential,** the tendency for an ion (or molecule) to gain electrons, is the value used to identify the relative ease of reduction for a half-reaction.

A **potentiometer** or **multimeter,** placed in the external circuit between the two electrodes, measures the cell potential, E_{cell}, a value that represents the *difference* between the tendencies of the metal ions in their respective half-cells to undergo reduction (i.e., the difference between the reduction potentials of the two redox couples).

For the copper and silver redox couples, we can represent their reduction potentials as $E_{Cu^{2+},Cu}$ and $E_{Ag^+,Ag}$, respectively. The cell potential being the difference of the two reduction potentials is therefore

$$E_{cell} = E_{Ag^+,Ag} - E_{Cu^{2+},Cu} \qquad (32.4)$$

Silver jewelry is longer lasting than copper jewelry; therefore silver has a higher tendency to be in the reduced state, a higher reduction potential

Experimentally, silver ion has a greater tendency than does copper ion to be in the reduced (metallic) state; therefore, Ag^+ has a greater (more positive) reduction potential. Since the cell potential, E_{cell}, is measured as a positive value, $E_{Ag^+,Ag}$ is placed before $E_{Cu^{2+},Cu}$ in Equation 32.4.

The measured cell potential corresponds to the **standard cell potential** when the concentrations of all ions are 1 mol/L and the temperature of the solutions is 25°C.

The *standard* reduction potential for the $Ag^+(1\ M)/Ag$ redox couple, $E°_{Ag^+,Ag}$, is +0.80 V, and the *standard* reduction potential for the $Cu^{2+}(1\ M)/Cu$ redox couple, $E°_{Cu^{2+},Cu}$, is +0.34 V. Theoretically, a potentiometer (or multimeter) would show the difference between these two potentials, or, at standard conditions,

$$E°_{cell} = E°_{Ag^+,Ag} - E°_{Cu^{2+},Cu} = +0.80\ V - (+0.34\ V) = +0.46 V \qquad (32.5)$$

Deviation from the theoretical value may be the result of surface activity at the electrodes or activity of the ions in solution.

Measure Cell Potentials

In Part A of this experiment, several cells are "built" from a selection of redox couples and data are collected. From an analysis of the data, the relative reduction potentials for the redox couples are determined and placed in an order of decreasing reduction potentials.

In Part B, the formations of the complex, $[Cu(NH_3)_4]^{2+}$, and the precipitate, CuS, are used to change the concentration of $Cu^{2+}(aq)$ in the Cu^{2+}/Cu redox couple. The observed changes in the cell potentials are interpreted.

Measure Non-standard Cell Potentials

The Nernst equation is applicable to redox systems that are *not* at standard conditions, most often when the concentrations of the ions in solution are *not* 1 mol/L. At 25°C, the measured cell potential, E_{cell}, is related to $E°_{cell}$ and ionic concentrations by

$$\text{Nernst equation: } E_{cell} = E°_{cell} - \frac{0.0592}{n} \log Q \qquad (32.6)$$

where n represents the moles of electrons exchanged according to the cell reaction. For the copper/silver cell, $n = 2$; two electrons are lost per copper atom and two electrons are gained per two silver ions (see Equations 32.1–32.3). For dilute ionic concentrations, the *reaction quotient, Q*, equals the **mass action expression** for the cell reaction. For the copper/silver cell (see Equation 32.1),

Mass action expression: the product of the molar concentrations of the products divided by the product of the molar concentrations of the reactants, each concentration raised to the power of its coefficient in the balanced cell equation

$$Q = \frac{[Cu^{2+}]}{[Ag^+]^2}$$

In Part C of this experiment, we study, in depth, the effect that changes in concentration of an ion have on the potential of the cell. The cell potentials for a number of zinc/copper redox couples are measured in which the copper ion concentrations are varied but the zinc ion concentration is maintained constant.

$$Zn(s) + Cu^{2+}(aq) \rightarrow Cu(s) + Zn^{2+}(aq)$$

The Nernst equation for this reaction is

$$E_{cell} = E°_{cell} - \frac{0.0592}{2} \log \frac{[Zn^{2+}]}{[Cu^{2+}]} \qquad (32.7)$$

Rearrangement of this equation (where $E°_{cell}$ and $[Zn^{2+}]$ are constants in the experiment) yields an equation for a straight line:

$$E_{cell} = \underbrace{E°_{cell} - \frac{0.0592}{2} \log [Zn^{2+}]}_{} + \frac{0.0592}{2} \log [Cu^{2+}] \qquad (32.8)$$
$$y = \qquad\qquad b \qquad\qquad + \quad m \qquad x$$

To simplify

$$pCu = -\log [Cu^{2+}]$$

$$E_{cell} = \text{constant} - \frac{0.0592}{2} pCu \qquad (32.9)$$

A plot of E_{cell} versus pCu for solutions of known copper ion concentrations has a negative slope of 0.0592/2 and an intercept b that not only includes the constants in Equation 32.8, but also the inherent characteristics of the cell and potentiometer (Figure 32.2).

The E_{cell} of a solution with an *unknown* copper ion concentration is then measured; from the linear plot, its concentration is determined.

EXPERIMENTAL PROCEDURE

Procedure Overview: The cell potentials for a number of galvanic cells are measured, and the redox couples are placed in order of decreasing reduction potentials. The effects of changes in ion concentrations on cell potentials are observed and analyzed.

Perform the experiment with a partner. At each circled superscript ①–⑫ in the procedure, *stop,* and record your observation on the Report Sheet. Discuss your observations with your lab partner and your instructor.

A. Reduction Potentials of Several Redox Couples

Chemists often use the "red, right, plus" rule in connecting the red wire to the right-side positive electrode (cathode) of the galvanic cell

1. **Collect the Electrodes, Solutions, and Equipment.** Obtain four, small ($\sim$50 mL) beakers and fill them three-fourths full of the 0.1 M solutions as shown in Figure 32.3. Share these solutions with other chemists/groups of chemists in the laboratory.

    Polish strips of copper, zinc, magnesium, and iron metal with steel wool or sandpaper, rinse briefly with dilute ($\sim$1 M) HNO_3 (**Caution!**), and rinse with deionized water. These polished metals, used as electrodes, should be bent to extend over the lip of their respective beakers. Check out a multimeter (Figure 32.4) (or a voltmeter) with two electrical wires (preferably a red and black wire) attached to "alligator" clips.

2. **Set Up the Copper/Zinc Cell.** Place a Cu strip (electrode) in the $CuSO_4$ solution and a Zn strip (electrode) in the $Zn(NO_3)_2$ solution. Roll and flatten a piece of filter paper; wet the filter paper with a 0.1 M KNO_3 solution. Fold and insert the ends of the filter paper into the solutions in the two beakers; this is the salt bridge. Set the multimeter to the 2000mV range or as appropriate. Connect one electrode to the negative terminal of the multimeter and the other to the positive terminal.[1]

3. **Determine the Copper/Zinc Cell Potential.** If the multimeter reads a negative potential, reverse the connections to the electrodes. Read and record the (positive) cell potential. Identify the metal strips that serve as the cathode (positive terminal) and the anode. Write an equation for the half-reaction occurring at each electrode. Combine the two half-reactions to write the equation for the cell reaction.①

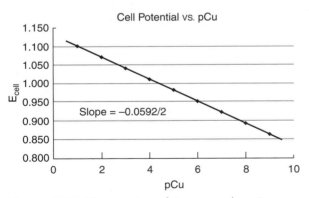

Figure 32.2 The variation of E_{cell} versus the pCu.

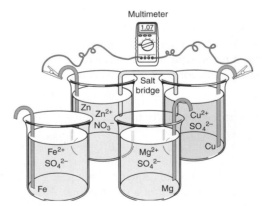

Figure 32.3 Setup for measuring the cell potentials of six galvanic cells.

[1]You have now combined two half-cells to form a galvanic cell.

4. **Repeat for the Remaining Cells.** Determine the cell potentials for all possible galvanic cells that can be constructed from the four redox couples. Refer to the Report Sheet for the various galvanic cells. Prepare a "new" salt bridge for each galvanic cell.[2]

5. **Determine the Relative Reduction Potentials.** Assuming the reduction potential of the $Zn^{2+}(0.1\ M)/Zn$ redox couple is -0.79 V, determine the reduction potentials of all other redox couples.[2][3]

6. **Determine the Reduction Potential of the Unknown Redox Couple.** Place a 0.1 M solution and electrode obtained from your instructor in a small beaker. Determine the reduction potential, relative to the $Zn^{2+}(0.1\ M)/Zn$ redox couple, for your unknown redox couple.[4]

---

1. **Effect of Different Molar Concentrations.** Set up the galvanic cell shown in Figure 32.5, using 1 M $CuSO_4$ and 0.001 M $CuSO_4$ solutions. Immerse a polished copper electrode in each solution. Prepare a salt bridge to connect the two half-cells. Measure the cell potential. Determine the anode and the cathode. Write an equation for the reaction occurring at each electrode.[5]

2. **Effect of Complex Formation.** Add 2–5 mL of 6 M NH_3 to the 0.001 M $CuSO_4$ solution, until any precipitate redissolves.[3] (**Caution:** *Do not inhale* NH_3.) Observe and record any changes in the half-cell and the cell potential.[6]

3. **Effect of Precipitate Formation.** Add 2–5 mL of 0.2 M Na_2S to the 0.001 M $CuSO_4$ solution, now containing the added NH_3. What is observed in the half-cell and what happens to the cell potential? Record your observations.[7]

B. Effect of Concentration Changes on Cell Potential

---

1. **Prepare the Diluted Solutions.** Prepare Solutions 1 through 4 as shown in the margin figure using a 1-mL pipet and 100-mL volumetric flasks.[4] Be sure to rinse the pipet with the more concentrated solution before making the transfer. Use deionized water for dilution "to the mark" in the volumetric flasks. Calculate the molar concentration of the Cu^{2+} ion for each solution and record.[8]

2. **Measure and Calculate the Cell Potential for *Solution 4*.** Set up the experiment as shown in Figure 32.6, using small (~50 mL) beakers.

   The Zn^{2+}/Zn redox couple is the reference half-cell for this part of the experiment. Connect the two half-cells with a "new" salt bridge. Reset the multimeter to the

C. The Nernst Equation and an Unknown Concentration

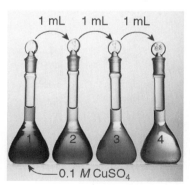

Successive quantitative dilution, starting with 0.1 M $CuSO_4$.

Figure 32.4 A modern multimeter.

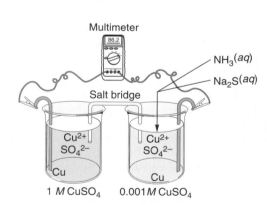

Figure 32.5 Setup for measuring the cell potential of a Cu^{2+} concentration cell.

---

[2]*Note:* These are *not* standard reduction potentials because 1 M concentrations of cations at 25°C are not used.
[3]Copper ion forms a complex with ammonia: $Cu^{2+}(aq) + 4\ NH_3(aq) \rightarrow [Cu(NH_3)_4]^{2+}(aq)$
[4]Share these prepared solutions with other chemists/groups of chemists in the laboratory.

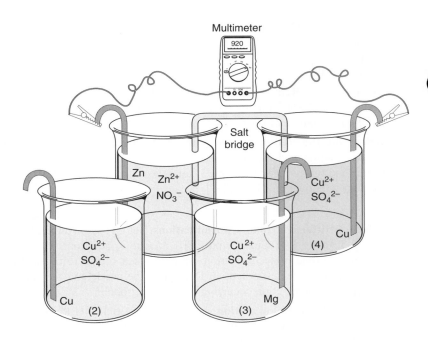

Multimeter

Salt bridge

Zn Zn²⁺
NO₃⁻

Cu²⁺
SO₄²⁻

Cu

(4)

Cu²⁺
SO₄²⁻

Cu

(2)

Cu²⁺
SO₄²⁻

Mg

(3)

Figure 32.6 Setup to measure the effect that diluted solutions have on cell potentials.

lowest range (~200 mV). Connect the electrodes to the multimeter and record the potential difference, $E_{cell, expt}$.⑨Calculate the *theoretical* cell potential $E_{cell, calc}$. (Use a table of standard reduction potentials and the Nernst equation.)⑩

3. **Measure and Calculate the Cell Potentials for Solutions 3 and 2.** Repeat Part C.2 with Solutions 3 and 2, respectively. A freshly prepared salt bridge is required for each cell.

4. **Plot the Data.** Plot $E_{cell, expt}$ *and* $E_{cell, calc}$ (ordinate) versus pCu (abscissa) on the *same* piece of linear graph paper (page 362) for the *four* concentrations of $CuSO_4$ (see data from Part A.3 for the potential of Solution 1). Have your instructor approve your graph.⑪

Appendix C

5. **Determine the Concentration of the Unknown.** Obtain a $CuSO_4$ solution with an "unknown" copper ion concentration from your instructor and set up a galvanic cell. Determine E_{cell} as in Part C.2. Using the graph, determine the unknown copper(II) ion concentration in the solution.⑫

Disposal: Dispose of the waste zinc, copper, magnesium, and iron solutions in the "Waste Metal Solutions" container. Return the metals to appropriately marked containers.

CLEANUP: Rinse the beakers twice with tap water and twice with deionized water. Discard the rinses in the "Waste Metal Solutions" container.

The Next Step

Galvanic cells are the basis for the design of specific ion electrodes, electrodes that sense the relative concentration of a specific ion, e.g., hydrogen ion, relative to the electrode that has a fixed concentration. Part C of this experiment could be the apparatus for measuring concentrations of Cu^{2+} in "other" samples. According to Equation 32.8, the pCu (negative log of $[Cu^{2+}]$) is proportional to the E_{cell}! Research specific ion electrodes, their design and their application. Design an experiment where a specific ion electrode, other than the pH electrode, can be used for a systematic study of an ion of interest.

Galvanic Cells, the Nernst Equation

Date _____ Lab Sec. _____ Name _____ Desk No. _____

1. In a galvanic cell
   a. oxidation occurs at the (name of electrode)   _____

   b. the cathode is the (sign) electrode   _____

   c. cations flow toward the (name of electrode)   _____

   d. electrons flow from the (name of electrode) to (name of electrode) _____ _____

2. a. What is the purpose of a salt bridge? Explain.

   b. How is the salt bridge prepared in this experiment?

3. For the cell reaction, $Cu^{2+}(aq) + Zn(s) \rightarrow Cu(s) + Zn^{2+}(aq)$, $E°_{cell} = 1.10$ V. Use the Nernst Equation to determine E_{cell} when the molar concentrations of Zn^{2+} and Cu^{2+} are 1.0 M and 1.0×10^{-7} M, respectively.

4. Refer to Figure 32.2.
   a. What is the value of the cell constant?

   b. What is the $[Cu^{2+}]$ if the measured cell potential is 0.96 V?

   c. What should be the cell potential if the $[Cu^{2+}]$ is 1.0×10^{-5} mol/L?

5. Consider a galvanic cell consisting of the following two redox couples:

$$Ag^+(0.010\ M) +\ e^- \rightarrow Ag \qquad E° = +0.80\ V$$

$$Bi^{3+}(0.0010\ M) + 3\ e^- \rightarrow Bi \qquad E° = +0.20\ V$$

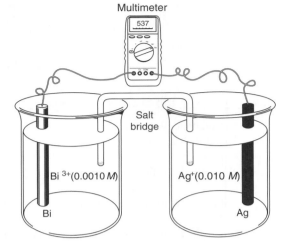

Multimeter

Salt bridge

Bi^{3+}(0.0010 M)   Ag$^+$(0.010 M)

Bi   Ag

a. Write the equation for the half-reaction occurring at the cathode.

b. Write the equation for the half-reaction occurring at the anode.

c. Write the equation for the cell reaction.

d. What is the *standard* cell potential, $E°_{cell}$, for the cell?

e. Realizing the nonstandard concentrations, what is the *actual* cell potential, E_{cell}, for the cell?

*6. The extent of corrosion in the steel reinforcing (rebar) of concrete is measured by the galvanic cell shown in the diagram of the instrument. The half-cell of the probe is usually a AgCl/Ag redox couple:

$$AgCl +\ e^- \rightarrow Ag + Cl^-\ (1.0\ M) \qquad E° = +0.23\ V$$

Corrosion is said to be "severe" if the cell potential is measured at greater than 0.41 V. Under these conditions, what is the iron(II) concentration on the rebar?

$$Fe^{2+} + 2\ e^- \rightarrow Fe \qquad E° = -0.44\ V$$

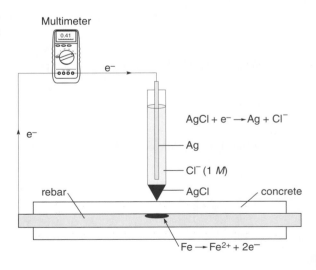

Multimeter

e^-

e^-

$AgCl + e^- \rightarrow Ag + Cl^-$

Ag

Cl$^-$ (1 M)

rebar

AgCl   concrete

$Fe \rightarrow Fe^{2+} + 2e^-$

Galvanic Cells, the Nernst Equation

Date _____ Lab Sec. _____ Name _____ Desk No. _____

A. Reduction Potentials of Several Redox Couples

Fill in the following table with your observations and interpretations from the galvanic cells.

Galvanic Cell	Measured E_{cell}	Anode	Equation for Anode Reaction	Cathode	Equation for Cathode Reaction
①Cu–Zn					
②Cu–Mg					
Cu–Fe					
Zn–Mg					
Fe–Mg					
Zn–Fe					

1. Write *balanced* equations for the six cell reactions.

2. What is the oxidizing agent in the Zn-Mg cell?

3. Compare the *sum* of the Zn–Mg and Cu–Mg cell potentials with the Cu–Mg cell potential. Explain.

4. Compare the *sum* of the Zn–Fe and Zn–Mg cell potentials with the Fe–Mg cell potential. Explain.

5. In the table below, list the "Measured E_{cell}" values from Part A (column 2). Setting the reduction potential for the Zn^{2+} (0.1 M)/Zn redox couple to –0.79V, calculate the "experimental reduction potential" for each redox couple (column 4). Using the Nernst equation, calculate the "theoretical reduction potential" for each redox couple, again setting the reduction potential for the Zn^{2+} (0.1 M)/Zn redox couple to –0.79V (column 5). Finally calculate the percent error of the experimental reduction potential for each redox couple (column 6).

③ Galvanic Cell	Measured E_{cell}	For the Redox Couple	Reduction Potential (expt'l, calculated)	Reduction Potential (calculated, theory)	% Error
Cu-Zn		Cu^{2+}/Cu			
Zn-Fe		Fe^{2+}/Fe			
Zn-Zn	0	Zn^{2+}/Zn	–0.79 V	–0.79	
Zn-Mg		Mg^{2+}/Mg			
Zn–unknown, X		X^{2+}/X			

6. ④Reduction potential of the unknown redox couple: _____

B. **Effect of Concentration Changes on Cell Potential**

1. ⑤Cell potential of "concentration cell": _____

   Anode reaction: _____

   Cathode reaction: _____
   Explain *why* a potential is recorded.

2. ⑥Cell potential from complex formation: _____
   Observation of solution in half-cell.

   Explain *why* the potential changes as it does with the addition of $NH_3(aq)$.

3. ⑦Cell potential from precipitate formation: _____
   Observation of solution in half-cell.

   Explain *why* the potential changes as it does with the addition of Na_2S.

C. The Nernst Equation and an Unknown Concentration

1. Complete the following table with the concentrations of the $Cu(NO_3)_2$ solutions and the measured cell potentials. Use a table of standard reduction potentials and the Nernst equation to *calculate* the E_{cell}.

Solution Number	[8] Concentration of CuSO$_4$	$-\log [Cu^{2+}]$, pCu	[9] E_{cell} (measured)	[10] E_{cell} (calculated)
1	0.1 mol/L	1	_____	
2	_____		_____	
3	_____		_____	
4	_____		_____	

2. [11] Instructor's approval of graph: _____

    Account for any significant difference between the measured and calculated E_{cell} values.

3. [12] E_{cell} for the solution of unknown concentration: _____

    Molar concentration of Cu^{2+} in the unknown: _____

Laboratory Questions

Circle the questions that have been assigned.

1. Part A.3. The filter paper salt bridge is *not* wetted with the 0.1 M KNO$_3$ solution. How will this affect the measured potential of the cell? Explain.

2. Part A.3. A positive potential is recorded when the copper electrode is the positive electrode. Is the copper electrode the cathode or the anode of the cell? Explain.

3. Part A.5. The measured reduction potentials are not equal to the calculated reduction potentials. Give two reasons why this might be observed.

4. Part B.2. How would the cell potential have been affected if the NH$_3(aq)$ had been added to the 1 M CuSO$_4$ solution instead of the 0.001 M CuSO$_4$ solution of the cell? Explain.

5. Part B.3. The cell potential increased (compared to Part B.2) with the addition of the Na$_2$S solution to the 0.001 M CuSO$_4$ solution. Explain.

6. Part C. As the concentration of the copper(II) ion increased from Solution 4 to Solution 1, what happened to the measured cell potentials? Explain.

7. Part C. Suppose the 0.1 M Zn^{2+} solution had been diluted (instead of the Cu^{2+} solution), what would have happened to the measured cell potentials? Explain.

8. Part C. How would you adjust the concentrations of Cu^{2+} and Zn^{2+} to maximize the cell potential? Explain.

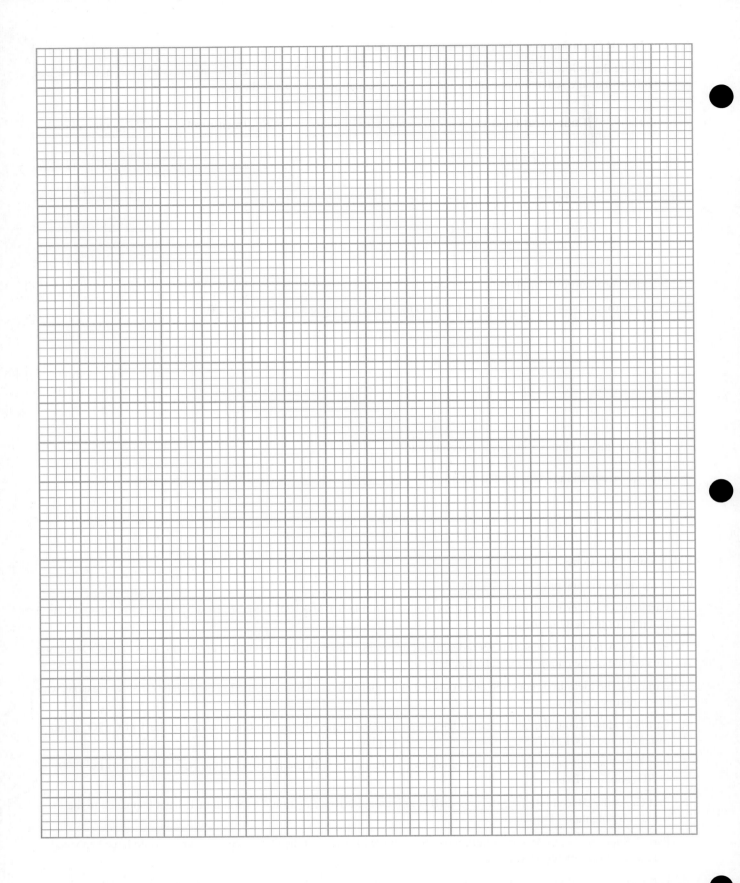

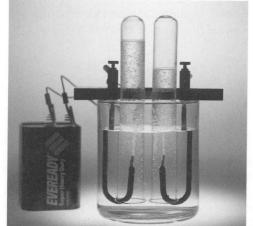

Electrolytic Cells, Avogadro's Number

A 1 : 2 mole (and volume) ratio of oxygen (left) to hydrogen (right) is produced from the electrolysis of water.

- To identify the reactions occurring at the anode and cathode during the electrolysis of various aqueous salt solutions
- To determine Avogadro's number and the Faraday constant

The following techniques are used in the Experimental Procedure

INTRODUCTION

Electrolysis processes are very important in achieving high standards of living. The industrial production of metals such as aluminum and magnesium, and nonmetals such as chlorine and fluorine, occurs in electrolytic cells. The highly refined copper metal required for electrical wiring is obtained through an **electroplating** process.

In an **electrolytic cell** the input of an electric current causes an otherwise nonspontaneous oxidation–reduction reaction, a nonspontaneous transfer of electrons, to occur. For example, sodium metal, a very active metal, and chlorine gas, a very toxic gas, are prepared industrially by the electrolysis of molten sodium chloride. Electrical energy is supplied to a molten NaCl system (Figure 33.1) by a direct current (dc) power source (set at an appropriate voltage) across the electrodes of an electrolytic cell.

The electrical energy causes the reduction of the sodium ion, Na^+, at the cathode and oxidation of the chloride ion, Cl^-, at the anode. Because cations migrate to the cathode and anions migrate to the anode, the cathode is the negative electrode (opposite charges attract) and the anode is designated the positive electrode.[1]

$$\text{cathode } (-) \text{ reaction:} \qquad Na^+ (l) + e^- \rightarrow Na(l)$$
$$\text{anode } (+) \text{ reaction:} \qquad 2\,Cl^-(l) \rightarrow Cl_2(g) + 2\,e^-$$

Electrolysis reactions also occur in aqueous solutions. For example, in the electrolysis of an aqueous copper(II) bromide, $CuBr_2$, solution, copper(II) ions, Cu^{2+}, are reduced at the cathode and bromide ions, Br^-, are oxidized at the anode (Figure 33.2).

$$\text{cathode } (-) \text{ reaction:} \qquad Cu^{2+} (aq) + 2\,e^- \rightarrow Cu(s)$$
$$\text{anode } (+) \text{ reaction:} \qquad 2\,Br^-(aq) \rightarrow Br_2(l) + 2\,e^-$$
$$\text{cell reaction:} \qquad Cu^{2+}(aq) + 2\,Br^-(aq) \rightarrow Cu(s) + Br_2(l)$$

Electrolysis: Use of electrical energy to cause a chemical reaction to occur

Electroplating: the use of electrical current to deposit a metal onto an electrode

Electrolytic cell: an apparatus used for an electrolysis reaction

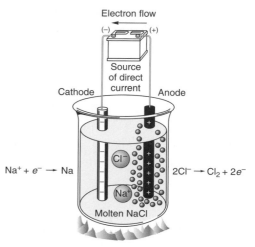

Figure 33.1 Schematic diagram of the electrolysis of molten sodium chloride.

[1]Note that the cathode is "−" in an electrolytic cell, but "+" in a galvanic cell; the anode is "+" in an electrolytic cell, but "−" in a galvanic cell.

Electrolysis of Aqueous Solution

In an aqueous solution, however, the reduction of water at the cathode (the negative electrode) and the oxidation of water at the anode (the positive electrode) are also possible reactions.

$$\text{cathode } (-) \text{ reaction for water: } \quad 2\,H_2O(l) + 2\,e^- \rightarrow H_2(g) + 2\,OH^-(aq) \quad (33.1)$$

$$\text{anode } (+) \text{ reaction for water: } \quad 2\,H_2O(l) \rightarrow O_2(g) + 4\,H^+(aq) + 4\,e^- \quad (33.2)$$

If the reduction of water occurs at the cathode, hydrogen gas is evolved and the solution near the cathode becomes basic as a result of the production of hydroxide ion. If oxidation of water occurs at the anode, oxygen gas is evolved and the solution near the anode becomes acidic. The acidity (or basicity) near the respective electrodes can be detected with pH paper or another acid–base indicator.

When two or more competing reduction reactions are possible at the cathode, the reaction that occurs most easily (the one with the *higher* reduction potential) is the one that usually occurs. Conversely, for two or more competing oxidation reactions at the anode, the reaction that takes place most easily (the one with the *higher* oxidation potential or the *lower* reduction potential) is the one that usually occurs.

In the electrolysis of the aqueous copper(II) bromide solution, Cu^{2+} has a higher reduction potential than H_2O and is therefore preferentially reduced at the cathode; Br^- has a greater tendency to be oxidized than water, and so Br^- is oxidized.

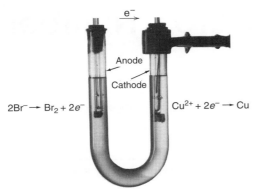

Figure 33.2 Electrolysis of a copper(II) bromide solution; the anode is on the left and the cathode is on the right.

Selective Electrolysis

Very few metals exist as metals in the natural state (gold being an exception), but rather most exist as cations in the earth's crust. Therefore, metal cations from the various ores must be reduced for the production of metals for commercial or industrial use. Some metals (such as iron and zinc) are reduced using a simple redox reaction, but others (such as aluminum and magnesium) require electrolytic processes. The molten salts (to avoid water interferences) of the latter (more active) metals are electrolyzed with the metals being electroplated at the cathode. By careful control of the applied potential in an electrolytic cell, metal cations of a mixture can be selectively electroplated, thus making electrolysis a process of refining the metal to a very high degree of purity.

In Part A of this experiment, a number of aqueous salt solutions, using different electrodes, are electrolyzed. The anode and cathode are identified and the products that are formed at each electrode are also identified.

Avogadro's Number and the Faraday Constant

In Part B, a quantitative investigation of the electrolytic oxidation of copper metal is used to determine Avogadro's number and the Faraday constant.

$$Cu(s) \rightarrow Cu^2(aq) + 2\,e^- \quad (33.3)$$

1 faraday = 1 mol e^- = 96,485 coulombs

Two moles of electrons (or 2 **faradays**) are released for one mole of $Cu(s)$; therefore, a mass measurement of the copper anode before and after the electrolysis determines the moles of copper that are oxidized. This in turn is used to calculate the moles of electrons that pass through the cell.

$$\text{moles of electrons} = \text{mass Cu} \times \frac{\text{mol Cu}}{63.54\ \text{g}} \times \frac{2\ \text{mol } e^-}{\text{mol Cu}} \quad (33.4)$$

Coulomb: SI base unit for electrical charge

The actual *number* of electrons that pass through the cell is calculated from the electrical current, measured in amperes (= coulombs/second), that passes through the cell for a recorded time period (seconds). The total charge (**coulombs, C**) that passes through the cell is

$$\text{number of coulombs} = \frac{\text{coulombs}}{\text{second}} \times \text{seconds} \quad (33.5)$$

As the charge of one electron equals 1.60×10^{-19} C, the number of electrons that pass through the cell can be calculated.

$$\text{number of electrons} = \text{number of coulombs} \times \frac{\text{electron}}{1.60 \times 10^{-19}\,\text{C}} \qquad (33.6)$$

Therefore, since the number of electrons (Equation 33.6) and the moles of electrons (Equation 33.4) can be separately determined, Avogadro's number is calculated as

$$\text{Avogadro's number} = \frac{\text{number of electrons}}{\text{mole of electrons}} \qquad (33.7)$$

In addition, the number of coulombs (Equation 33.5) per mole of electrons (Equation 33.4) equals the Faraday constant. With the available data, the Faraday constant can also be calculated:

$$\text{Faraday constant} = \frac{\text{number of coulombs}}{\text{mole of electrons}} \qquad (33.8)$$

Procedure Overview: The products that result from the electrolysis of various salt solutions are observed and identified; these are qualitative measurements. An experimental setup is designed to measure quantitatively the flow of current and consequent changes in mass of the electrodes in an electrolytic cell; from these data, experimental "constants" are calculated.

The electrolysis apparatus may be designed differently than the one described in this experiment. Ask your instructor.

1. **Set Up the Electrolysis Apparatus.** Connect two wire leads (different colors) attached to alligator clips to a direct current (dc) power supply.[2] Clean and mount the glass U-tube on a ring stand (see Figure 33.3). Connect the alligator clips to the corresponding electrodes, listed in Table 33.1.

2. **Electrolyze the Solutions.** Fill the U-tube three-fourths full with Solution #1 from Table 33.1. Insert the corresponding electrodes into the solution and electrolyze for ~5 minutes. During the electrolysis, watch for any evidence of a reaction in the anode and cathode chambers.

   - Does the pH of the solution change at each electrode? Test each chamber with litmus or pH paper.[3] Compare the color with a pH test on the original solution.
   - Is a gas evolved at either or both electrodes?
   - Look closely at each electrode. Is a metal depositing on the electrode or is the metal electrode slowly disappearing?

3. **Account for Your Observations.** Write the equations for the reactions occurring at the anode and cathode and for each cell reaction. Repeat for Solutions 2–5.

A. Electrolysis of Aqueous Salt Solutions

Disposal: Discard the salt solutions into the "Waste Salts" container.

Table 33.1 Electrolytic Cells for Study

Solution No.	Solution*	Electrodes
1	2 g NaCl/100 mL	Carbon (graphite)
2	2 g NaBr/100 mL	Carbon (graphite)
3	2 g KI/100 mL	Carbon (graphite)
4	0.1 M CuSO$_4$	Carbon (graphite)
5	0.1 M CuSO$_4$	Polished copper metal strips

*Try other solutions and electrodes as suggested by your laboratory instructor.

---

[2]The dc power supply can be a 9-V transistor battery.
[3]Several drops of universal indicator can be added to the solution in both chambers to detect pH changes.

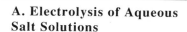

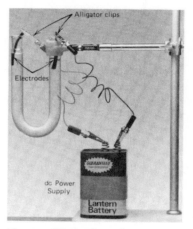

Figure 33.3 Electrolysis apparatus.

CLEANUP: Rinse the U-tube twice with tap water and twice with deionized water before preparing the next solution for electrolysis. Discard each rinse in the sink.

B. Determination of Avogadro's Number and the Faraday Constant

1. **Set up the Apparatus.** Refer to Figure 33.4. The U-tube from Part A can again be used. The dc power supply must provide 3–5 volts (two or three flashlight batteries in series or a lantern battery); the ammeter or multimeter must read from 0.2 to 1.0 A. Polish two copper metal strips (to be used as the electrodes) with steel wool or sandpaper. Briefly dip each electrode (use the fume hood) into 6 M HNO_3 (**Caution:** *do not allow skin contact*), and then rinse with deionized water.

   Add 100 mL of 1.0 M $CuSO_4$ (in 0.1 M H_2SO_4) to the 150-mL beaker (or fill the U-tube).

2. **Set the Electrodes.** Rinse the electrodes with ethanol if available. When dry, label the two electrodes because the mass of each will be determined before and after the electrolysis. Measure the mass (± 0.001 g, preferably ± 0.0001 g) of each labeled electrode. The copper electrode with the lesser mass is to serve as the anode (+ terminal), and the other is to serve as the cathode (− terminal) for the electrolytic cell. Connect the cathode (through the variable resistor and ammeter/multimeter) to the negative terminal of the dc power supply.

   Before electrolysis begins, obtain your instructor's approval of the complete apparatus.

3. **Electrolyze the CuSO$_4$ Solution.** Adjust the variable resistance to its maximum value.[4] Be ready to start timing (a stopwatch is ideal). Attach the anode to the positive terminal of the dc power supply and START TIME. During the electrolysis do *not* move the electrodes; this changes current flow. Adjust the current with the variable resistor to about 0.5 A and, periodically during the course of the electrolysis, readjust the current to 0.5 A.[5]

   Discontinue the electrolysis after 15–20 minutes. Record the exact time (minutes and seconds) of the electrolysis process.

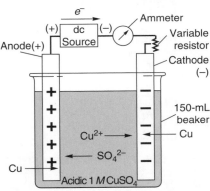

Figure 33.4 Setup for determining Avogadro's number and the Faraday constant.

4. **Dry and Measure the Mass.** Carefully remove the electrodes (be careful not to loosen the electroplated copper metal from the cathode); carefully dip each electrode into a 400-mL beaker of deionized water to rinse the electrodes (followed by ethanol if available). Air-dry, measure the mass (± 0.001 g, preferably ± 0.0001 g) of each electrode, and record.

5. **Repeat the Electrolysis.** If time allows, repeat Part B using the same copper electrodes (with new mass measurements!) and 1.0 M $CuSO_4$ solution.

Disposal: Discard the copper(II) sulfate solution into the "Waste Salts" container.

CLEANUP: Rinse the beaker or U-tube twice with tap water and twice with deionized water. Discard each rinse as directed by your instructor.

The Next Step

Electroplating of metals, such as nickel, chromium, silver, copper, etc., is a common industrial process. Research a specific process and design an apparatus and procedure of depositing quantitative amounts of metal to a cathode.

---

[4]If a variable resistor is unavailable, record the current at 1-minute intervals and then calculate an *average* current over the entire electrolysis period.
[5]If the current is greater or less than 0.5 A, vary the time of electrolysis proportionally.

Electrolytic Cells, Avogadro's Number

Date _____ Lab Sec. _____ Name _____ Desk No. _____

1. The standard reduction potential for the Cu^{2+}/Cu redox couple is $+0.34$ V; that for H_2O/H_2, OH^- at a pH of 7 is -0.41 V. For the electrolysis of a neutral 1.0 M $CuSO_4$ solution, write the equation for the half-reaction occurring at the cathode at standard conditions.

2. In an electrolytic cell
   a. reduction occurs at the (name of electrode)

   _____

   b. the anode is the (sign) electrode

   _____

   c. anions flow toward the (name of electrode)

   _____

   d. electrons flow from the (name of electrode) to (name of electrode)

   _____   _____

   e. the cathode should be connected to the (positive/negative) terminal of
      the dc power supply

   _____

3. Experimental Procedure, Part A.2. Describe the proper technique for testing the pH of a solution with litmus or pH paper.

   17b

4. a. Identify a chemical test(s) to determine if water is oxidized at the anode in an electrolytic cell.

   b. Similarly, identify a chemical test(s) to determine if water is reduced at the cathode in an electrolytic cell.

5. Very pure copper metal is produced by the electrolytic refining of blister (impure) copper. In the cell at right, label the anode, the cathode, and the polarity $(+, -)$ of each.

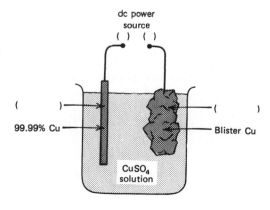

6. a. When a solution of $MgCl_2$ is electrolyzed, red litmus turns blue and a gas is evolved in the cathode chamber. Write the equation for the half-reaction occurring at the cathode.

   b. When a solution of nickel(II) sulfate adjusted to a pH of 7 is electrolyzed, the green color of the solution becomes less intense in the cathodic chamber and gas bubbles are detected in the anodic chamber. Write the equations for the half-reactions occurring in the cathodic and the anodic chambers.

7. A 1.0 M $CuSO_4$ solution was electrolyzed for 28 minutes and 22 seconds using copper electrodes. The average current flowing through the cell over the time period was 0.622 A. The mass of the copper anode before the electrolysis was 2.4852 g and 2.1335 g afterwards.
   a. Calculate the number of moles of copper oxidized *and* moles electrons that passed through the cell (see Equation 33.4).

   b. Calculate the total charge (coulombs) that passed through the cell (see Equation 33.5).

   c. How many electrons passed through the cell during the 28.0 minute, 22 second time period (see Equation 33.6)?

   d. From the data, calculate Avogadro's Number (see Equation 33.7).

Experiment 33 *Report Sheet*

Electrolytic Cells, Avogadro's Number

Date __________ Lab Sec. ______ Name ______________________________ Desk No. __________

A. Electrolysis of Aqueous Salt Solutions

Solution	Electrodes	Litmus Test	Gas Evolved?	Balanced Equations for Reactions
NaCl	C(*gr*)			Anode ______________________ Cathode ______________________ Cell ______________________
NaBr	C(*gr*)			Anode ______________________ Cathode ______________________ Cell ______________________
KI	C(*gr*)			Anode ______________________ Cathode ______________________ Cell ______________________
$CuSO_4$	C(*gr*)			Anode ______________________ Cathode ______________________ Cell ______________________
$CuSO_4$	Cu(*s*)			Anode ______________________ Cathode ______________________ Cell ______________________

B. Determination of Avogadro's Number and the Faraday Constant

Data	Trial 1	Trial 2
1. *Initial* mass of copper anode (g)	_____	_____
2. *Initial* mass of copper cathode (g)	_____	_____
3. Instructor's Approval of Apparatus	_____	
4. Time of electrolysis (s)	_____	_____
5. Current (or average current) (A)	_____	_____
6. *Final* mass of copper anode (g)	_____	_____
7. *Final* mass of copper cathode (g)	_____	_____

Data Analysis

	Trial 1	Trial 2
1. Mass of copper oxidized at anode (g)		
2. Moles of copper oxidized (*mol*)		
3. Moles of electrons transferred (*mol e$^-$*)		
4. Coulombs passed through cell (C)		
5. Electrons passed through cell (*e$^-$*)		
6. Avogadro's number (*e$^-$/mol e$^-$*)		
7. Average value of Avogadro's number		
8. Literature value of Avogadro's number		
9. Percent error		
10. Faraday constant (*C/mol e$^-$*)		
11. Average Faraday constant (*C/mol e$^-$*)		
12. Literature value of Faraday constant		
13. Percent error		

Laboratory Questions

Circle the questions that have been assigned.

1. Part A.2. If zinc electrodes are used instead of the graphite electrodes, the reaction occurring at the anode may be different but the reaction occurring at the cathode would remain unchanged. Explain.

2. Part B. Repeat the calculation of Avogadro's number, using the mass gain of the cathode *instead* of the mass loss of the anode. Account for any difference in the calculated values.

3. Part B.2. If the current is recorded as being greater than it actually is, would Avogadro's number be calculated too high, too low, or be unaffected? Explain.

4. Account for the percent errors in determining Avogadro's number and the Faraday constant.

5. The electrolytic refining of copper involves the oxidation of impure copper, containing such metals as iron and nickel (oxidized to copper(II), iron(II), and nickel(II) ions), at the anode and then reduction of the copper(II) ion to copper metal at the cathode. Explain why the iron(II) and nickel(II) ions are not deposited on the cathode.

An Equilibrium Constant

The nearly colorless iron(III) ion (left) forms an intensely colored complex (right) in the presence of the thiocyanate ion.

OBJECTIVES

- To use a **spectrophotometer** to determine the equilibrium constant of a chemical system
- To use graphing techniques and data analysis to evaluate data
- To determine the equilibrium constant for a soluble equilibrium

Spectrophotometer: a laboratory instrument that measures the amount of light transmitted through a sample

TECHNIQUES

The following techniques are used in the Experimental Procedure

INTRODUCTION

A spectrophotometric method of analysis involves the interaction of electromagnetic (EM) radiation with matter. The most common regions of the EM spectrum used for analyses are the ultraviolet, visible, and the infrared regions. We are most familiar with the visible region of the spectrum, having a wavelength range from 400 to 700 nm.

The visible spectra of ions and molecules in solution arise from *electronic* transitions within their respective structures. The greater the concentration of the absorbing ions/molecules in solution the greater is the absorption of the visible EM radiation (and the greater is the transmittance of the *complimentary* radiation). The degree of absorbed radiation (or the intensity of the transmitted radiation) is measured using an instrument called a **spectrophotometer**, an instrument that measures transmitted light intensities with a photosensitive detector at specific (but variable) visible wavelengths (Figure 34.1). The wavelength where the absorbing ions or molecules has a maximum absorption of visible radiation is determined and set on the spectrophotometer for the analysis.

The visible light path through the spectrophotometer from the light source through the sample to the photosensitive detector is shown in Figure 34.2.

Several factors control the amount of EM radiation (light energy) that a sample absorbs:

- Concentration of the absorbing substance.
- Thickness of the sample containing the absorbing substance (determined by the width of the **cuvet**)
- Probability of light absorption by the absorbing substance (called the **molar absorptivity coefficient** or **extinction coefficient**)

The Introduction to Dry Lab 3 discusses in more detail the interaction of electromagnetic radiation with atoms, ions, and molecules in terms of energy states, excited states, wavelengths, and spectra.

Cuvet: a special piece of glassware to hold solutions for measurement in the spectrophotometer

Figure 34.1 Common laboratory visible spectrophotometer.

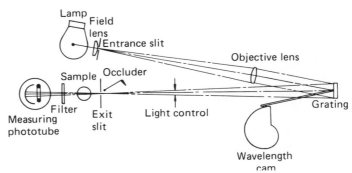

Figure 34.2 The light path through a visible spectrophotometer.

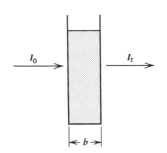

Figure 34.3 Incident light, I_0, and transmitted light, I_t, for a sample of thickness b.

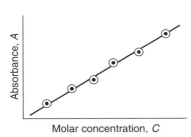

Molar concentration, C

Figure 34.4 A data plot of absorbance vs. concentration.

The ratio of the intensity of the transmitted light, I_t, to that of the incident light, I_0 (Figure 34.3), is called the transmittance, T, of the EM radiation by the sample. This ratio, expressed as percent, is

$$\frac{I_t}{I_0} \times 100 = \%T \tag{34.1}$$

Most spectrophotometers have a $\%T$ (percent transmittance of light) scale. Because it is linear, the $\%T$ scale is easy to read and interpolate. However, chemists often perform calculations based on the amount of light *absorbed* by the sample, rather than the amount of light transmitted, because absorption is directly proportional to the concentration of the absorbing substance. The absorbance, A, of the substance is related to the intensity of the incident and transmitted light (and the percent transmittance) by the equations

$$A = \log\frac{I_0}{I_t} = \log\frac{1}{T} = \log\frac{100}{\%T} = a \cdot b \cdot c \tag{34.2}$$

a, the molar absorptivity coefficient, is a constant at any given wavelength for a particular absorbing substance, b is the thickness of the absorbing substance in centimeters, and c is the molar concentration of the absorbing substance.[1]

The absorbance value is directly proportional to the molar concentration of the absorbing substance, *if* the same (or a matched) cuvet and a *set* wavelength are used for all measurements. A plot of absorbance versus concentration data is linear; a calculated slope and absorbance data can be used to determine the molar concentration of the same absorbing species in a solution of unknown concentration (Figure 34.4) from the linear relationship.

Measuring an Equilibrium Constant

The magnitude of an equilibrium constant, K_c, expresses the equilibrium position for a chemical system. For the reaction, $a\mathrm{A} + b\mathrm{B} \rightleftharpoons x\mathrm{X} + y\mathrm{Y}$, the mass action expression, $\dfrac{[\mathrm{X}]^x[\mathrm{Y}]^y}{[\mathrm{A}]^a[\mathrm{B}]^b}$, equals the equilibrium constant, K_c, when a dynamic equilibrium has been established between reactants and products. The brackets in the mass action expression denote the equilibrium molar concentration of the respective substance.

The magnitude of the equilibrium constant indicates the principal species, products or reactants, that exist in the chemical system at equilibrium. For example, a large equilibrium constant indicates that the equilibrium lies to the right with a high concentration of products and correspondingly low concentration of reactants. The value of K_c is constant for a chemical system at a given temperature.

---

[1]Since the quantity log (I_0/I_t) is generally referred to as absorbance, Equation 34.2 becomes $A = abc$. This equation is commonly referred to as **Beer's law.**

This experiment determines K_c for a chemical system in which all species are soluble. The chemical system involves the equilibrium between iron(III) ion, Fe^{3+}, thiocyanate ion, SCN^-, and thiocyanatoiron(III) ion, $FeNCS^{2+}$:

$$[Fe(H_2O)_6]^{3+}(aq) + SCN^-(aq) \rightleftharpoons [Fe(H_2O)_5NCS]^{2+}(aq) + H_2O(l) \quad (34.3)$$

The "free" thiocyanate ion is commonly written as SCN^-; however its bond to the ferric ion is through the nitrogen atom, thus the formula of the complex is written $FeNCS^{2+}$

Because the concentration of water is essentially constant in dilute aqueous solutions, we omit the waters of hydration and simplify the equation to read

$$Fe^{3+}(aq) + SCN^-(aq) \rightleftharpoons FeNCS^{2+}(aq) \quad (34.4)$$

The mass action expression for the equilibrium system, equal to the equilibrium constant, is

$$K_c = \frac{[FeNCS^{2+}]}{[Fe^{3+}][SCN^-]} \quad (34.5)$$

In Part A you will prepare a set of five **standard solutions** of the $FeNCS^{2+}$ ion. As $FeNCS^{2+}$ is a deep, blood-red complex, its absorption maximum occurs at about 447 nm. The absorbance at 447 nm for each solution is plotted versus the molar concentration of $FeNCS^{2+}$; this establishes a **calibration curve** from which the concentrations of $FeNCS^{2+}$ are determined for the chemical systems in Part B.

Standard solution: a solution with a very well known concentration of solute

Calibration curve: a plot of known data from which further interpretations can be made

In preparing the standard solutions of $FeNCS^{2+}$, the Fe^{3+} concentration is set to *far* exceed the SCN^- concentration. This huge excess of Fe^{3+} pushes the equilibrium (Equation 34.4) *far* to the right, consuming nearly all of the SCN^- placed in the system. As a result the $FeNCS^{2+}$ concentration at equilibrium approximates the original SCN^- concentration. In other words, we assume that the position of the equilibrium is driven so far to the right by the excess Fe^{3+} that all of the SCN^- is **complexed,** forming $FeNCS^{2+}$ (Figure 34.5).

Complexed: the formation of a bond between the Lewis base, SCN^-, and the Lewis acid, Fe^{3+}

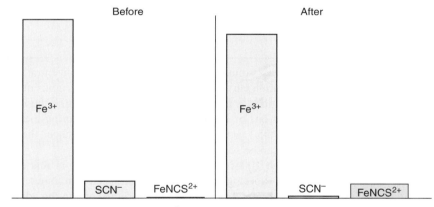

Figure 34.5 A large excess of Fe^{3+} consumes nearly all of the SCN^- to form $FeNCS^{2+}$. The amount of Fe^{3+} remains essentially unchanged in solution.

In Part B, the concentrations of the Fe^{3+} and SCN^- ions in the various test solutions are nearly the same, thus creating equilibrium systems in which there is an appreciable amount of each of the species after equilibrium is established (Figure 34.6).

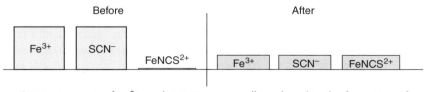

Figure 34.6 Amounts of Fe^{3+} and SCN^- are equally reduced in the formation of $FeNCS^{2+}$.

| **Measurements for K_c** | In Part B, precise volumes of known molar concentrations of Fe^{3+} and SCN^- are mixed. The equilibrium molar concentration of $FeNCS^{2+}$ of the system is determined by measuring its absorbance and then using the calibration curve from Part A. Since the total volume of the mixed solution is measured precisely, the *initial* moles of Fe^{3+} and SCN^- and the *equilibrium* moles of $FeNCS^{2+}$ are easily calculated from known molar concentrations. |

From Equation 34.4, for every mole of $FeNCS^{2+}$ that exists at equilibrium, an equal number of moles of Fe^{3+} and SCN^- have reacted to reach equilibrium:

$$\text{mol FeNCS}^{2+}_{\text{equilibrium}} = \text{mol Fe}^{3+}_{\text{reacted}} = \text{mol SCN}^-_{\text{reacted}} \tag{34.6}$$

Therefore, the moles of Fe^{3+} at equilibrium (unreacted) is

$$\text{mol Fe}^{3+}_{\text{equilibrium}} = \text{mol Fe}^{3+}_{\text{initial}} - \text{mol Fe}^{3+}_{\text{reacted}} \tag{34.7}$$

Similarly, the moles of SCN^- at equilibrium (unreacted) is

$$\text{mol SCN}^-_{\text{equilibrium}} = \text{mol SCN}^-_{\text{initial}} - \text{mol SCN}^-_{\text{reacted}} \tag{34.8}$$

Again, since the total volume of the reaction mixture is known precisely, the equilibrium molar concentrations of Fe^{3+} and SCN^- (their *equilibrium* concentrations) can be calculated. Knowing the measured equilibrium molar concentration of $FeNCS^{2+}$ from the calibration curve, substitution of the three equilibrium molar concentrations into the mass action expression provides the value of the equilibrium constant, K_c.

| **Calculations for K_c** | The calculations for K_c are involved, but completion of the Prelaboratory Assignment should clarify most of the steps. The Report Sheet is also outlined in such detail as to help with the calculations. |

| **EXPERIMENTAL PROCEDURE** | **Procedure Overview:** One set of solutions having known molar concentrations of $FeNCS^{2+}$ is prepared for a calibration curve, a plot of absorbance versus concentration. A second set of equilibrium solutions is prepared and mixed to determine the respective equilibrium molar concentrations of $FeNCS^{2+}$. By carefully measuring the initial amounts of reactants placed in the reaction systems and the absorbance, the mass action expression at equilibrium can be solved; this equals K_c. |

A large number of pipets and 100-mL volumetric flasks are used in this experiment. Ask your instructor about working with a partner. A spectrophotometer is an expensive, delicate analytical instrument. Operate it with care, following the advice of your instructor, and it will give you good data.

A. A Set of Standard Solutions to Establish a Calibration Curve

The set of standard solutions is used to determine the absorbance of known molar concentrations of $FeNCS^{2+}$. A plot of the data, known as a calibration curve, is used to determine the equilibrium molar concentrations of $FeNCS^{2+}$ in Part B.

Once the standard solutions are prepared, proceed smoothly and methodically through Part A.4. Therefore, read through all of Part A before proceeding.

1. **Prepare a Set of the Standard Solutions.** Prepare the solutions in Table 34.1. Pipet 0, 1, 2, 3, 4 and 5 mL of 0.001 *M* NaSCN into separate, labeled, and clean 25-mL volumetric flasks (or 200-mm test tubes). Pipet 10.0 mL of 0.2 *M* $Fe(NO_3)_3$ into each flask (or test tube) and *quantitatively* dilute to 25 mL (the "mark" on the volumetric flask) with 0.1 *M* HNO_3. Stir/agitate each solution thoroughly to ensure that equilibrium is established.

Table 34.1 Composition of the Set of Standard FeNCS^{2+} Solutions for Preparing the Calibration Curve

Standard Solution	0.2 M Fe(NO$_3$)$_3$ (in 0.1 M HNO$_3$)	0.001 M NaSCN (in 0.1 M HNO$_3$)	0.1 M HNO$_3$
Blank	10.0 mL	0 mL	dilute to 25 mL
1	10.0 mL	1 mL	dilute to 25 mL
2	10.0 mL	2 mL	dilute to 25 mL
3	10.0 mL	3 mL	dilute to 25 mL
4	10.0 mL	4 mL	dilute to 25 mL
5	10.0 mL	5 mL	dilute to 25 mL

Record on the Report Sheet the *exact* molar concentrations of the Fe(NO$_3$)$_3$ and NaSCN reagent solutions.

2. **Prepare the Blank Solution.** After the spectrophotometer has been turned on for 10 minutes and the wavelength scale has been set at 447 nm, rinse a cuvet with several portions of the **blank solution.** Dry the outside of the cuvet with a clean Kimwipe, removing water and fingerprints.[2] Handle the lip of the cuvet thereafter. If a cuvet has two clear and two cloudy sides, be sure light passes through the clear sides and handle the cuvet on the cloudy sides.

3. **Calibrate the Spectrophotometer.** Place the cuvet, three-fourths filled with the blank solution, into the sample compartment, align the mark on the cuvet with that on the sample holder, and close the cover. Set the meter on the spectrophotometer to read zero absorbance (or 100%T).[3] Remove the cuvet. Consult with your instructor for any further calibration procedures. Once the instrument is set, *do not* perform any additional adjustments for the remainder of the experiment. If you accidentally do, merely repeat the calibration procedure.

4. **Record the Absorbance of the Standard Solutions.** Empty the cuvet and rinse it *thoroughly* with several small portions of Solution 1.[4] Fill it approximately three-fourths full. Again, carefully dry the outside of the cuvet with a clean Kimwipe. Remember, handle only the lip of the cuvet. Place the cuvet into the sample compartment and align the cuvet and sample holder marks; read the absorbance (or percent transmittance if the spectrophotometer has a meter readout) and record. Repeat for Solutions 2, 3, 4, and 5.

    Share the set of standard solutions with other chemists in the laboratory.

5. **Graph the Data.** Plot absorbance, A (ordinate), versus [FeNCS^{2+}] (abscissa) for the six solutions on linear graph paper or by using appropriate software. Draw the *best straight line* through the six points (see Figure 34.4) to establish the calibration curve. Ask your instructor to approve your graph.

1. **Prepare the Test Solutions.** In *clean* 150-mm test tubes (or 10-mL volumetric flasks)[5] prepare the test solutions in Table 34.2. Use pipets for the volumetric measurements. Be careful not to "mix" pipets to avoid contamination of the reagents prior to the preparation. Also note that the molar concentration of Fe(NO$_3$)$_3$ for

Blank 1 2 3 4 5

A set of labeled standard solutions.

Blank solution: a solution that contains all light-absorbing species except the one being investigated in the experiment

A box of lint-free tissue.

Appendix C

B. Absorbance for the Set of Test Solutions

---

[2]Water and fingerprints (or any foreign material) on the outside of the cuvet reduce the intensity of the light transmitted to the detector.

[3]For spectrophotometers with a meter readout (as opposed to digital), record the percent transmittance and then calculate the absorbance (Equation 34.2). This procedure is more accurate because %T is a linear scale (whereas absorbance is logarithmic) and because it is easier to estimate the linear %T values more accurately and consistently.

[4]If possible, prepare six matched cuvets, one for each standard solution, and successively measure the absorbance of each, remembering that the first solution is the blank solution.

[5]If 10-mL volumetric flasks are used, use pipets to dispense the volumes of the NaSCN and Fe(NO$_3$)$_3$ solutions and then dilute to the mark with 0.1 M HNO$_3$.

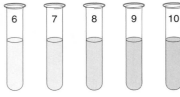

A set of labeled test solutions.

Table 34.2 Composition of the Set of Equilibrium Test Solutions for the Determination of K_c

Test Solution	0.002 M Fe(NO₃)₃* (in 0.1 M HNO₃)	0.002 M NaSCN (in 0.1 M HNO₃)	0.1 M HNO₃
6	5 mL	1 mL	4 mL
7	5 mL	2 mL	3 mL
8	5 mL	3 mL	2 mL
9	5 mL	4 mL	1 mL
10	5 mL	5 mL	—

*If 0.002 M Fe(NO₃)₃ is not available, dilute 1.0 mL (measure with a 1.0-mL pipet) of the 0.2 M Fe(NO₃)₃ used in Part A with 0.1 M HNO₃ in a 100-mL volumetric flask and then share the remaining solution with other students.

this set of solutions is 0.002 mol/L, *not* the 0.2 mol/L solution used in Part A and the molar concentration of NaSCN in 0.002 *M*, not 0.001 *M*.

Record the *exact* molar concentrations of the Fe(NO₃)₃ and NaSCN reagent solutions on the Report Sheet.

Once the test solutions are prepared, proceed smoothly and methodically (you need not hurry!) through Part B.3.

2. **Recalibrate the Spectrophotometer.** Use the blank solution from Part A to check the calibration of the spectrophotometer. See Part A.3.

3. **Determine the Absorbance of the Test Solutions.** Stir or agitate each test solution until equilibrium is reached (approximately 30 seconds). Rinse the cuvet thoroughly with several portions of the test solution and fill it three-fourths full. Clean and dry the outside of the cuvet. Be cautious in handling the cuvets. Record the absorbance of each test solution as was done in Part A.4.

> *Disposal:* Dispose of all waste thiocyanatoiron(III) ion solutions from Parts A and B in the "Waste Salts" container.

CLEANUP: Rinse the volumetric flasks, the pipets, and the cuvets twice with tap water and twice with deionized water. Discard each rinse in the sink.

4. **Use Data to Determine Equilibrium Concentrations.** From the calibration curve prepared in Part A.5, use the recorded absorbance value for each test solution to determine the equilibrium molar concentration of FeNCS²⁺.

C. Calculation of K_c

Appendix B

1. **Data Analysis.** Complete the calculations as outlined on the Report Sheet and described in the Introduction. Complete an entire K_c calculation for Test Solution 6 before attempting the calculations for the remaining solutions.

The equilibrium constant will vary from solution to solution and from chemist to chemist in this experiment, depending on chemical technique and the accumulation and interpretation of the data. Consequently, it is beneficial to work through your own calculations with other colleagues and then "pool" your final, experimental K_c values to determine an accumulated "probable" value and a standard deviation for K_c. The instructor may offer to assist in the calculations for K_c.

The Next Step

Spectrophotometry is a powerful tool for analyzing substances that have color, such as an aspirin (Experiment 19) complex, transition metal ions (Experiments 34, 35, 36), and anions (Experiment 3 and 37) to mention only a few. Research the spectrophotometric analysis of a specific substance and design a systematic study for its presence.

An Equilibrium Constant

Date _____ Lab Sec. _____ Name _____ Desk No. _____

1. a. Why do chemists prefer to read/record the absorbance, rather than the percent transmittance, of light (electromagnetic radiation) when analyzing a sample having a visible color?

   b. Three parameters affect the absorbance of a sample. Which one is the focus of this experiment?

2. Experimental Procedure, Part A.1, Table 34.1. A 3.00 mL aliquot of 0.001 M NaSCN is diluted to 25.0 mL with 0.1 M HNO_3.
   a. How many moles of SCN^- are present?

   b. If all of the SCN^- is complexed with Fe^{3+} to $FeNCS^{2+}$, what is the molar concentration of $FeNCS^{2+}$?

3. Experimental Procedure, Part A.1. For preparing a set of standard solutions of $FeNCS^{2+}$, the equilibrium molar concentration of $FeNCS^{2+}$ is assumed to equal the initial molar concentration of the SCN^- in the reaction mixture. Why is this assumption valid?

4. Experimental Procedure, Part A.3. The blank solution used to calibrate the spectrophotometer is 10.0 mL of 0.2 M $Fe(NO_3)_3$ diluted to 25.0 mL with 0.1 M HNO_3. Why is this solution preferred to simply using de-ionized water for the calibration?

5. Plot the following data as absorbance versus $[M^{n+}]$ as a calibration curve:

Absorbance, A	Molar Concentration of M^{n+}
0.045	3.0×10^{-4} mol/L
0.097	6.2×10^{-4} mol/L
0.14	9.0×10^{-4} mol/L
0.35	2.2×10^{-3} mol/L
0.51	3.2×10^{-3} mol/L

A "test" solution showed a percent transmittance ($\%T$) reading of 34.2%T. Interpret the calibration curve to determine the molar concentration of M^{n+} in the test solution.

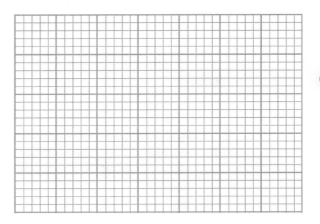

6. A 5.0 mL volume of 0.00200 M SCN$^-$ is mixed with 5.0 mL of 0.0200 M Fe^{3+} to form the blood-red FeNCS^{2+} complex. The equilibrium molar concentration of the FeNCS^{2+} determined from a calibration curve, is 7.0×10^{-4} mol/L. Calculate, in sequence, each of the following quantities in the aqueous solution to determine the equilibrium constant for the reaction,

$$Fe^{3+}(aq) + SCN^-(aq) \rightleftharpoons FeNCS^{2+}(aq)$$

a. *moles* of FeNCS^{2+} that form in reaching equilibrium _____

b. moles of Fe^{3+} that react to form the FeNCS^{2+} at equilibrium _____

c. moles of SCN$^-$ that react to form the FeNCS^{2+} at equilibrium _____

d. moles of Fe^{3+} initially placed in the reaction system _____

e. moles of SCN$^-$ initially placed in the reaction system _____

f. moles of Fe^{3+} (*un*reacted) at equilibrium (Eq. 34.7) _____

g. moles of SCN$^-$ (*un*reacted) at equilibrium (Eq. 34.8) _____

h. molar concentration of Fe^{3+} (*un*reacted) at equilibrium _____

i. molar concentration of SCN$^-$ (*un*reacted) at equilibrium _____

j. molar concentration of FeNCS^{2+} at equilibrium _____7.0×10^{-4} mol/L_____

k. $K_c = \dfrac{[FeNCS^{2+}]}{[Fe^{3+}][SCN^-]}$ _____

7. When 90.0 mL of 0.10 M Fe^{3+} is added to 10.0 mL of an SCN$^-$ solution, the equilibrium molar concentration of FeNCS^{2+}, as determined from a calibration curve, is 1.0×10^{-6} mol/L. Using the value of K_c from Question #6, determine the equilibrium molar concentration of SCN$^-$ in the system.

Experiment 34 *Report Sheet*

An Equilibrium Constant

Date ________ Lab Sec. ______ Name ______________________________ Desk No. ________

A. A Set of Standard Solutions to Establish a Standardization Curve

Molar concentration of $Fe(NO_3)_3$ ______________; Molar concentration of NaSCN ______________

Standard Solutions	***Blank***	***1****	***2***	***3***	***4***	***5***
A.1. Volume of NaSCN (*mL*)						
A.2. Moles of SCN^- (*mol*)						
A.3. $[SCN^-]$ (25.0 *mL*)						
A.4. $[FeNCS^{2+}]$ (*mol/L*)						
A.5. Percent transmittance, %*T* (for meter readings only)						
A.6. Absorbance, *A*						

*Calculation for Standard Solution 1.

A7. Plot data of *A* versus $[FeNCS^{2+}]$. Instructor's approval ________________________________

B. Absorbance for the Set of Test Solutions

Molar concentration of $Fe(NO_3)_3$ _____ ; Molar concentration of NaSCN _____

Test Solutions	6	7	8	9	10
B.1. Volume of $Fe(NO_3)_3$ (mL)					
B.2. Moles of Fe^{3+}, initial (mol)	*				
B.3. Volume of NaSCN (mL)					
B.4. Moles of SCN^-, initial (mol)	*				
B.5. Percent transmittance, %T (for meter readings only)					
B.6. Absorbance, A					

*Calculation for Test Solution 6.

C. Calculation of K_c

	6	7	8	9	10
C.1. $[FeNCS^{2+}]$, equilibrium, from calibration curve (mol/L)					
C.2. Moles $FeNCS^{2+}$ at equilibrium (10 mL) (mol)					

	6	7	8	9	10

[Fe^{3+}], equilibrium

C.3. Moles Fe^{3+}, reacted (*mol*)

C.4. Moles Fe^{3+}, equilibrium (*mol*)

C.5. [Fe^{3+}], equilibrium (*unreacted*). (10 mL) (*mol/L*) *

*Calculation for Test Solution 6.

[SCN$^-$], equilibrium

C.6. Moles SCN$^-$, reacted (*mol*)

C.7. Moles SCN$^-$, equilibrium (*mol*)

C.8. [SCN$^-$], equilibrium (*unreacted*). (10 mL) (*mol/L*) *

*Calculation for Test Solution 6.

C.9. $K_c = \dfrac{[\text{FeNCS}^{2+}]}{[\text{Fe}^{3+}][\text{SCN}^-]} = \dfrac{[\text{C.1.}]}{[\text{C.5}][\text{C.8}]}$ *

*Calculation for Test Solution 6.

C.10 Average K_c _____

C.11 Standard deviation of K_c (Appendix B) _____

C.12 Relative standard deviation of K_c (%RSD) _____

Calculation for standard deviation.

Class Data/Group	1	2	3	4	5
Average K_c					

Calculate the standard deviation of K_c from class data (Appendix B).

Calculate the relative standard deviation of K_c (%RSD) from class data

Laboratory Questions

Circle the questions that have been assigned.

1. Part A.2. All spectrophotometers are different. The spectrophotometer is to be set at 447 nm. What experiment could you do, what data would you collect, and how would you analyze the data to ensure that 447 nm is the "best" setting for measuring the absorbance of $FeNCS^{2+}$ in this experiment?

2. Part A.5. One of the standard solutions had an abnormally "high" absorbance reading causing a more positive slope for the data plot.
   a. Will the equilibrium concentrations of $FeNCS^{2+}$ in the Test Solutions (Part B) be too high or too low? Explain.
   b. Will the calculated K_c for the equilibrium be too high, too low, or unaffected by the erred data plot? Explain.

3. Part B.3. Fingerprint smudges are present on the cuvet containing the solution placed into the spectrophotometer for analysis.
   a. How does this technique error affect the absorbance reading for $FeNCS^{2+}$ in the analysis? Explain.
   b. Will the equilibrium concentration of $FeNCS^{2+}$ be recorded as being too high or too low? Explain.
   c. Will the equilibrium concentration of SCN^- be too high, too low, or unaffected by the technique error? Explain.
   d. Will the K_c for the equilibrium be too high, too low, or unaffected by the technique error? Explain.

4. Part B.3. For the preparation of Test Solution 8 (Table 34.2), the 2.0 mL of 0.1 M HNO_3 is omitted.
   a. How does this technique error affect the absorbance reading for $FeNCS^{2+}$? Explain.
   b. Will the K_c for the equilibrium be too high, too low or unaffected by the technique error? Explain.

*5. The equation, $A = a \cdot b \cdot c$ (see footnote 1), becomes nonlinear at high concentrations of the absorbing substance. Suppose you prepare a solution with a very high absorbance that is suspect in *not* following the linear relationship. How might you still use the sample for your analysis, rather than discarding the sample and the data?

6. Glass cuvets, used for precision absorbance measurements in a spectrophotometer, are marked so that they always have the *same* orientation in the sample compartment. What error does this minimize and why?

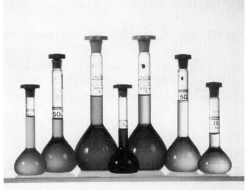

Experiment 35

Spectrophotometric Metal Ion Analysis

The absorbance of light indicates the relative concentrations of a substance in solution.

• To use a spectrophotometer to measure the concentration of a metal ion
• To use graphing techniques for data analysis
• To learn of the adaptability of spectrophotometric analyses

OBJECTIVES

The following techniques are used in the Experimental Procedure

TECHNIQUES

INTRODUCTION

Many transition metal cations have color, but in respective concentration ranges may also be considered hazardous wastes. Disposal of electroplating baths containing e.g., copper, nickel, or chromium ions cannot be simply discarded without some type of treatment. Often the treatment is precipitation, but also changes in oxidation states, acidification, complexation, or simple dilution procedures are used prior to disposal.

Because many transition metal ions do have color, their concentrations can be determined by a visible spectroscopic analysis. Those metal ions that do not have color may be analyzed by using ultraviolet radiation; however, a flame atomic absorption spectrophotometric (FAAS) analysis is by far the more common technique. The basic principles for either of the analyses are the same—standard solutions of the metal ion of interest are prepared, the absorption of each is determined to prepare a calibration curve. The absorption of the unknown is then determined and its concentration is determined by referring to the calibration curve.

Before beginning this experiment, read closely the Introduction to Experiment 34. An understanding of the absorbance and transmission of electromagnetic radiation through samples of varying concentrations in a spectrophotometer is imperative for an appreciation of the underlying chemical principles of this experiment.

Mixtures of ions can be problematic. The absorption spectrum for one ion may overlap and interfere with that of another ion. For that reason, FAAS or an induced couple plasma-atomic emission spectroscopy (ICP-AES) technique minimizes *some* of the interferences.

Procedure Overview: A set of standard solutions is prepared. First, the wavelength for maximum absorption, λ_{max}, of the metal ion in the visible region is determined from an absorbance, A, versus λ data plot. Next, the absorbance, A, of the standard solutions is determined at λ_{max} in order to construct a calibration curve of A versus [*conc*]. The two data plots can be constructed with the use of suitable software (e.g.,

EXPERIMENTAL PROCEDURE

Excel). The concentration of the metal ion in the unknown sample is then determined from the calibration curve.

The metal ion for your analysis will be selected by the instructor (or chosen by you) from one of the following cations: Cu^{2+}, Ni^{2+}, Co^{2+}, Cr^{3+}, or Fe^{3+}. You should also be aware that the color of some of these cations may be enhanced with the addition of a complexing agent. If a complexing agent is to be used in the analysis, it must be added to the stock solution in Part A and to the unknown in Part D. See Experiment 36. Consult your instructor.

A. A Set of Standard Solutions

Identify the metal ion for analysis on the Report Sheet.

1. **Prepare a Stock Solution.** Use a clean 100-mL volumetric flask to prepare a 0.20 *M* solution of the metal ion selected for analysis using 0.1 *M* HNO_3 as a diluent.[1] To do this, you will need to know the formula of the salt and calculate the mass of the salt for the solution. See Prelaboratory Assignment, Exercise 2. The mass of the salt is to be measured ± 0.001 g or as accurately as possible. Record on the Report Sheet the exact molar concentration of the stock solution.

2. **Prepare the Standard Solutions.** Prepare the solutions in Table 35.1 using clean, labeled 25-mL volumetric flasks (or 200-mm test tubes). Use pipets to dispense the solutions.

   Additional standard solutions may need to be prepared—the absorbance values for the set of standard solutions should range from 0 to ~1.1. See Part C.1. Calculate the molar concentrations of the standard solutions and record them in Part C of the Report Sheet.

B. Determination of λ_{max}

1. **Calibrate the Spectrophotometer.** After the spectrophotometer has been turned on for at least 10 minutes, set the wavelength scale to its minimum (~350 nm), and set the zero (0%T) on the spectrophotometer. Rinse twice a cuvet with the blank solution. Fill the cuvet at least three-fourths full with the *blank solution* and dry the outside of the cuvet with a clean Kimwipe, removing fingerprints and water. Place the cuvet into the sample compartment, aligning the mark on the cuvet with that on the sample holder (or the clear sides of the cuvet in the path of EM radiation). The meter on the spectrophotometer should now read zero absorbance or 100%T. If not, consult with your instructor.

Appendix B

2. **Wavelength Scan.** Set the wavelength of the spectrophotometer to the shortest wavelength setting (~350 nm). Place the stock solution (*Solution 5* in Table 35.1) in a cuvet following the "blank solution procedure" in Part B.1. Measure and record the wavelength and absorbance.

   Scan the wavelength range of the spectrophotometer while recording absorbance values. The spectrophotometer may do this continuously/automatically or you may need to manually scan every ~10 nm. If you scan manually, set the

Table 35.1 A Set of Standard Solutions for Metal Ion Analysis

Standard Solution	0.20 *M* Metal Ion	0.1 *M* HNO_3
Blank	0 mL	dilute to 25 mL
1	1 mL	dilute to 25 mL
2	5 mL	dilute to 25 mL
3	10 mL	dilute to 25 mL
4	20 mL	dilute to 25 mL
5	25 mL	—

[1]Add the complexing agent, if applicable, prior to the dilution in the 100-mL volumetric flask. The amount of complexing agent can be determined from a literature search.

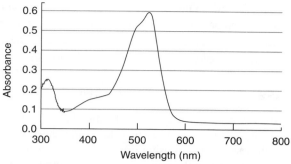

Figure 35.1 A representative absorption spectrum of a laboratory solution. Provided by Mike Schuder, Carroll College, Minnesota.

absorbance at zero (100% T) with the blank solution (0.1 M HNO$_3$) at *each* chosen wavelength before recording the absorbance of the stock solution.

Plot the data, absorption, A (ordinate) versus wavelength, λ (abscissa), manually or with the appropriate software to determine the λ_{max}, the wavelength where maximum absorption occurs for your metal ion (see Figure 35.1). Ask your instructor to approve your graph.

---

1. **Absorbance of Standard Solutions.** Set the spectrophotometer at λ_{max}. Record the absorbance of the stock solution (#5 in Table 35.1) at λ_{max}. Measure the absorbance for the other standard solutions in Table 35.1, starting with the most dilute (the blank).

    If *at least four* of the standard solutions are not in the absorbance range of 0 to ~1.1, prepare additional standard solutions.

2. **Plot the Data for the Calibration Curve.** Plot absorbance, A (ordinate), versus molar concentration (abscissa) for the six solutions in Table 35.1. Draw the best straight line through the data points to establish the calibration curve. Use Excel or similar graphing software to obtain values of the slope and y-intercept for the data plot. Ask your instructor to approve your graph.

C. Plot the Calibration Curve

Appendix B

---

1. **Prepare the Unknown.** Filter the solution if it is cloudy (for example, if it is a sample from an unknown source or sample from an electroplating bath).[2] Record the volume of the sample.[3] The solution may need to be diluted quantitatively (using 0.1 M HNO$_3$) with a pipet and 25-mL volumetric flask to have an absorbance measurement that falls within the range of the standard solutions (<1.1) on the calibration curve.

2. **Concentration of Metal Ion.** Once the sample is "prepared," measure its absorbance in the same manner as were the absorbance values for the standard solutions in Part C.1. Read the calibration curve to determine the molar concentration of the metal ion in the sample. Account for any dilution to determine the molar concentration of the metal ion in the original sample.

3. **Expressing Concentration.** Conventionally, in "real-world" samples, the metal ion concentrations are expressed in units of ppm (mg/L) or even smaller units depending upon the concentration of the metal. Express the concentration of your

D. Unknown Metal Ion Concentration

---

[2]Turbidity in the sample can dramatically affect the absorbance reading of a sample and therefore affect the presumed concentration of the metal ion in solution.
[3]Add the complexing agent, if applicable, to unknown sample at this point.

metal ion "appropriately" in units of mass/volume. Mass per volume units may be parts per hundred (pph or %), parts per million (ppm), parts per billion (ppb), etc.

Dispose of all prepared solutions in the "Waste Metal Salts" container.

The Next Step

(1) Research the spectrophotometric analysis of a metal ion of interest. Design a procedure(s) for its analysis as a function of a few select parameters. (2) Read to determine the advantages (or disadvantages) of using FAAS or ICP-AES for the analysis of metal ions. What are the similarities in the procedure for FAAS and visible spectrophometric analyses? (3) Other metal ion analysis methods include the use of graphite furnace atomic absorption (GFAA), microwave induced plasma (MIP), and DC arc plasma . . . where are these methods of analysis most applicable?

Spectrophotometric Metal Ion Analysis

Date _____ Lab Sec. _____ Name _____ Desk No. _____

1. Of the 100 mL of stock solution that is to be prepared for Part A.1, how many milliliters will be used for preparing the standard solutions in Part A.2?

2. A 100.0-mL volume of a 0.20 M stock solution of Cu^{2+} is to be prepared using $CuSO_4 \cdot 5H_2O$ (molar mass = 249.68 g/mol).[2]
   a. How many grams of $CuSO_4 \cdot 5H_2O$ must be measured for the preparation?

   b. Describe the procedure for the preparation of the solution using 0.1 M HNO_3 as a diluent.

   c. A 2.0-mL pipet transfers the Cu^{2+} stock solution to a 25.0-mL volumetric flask that is then diluted "to the mark" of the volumetric flask with 0.1 M HNO_3. What is the molar concentration of the diluted Cu^{2+} solution?

3. Refer to Figure 35.1.
   a. Identify the color of the visible spectrum where maximum absorption occurs (see Dry Lab 3).

   b. What is the predicted color of the solution?

   c. The Co^{2+} ion has a λ_{max} for absorption at 510 nm. Will a Co^{2+} solution have the same exact color as the solution of Figure 35.1? Explain.

---

[2] $CuSO_4 \cdot 5H_2O$ is used as an agricultural fungicide, bactericide, and herbicide.

4. A calibration curve for a common metal(II) ion is shown:

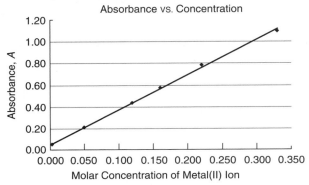

a. A metal(II) ion solution of unknown concentration shows an absorbance of 0.71. What is the molar concentration of the metal(II) ion in the sample?

b. For a metal(II) ion solution having a 0.288 M concentration, what would be its predicted absorbance?

c. Assuming a 1-cm cell, what is the value of the absorptivity coefficient for this solution?

5. Briefly describe the procedure of setting λ_{max} for a metal ion solution on the spectrophotometer.

6. The unknown metal ion concentration in your sample has an absorbance that is outside the range of the absorbance values for the standard solutions. What procedure(s) should be taken to rectify the discrepancy? Explain.

Spectrophotometric Metal Ion Analysis

Date _____ Lab Sec. _____ Name _____ Desk No. _____

Metal Ion for Analysis _____

A. A Set of Standard Solutions

1. **Prepare a Stock Solution.** Show the calculation for the mass of metal ion salt in the preparation of the stock solution.

Measured tared mass of metal ion salt (g) _____
Describe the preparation of the 0.20 M stock solution.

Concentration of stock solution (mol/L) _____

B. Determination of λ_{max}

2. **Wavelength scan.** Use the following table to record wavelength/absorbance data.

λ	Abs	λ	Abs	λ	Abs	λ	Abs	λ	Abs	λ	Abs	λ	Abs

Plot the data of absorbance vs. wavelength to set λ_{max}. From the data plot, λ_{max} = _____ nm

Have the instructor approve your graph. _____

C. Plot the Calibration Curve

1. **Absorbance of Standard Solutions.** Read and record the absorbance values for the standard solutions.

Standard Solution	Volume of Standard Solution (mL)	Absorbance	Calculated Molar Concentration
Blank	0		
1	1		
2	5		
3	10		
4	20		
5	25		
others as needed			

Plot the calibration curve of absorbance vs. molar concentration for the standard solutions.

Calculate the absorptivity coefficient for the metal ion.

Have the instructor approve your graph. _____

D. Unknown Metal Ion Concentration

1. **Prepare the Unknown.** Describe the preparation for the solution containing the unknown concentration of the metal ion.

   Volume of unknown sample solution (*mL*) _____

   Volume of diluted sample (as needed) (*mL*) _____

2. **Concentration of Metal Ion.** From the calibration curve determine the molar concentration of the metal ion in the unknown solution.

   Concentration of metal ion from the calibration curve (*mol/L*) _____

   Concentration of metal ion in the original sample if corrected for dilution (*mol/L*) _____

   Show your calculations to account for the dilution of the original sample.

3. **Expressing Concentration.** Express the concentration of the metal ion in the sample in appropriate units of mass/volume, i.e., pph (%), ppm, ppb, etc.

   Concentration of metal ion in the original sample (*mass/volume*) _____

   Show calculations.

Laboratory Questions

Circle the questions that have been assigned.

1. Part A.1. For the preparation of the stock solution, 0.1 *M* HNO_3 is used as a diluent rather than deionized water. Explain why. Hint: Review the solubility rules of transition metal ions in Appendix G.

2. Part A.2. In diluting the standard solutions, 0.1 *M* HNO_3 is used. In the dilution, is it more important to use the correct volume or the correct concentration of the HNO_3 solution for the dilution? Explain.

3. Part B.2. Will $Co^{2+}(aq)$ or $Cu^{2+}(aq)$ have the shorter λ_{max} for absorption in the visible spectrum? Explain. Hint: $Co^{2+}(aq)$ has a pink/rose color and $Cu^{2+}(aq)$ has a blue color. See Dry Lab 3.

4. Part D.1. An electrolytic-bath sample containing the metal ion of unknown concentration contains suspended matter. Because of a lack of time the absorbance of the sample was measured and recorded as is. As a result, will the concentration of the metal ion in the unknown be reported too high, too low, or remain unchanged as a result of chemist's hasty decision? Explain.

5. Part D.2. The cuvet used for the absorbance measurements is not wiped clean with a Kimwipe before the absorbance measurement. As a result, will the concentration of the metal ion in the unknown be reported too high, too low, or remain unchanged as a result of this poor technique? Explain.

The appearance of a nickel(II) ion complex depends on the ligands; at left is $[Ni(H_2O)_6]^{2+}$ and at right is $[Ni(NH_3)_6]^{2+}$.

OBJECTIVES

- To observe the various colors associated with transition metal ions
- To determine the relative strengths of ligands
- To compare the stability of complexes
- To synthesize a coordination compound

TECHNIQUES

The following techniques are used in the Experimental Procedure

INTRODUCTION

One of the most intriguing features of the transition metal ions is their vast array of colors. The blues, greens, and reds that we associate with chemicals are oftentimes due to the presence of transition metal ions. We observed these colors in Experiments 6, 27, 28, 34 and 35, where it was noted that some colors are characteristic of certain hydrated transition metal ions; for example, hydrated Cu^{2+} salts are blue, Ni^{2+} salts are green, and Fe^{3+} salts are rust-colored.

A second interesting feature of the transition metal ions is that subtle, and on occasion very significant, color changes occur when molecules or ions other than water bond to the metal ion to form a **complex.** These molecules or ions including water, called **ligands,** are Lewis bases (electron pair donors) which bond directly to the metal ion, producing a change in the electronic energy levels of the metal ion. As a result, the energy (and also the wavelengths) of light absorbed by the electrons in the transition metal ion and, consequently, the energy (and wavelengths) of light transmitted change.[1] The solution has a new color.

The complex has a number of ligands bonded to the transition metal ion which form a **coordination sphere.** The complex along with its neutralizing ion is called a **coordination compound.** For example, $K_4[Fe(CN)_6]$ is a coordination compound: the six CN^- ions are the ligands, $Fe(CN)_6$ is the coordination sphere, and $[Fe(CN)_6]^{4-}$ is the complex.

Complex: an ion or molecule formed between a metal ion (a Lewis acid) and anions or molecules (Lewis bases)

Ligand: a Lewis base (an electron-pair donor) that combines with a metal ion to form a complex

Coordination sphere: the metal ion and all ligands of the complex

Coordination compound: a neutral compound which has a cationic and/or an anionic complex ion

[1] See Dry Lab 3 for a discussion of the theory of the origin of color in substances.

Figure 36.1 A photographic negative and a positive print made from it.

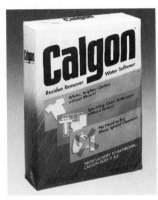

Sodium tripolyphosphate, a component of Calgon®, complexes and solubilizes $Ca^{2+}(aq)$—thus the undesirable "calcium is gone" from the wash water!

Figure 36.2 The complex shows three ethylenediamine ligands, each as a bidentate ligand. Ethylenediamine is a chelating agent.

More than the intrigue of the color and color changes are the varied uses that these complexes have in our society. A few examples of the applicability of complex formation follow:

- For photographic film development (Figure 36.1), sodium thiosulfate, $Na_2S_2O_3$, called **hypo**, removes the unsensitized silver ion from the film in the form of the soluble silver complex, $[Ag(S_2O_3)_2]^{3-}$:

$$AgBr(s) + 2\,S_2O_3{}^{2-}(aq) \rightarrow [Ag(S_2O_3)_2]^{3-}(aq) + Br^-(aq)$$

- For the removal of calcium ion hardness from water, soluble polyphosphates, such as sodium tripolyphosphate, $Na_5P_3O_{10}$, are added to detergents to form a soluble calcium complex:

$$Ca^{2+}(aq) + P_3O_{10}{}^{5-}(aq) \rightarrow [CaP_3O_{10}]^{3-}(aq)$$

- For mercury and lead poisoning by ingestion, mercaptol (British antilewisite), $C_3H_8OS_2$, is swallowed for purposes of forming the complex with the metal ion and rendering the "free" ion ineffective:

$$Hg^{2+}(aq)\ [\text{or } Pb^{2+}\ (aq)] + 2\,CH_2{-}CH{-}CH_2(aq) \rightarrow [Hg(C_3H_8OS_2)_2]^{2+}(aq)$$
$$\qquad\qquad\qquad\qquad\quad |\qquad |\qquad |$$
$$\qquad\qquad\qquad\qquad\ SH\quad SH\quad OH$$

- In the qualitative identification of transition metal ions (Experiments 38 and 39), many of the transition metal ions are confirmed present as a result of complex formation: copper ion as $[Cu(NH_3)_4]^{2+}$, nickel ion as $Ni(HDMG)_2$, iron(III) ion as $[FeNCS]^{2+}$, and zinc ion as $[Zn_3[Fe(CN)_6]_2]^{2-}$.

- Trace amounts of metal ions, such as zinc, aluminum, and iron, tend to catalyze the air oxidation (the spoilage) of various foods. To retard spoilage and extend the shelf life, some food companies add a small amount of calcium disodium ethylenediaminetetraacetate, abbreviated $CaNa_2Y$, to their product. The Y^{4-} ion (Table 36.1) complexes (or sequesters) the metal ions and nullifies their catalytic activity. The H_2Y^{2-} ion is also used to analyze for water hardness, complexing the Ca^{2+} and Mg^{2+} as water-soluble CaH_2Y and MgH_2Y (Experiment 21) in a titration procedure.

The bond strength between the transition metal ion and its ligands varies, depending on the electron pair donor (Lewis base) strength of the ligand and the electron pair acceptor (Lewis acid) strength of the transition metal ion. Ligands may be neutral (e.g., H_2O, NH_3, $H_2NCH_2CH_2NH_2$) or anionic (e.g., CN^-, SCN^-, Cl^-).

A single ligand may form one bond to the metal ion (a **monodentate** ligand), two bonds to the metal ion (a **bidentate** ligand), three bonds to the metal ion (a **tridentate** ligand), and so on. Ligands that form two or more bonds to a transition metal ion are also called **chelating agents** and **sequestering agents** (Figure 36.2). In Table 36.1 note that a nonbonding electron pair (a Lewis base) is positioned on each atom of the ligand that serves as a bonding site to the transition metal ion.

The complex formed between a chelating agent (a *poly*dentate ligand) and a metal ion is generally *more* stable than that formed between a monodentate ligand and a metal ion. The explanation is that the several ligand–metal bonds between a polydentate ligand and a metal ion are more difficult to break than a single bond between a monodentate ligand and a metal ion. The stability of complexes having polydentate ligands will be compared with the stability of those having only monodentate ligands in this experiment.

The *number of bonds* between a metal ion and its ligands is called the **coordination number** of the complex. If four monodentate ligands or if two bidentate ligands bond to a metal ion, the coordination number for the complex is 4. Six water molecules, or six cyanide ions, or three ethylenediamine (a bidentate ligand) molecules (see Figure 36.2), or two diethylenetriamine (a tridentate ligand) molecules

Table 36.1 Common Ligands

Monodentate Ligands

$H_2O:$, $:NH_3$, $:CN^-$, $:SCN^-$, $:S_2O_3{}^{2-}$, $:F^-$, $:Cl^-$, $:Br^-$, $:I^-$ $:OH^-$

Bidentate Ligands (chelating agents)

Ethylenediamine (en)

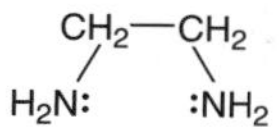

Oxalate ion

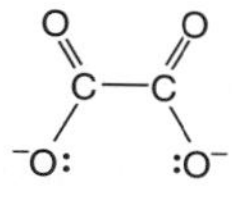

Mercaptol

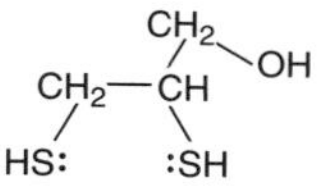

Tartrate ion

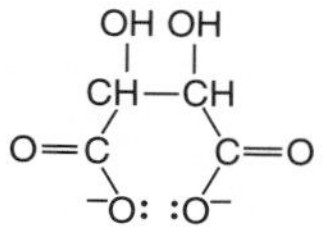

Polydentate Ligands (chelating agents)

Diethylenetriamine

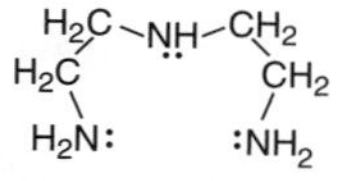

Tripolyphosphate ion, $P_3O_{10}{}^{5-}$

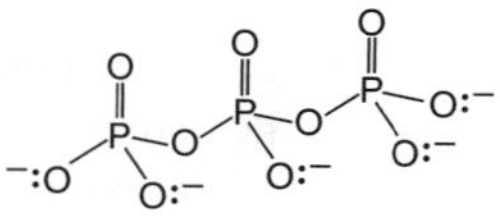

Ethylenediaminetetraacetate ion, Y^{4-}

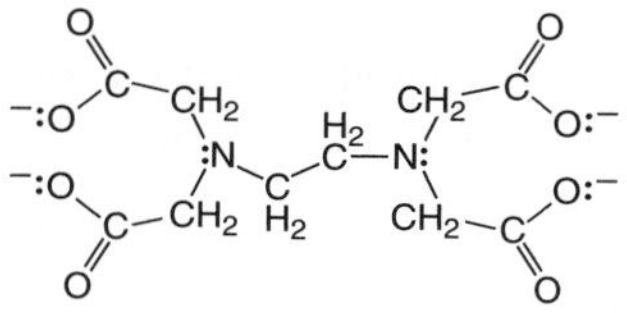

Table 36.2 Common Coordination Numbers of Some Transition Metal Ions

Coordination Number	Transition Metal Ions
2	Ag^+, Au^+
4	Hg^{2+}, Cu^{2+}, Ni^{2+}, Co^{2+}, Zn^{2+}, Pd^{2+}, Pt^{2+}, Au^{3+}
6	Co^{2+}, Co^{3+}, Fe^{2+}, Fe^{3+}, Cr^{3+}, Ni^{2+}, Cu^{2+}

bonded to a given metal ion all form a complex with a coordination number of 6 (Figure 36.3). Coordination numbers of 2, 4, and 6 are most common among the transition metal ions.

Table 36.2 lists the common coordination numbers for some transition metal ions.

In Parts A, B, C, and D of this experiment we will observe the formation of complexes between Cu^{2+}, Ni^{2+}, and Co^{2+} ions and the ligands Cl^-, H_2O, NH_3, ethylenediamine (en), and the thiocyanate ion, SCN^-. In specified examples, we will determine the stability of the complex.

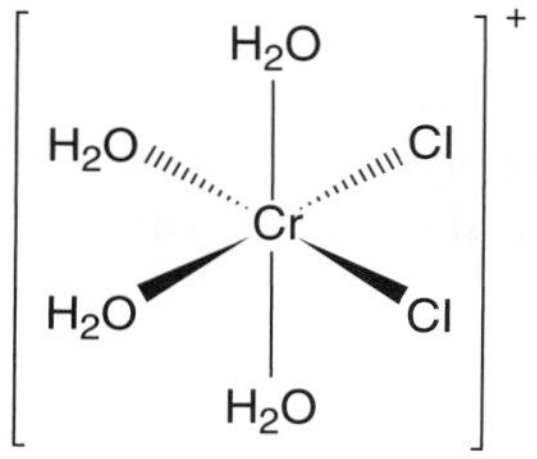

Figure 36.3 The complex has a coordination number of six, each ligand is monodentate.

Stablity of a Complex

The stability of a complex can be determined by mixing it with an anion known to form a precipitate with the cation of the complex. The anion most commonly used to measure the stability of a complex is the hydroxide ion. For example, consider the copper(II) ion and the generic mondentate ligand, X^-, forming the complex $[CuX_4]^{2-}$:

$$Cu^{2+}(aq) + 4\,X^-(aq) \rightleftharpoons [CuX_4]^{2-}(aq)$$

When hydroxide ion is added to this system (in a state of dynamic equilibrium), the "free" copper(II) ion has a choice of now combining with the anion that forms the stronger bond—in other words, a competition for the copper(II) ion between the two anions, X^- and OH^-, exists in solution. If the ligand forms the stronger bond to the copper(II) ion, the complex remains in solution and no change is observed; if, on the other hand, the hydroxide forms a stronger bond to the copper(II) ion, $Cu(OH)_2$ precipitates from solution.

$$Cu^{2+}(aq) + 2\ OH^-(aq) \rightleftharpoons Cu(OH)_2(s)$$

Therefore, a measure of the stability of the complex is determined.

To write the formulas of the complexes in this experiment, we will assume that the coordination number of Cu^{2+} is always 4 and that of Ni^{2+} and Co^{2+} is always 6.

Parts E, F, and G outline the syntheses of several coordination compounds:

- tetraamminecopper(II) sulfate monohydrate, $[Cu(NH_3)_4]SO_4 \cdot H_2O$
- hexaamminenickel(II) chloride, $[Ni(NH_3)_6]Cl_2$
- *tris*(ethylenediamine)nickel(II) chloride dihydrate, $[Ni(en)_3]Cl_2 \cdot 2H_2O$

Tetraamminecopper(II) Sulfate Monohydrate

Ammonia is added to an aqueous solution of copper(II) sulfate pentahydrate, $[Cu(H_2O)_4]SO_4 \cdot H_2O$ (molar mass = 249.68 g/mol). Ammonia displaces the four water ligands from the coordination sphere:

$$[Cu(H_2O)_4]SO_4 \cdot H_2O + 4\ NH_3 \xrightarrow{\text{aqueous}} [Cu(NH_3)_4]SO_4 \cdot H_2O + 4\ H_2O \qquad (36.1)$$

The solution is cooled and 95% ethanol is added to reduce the solubility of the *blue* tetraamminecopper(II) sulfate (molar mass = 245.61 g/mol) salt which then crystallizes.

Hexaamminenickel(II) Chloride

For the preparation of hexaamminenickel(II) chloride, six ammonia molecules displace the six water ligands in $[Ni(H_2O)_6]Cl_2$ (molar mass = 237.71 g/mol).

$$[Ni(H_2O)_6]Cl_2 + 6\ NH_3 \xrightarrow{\text{aqueous}} [Ni(NH_3)_6]Cl_2 + 6\ H_2O \qquad (36.2)$$

The $[Ni(NH_3)_6]Cl_2$ coordination compound (molar mass = 231.78 g/mol) is cooled and precipitated with 95% ethanol; the lower polarity of the 95% ethanol causes the *lavender* salt to become less soluble and precipitate.

***Tris*(ethylenediamine)nickel (II) Chloride Dihydrate**

Three ethylenediamine (en) molecules displace the six water ligands of $[Ni(H_2O)_6]Cl_2$ (molar mass = 237.71 g/mol) for the preparation of the *violet tris*(ethylenediamine) nickel(II) chloride dihydrate (molar mass = 345.92 g/mol).

$$[Ni(H_2O)_6]Cl_2 + 3\ en \xrightarrow{\text{aqueous}} [Ni(en)_3]Cl_2 \cdot 2H_2O + 4\ H_2O \qquad (36.3)$$

EXPERIMENTAL PROCEDURE

Procedure Overview: Several complexes of Cu^{2+}, Ni^{2+}, and Co^{2+} are formed and studied. The observations of color change that result from the addition of a ligand are used to understand the relative stability of the various complexes that form. One or more coordination compounds are synthesized and isolated.

Notes on observation. Some colors may be difficult to distinguish. If a precipitate initially forms when ligand is added, add more of the ligand. *Always* compare the test solution with the original aqueous solution. On occasion you may need to discard some of the test solution if too much of the original solution was used. If a color change occurs (*not* a change in color intensity) then a new complex has formed. Also, after each addition of the ligand-containing solution, tap the test tube to agitate the mixture (Figure 36.4) and view the solution at various angles to note the color (change) (Figure 36.5).

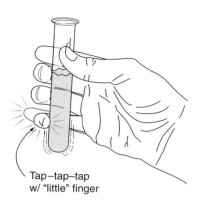

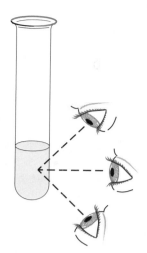

Figure 36.4 Shake the contents of the test tube with the "little" finger.

Figure 36.5 View solution from all angles.

Caution: *Several of the ligand-containing solutions should be handled with care. The conc HCl, conc NH₃, and ethylenediamine reagents produce characteristic odors that are skin and respiratory irritants. Use "drops" as suggested and avoid inhalation and skin contact.*

Most complex ion formation reactions are completed in small test tubes. While volumes of solutions need not be exact, a volume close to the suggested amount should be used. A 75-mm test tube has a volume of 3 mL—proportional fractions should be used in estimating the volume suggested in the procedure.

From three to six small test tubes are used for Parts A, B, C, and D of the experiment. Plan to keep them clean, rinsing thoroughly after each use. However, do *not* discard any test solutions until the entire "Part" of the Experimental Procedure has been completed.

For Parts A–D, perform the experiment with a partner. At each circled superscript①-⑨ in the procedure, *stop,* and record your observation on the Report Sheet. Discuss your observations with your lab partner and your instructor. Consult with your laboratory instructor to determine which of the coordination compounds outlined in Parts E, F, and G you are to synthesize.

---

1. **Form the Complexes.** Place about 0.5 mL of 0.1 *M* $CuSO_4$, 0.1 *M* $Ni(NO_3)_2$, and 0.1 *M* $CoCl_2$ into each of three separate small test tubes.①Add 1 mL (20 drops) of conc HCl to each. (**Caution:** *Do not allow conc HCl to contact skin or clothing. Flush immediately the affected area with water.*) Tap the test tube to agitate. Record your observations on the Report Sheet.②

2. **Dilute the Complexes.** Slowly add ~1 mL of water to each test tube. Compare the colors of the solutions to about 2.5 mL of the original solutions. Does the original color return? Record.③

A. Chloro Complexes of the Copper(II), Nickel(II), and Cobalt(II) Ions

---

1. **Form the Complexes.** Place about 0.5 mL of 0.1 *M* $CuSO_4$ in each of five test tubes and transfer them to the fume hood. Add 5 drops of conc NH₃ (**Caution:** *avoid breathing vapors*) to the first test tube, 5 drops of ethylenediamine (**Caution:** *avoid breathing its vapors, too*) to the second, 5 drops of 0.1 *M* KSCN to the third test tube, and nothing more to a fourth test tube (Figure 36.6).

The fifth test tube can be used to form a complex with a ligand selected by your instructor—the thiosulfate anion, $S_2O_3^{2-}$, the oxalate anion, $C_2O_4^{2-}$, the

B. Complexes of the Copper(II) Ion

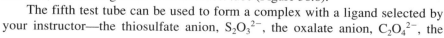

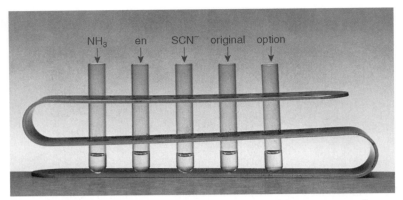

Figure 36.6 Set up of five test tubes for ligand addition to a metal ion.

tartrate anion, $(CHOH)_2(CO_2)_2^{2-}$, and the ethylenediaminetetraacetate anion, Y^{4-}, readily form complexes and are suggested.

If a precipitate forms in any of the solutions, *add an excess* of the ligand-containing solution. Compare the appearance of the test solutions with the 1 mL of original $CuSO_4$ solution in the fourth test tube.[4]

2. **Test for Stability.** Add 3–5 drops of 1 M NaOH to each of the three (or four) solutions. Account for your observations.[5]

C. Complexes of the Nickel(II) Ion

1. **Form the Complexes.** Repeat Part B.1, substituting 0.1 M $NiCl_2$ for 0.1 M $CuSO_4$. Record your observations.[6]

2. **Test for Stability.** Add 3–5 drops of 1 M NaOH to each test solution. Account for your observations.[7]

D. Complexes of the Cobalt(II) Ion

1. **Form the Complexes.** Repeat Part B.1, substituting 0.1 M $Co(NO_3)_2$ for 0.1 M $CuSO_4$. Record your observations.[8]

2. **Test for Stability.** Add 3–5 drops of 1 M NaOH to each test solution. Account for your observations.[9]

 Disposal: Dispose of the waste solutions from Parts A, B, C, and D in the "Waste Metal Ion Solutions" container.

 CLEANUP: Rinse the test tubes twice with tap water and twice with deionized water and discard in the "Waste Metal Ion Solutions" container. Additional rinses can be discarded in the sink, followed by a generous amount of tap water.

E. Synthesis of Tetraamminecopper(II) Sulfate Monohydrate

1. **Dissolve the Starting Material.** Measure 6 g (± 0.01 g) of copper(II) sulfate pentahydrate, $[Cu(H_2O)_4]SO_4 \cdot H_2O$, in a clean, dry previously mass-measured 125-mL Erlenmeyer flask. Dissolve the sample in 15 mL of deionized water. Heating may be necessary. Transfer the beaker to the fume hood.

2. **Precipitate the Complex Ion.** Add conc NH_3 (**Caution:** *do not inhale*) until the precipitate that initially forms has dissolved. Cool the deep-blue solution in an ice bath. Cool 20 mL of 95% ethanol (**Caution:** *flammable, extinguish all flames*) to ice bath temperature and then slowly add it to the solution. The blue, solid complex should form.

3. **Isolate the Product.** Premeasure the mass (± 0.01 g) of a piece of filter paper and fit it in a Büchner funnel. Vacuum filter the solution (Figure 36.7); wash the solid with two 5-mL portions of cold 95% ethanol (*not* water!). Place the filter and sam-

Figure 36.7 A vacuum filtration setup for filtering the coordination compound.

ple on a watchglass and allow the sample to air-dry. Determine the mass of the filter paper and product. Calculate the percent yield.

4. **Obtain the Instructor's Approval.** Transfer your product to a *clean, dry* test tube, stopper and (properly) label the test tube, and submit it to your instructor for approval.

1. **Form the Complex Ion.** In a previously mass-measured 125-mL Erlenmeyer flask, measure and dissolve 6 g (±0.01 g) of nickel(II) chloride hexahydrate, $[Ni(H_2O)_6]Cl_2$, in (no more than) 10 mL of warm (~50°C) deionized water. In a fume hood, slowly add 20 mL of conc NH_3 (**Caution!**).

2. **Precipitate the Complex Ion.** Cool the mixture in an ice bath. Cool 15 mL of 95% ethanol (**Caution:** *flammable, extinguish all flames*) to ice bath temperature and then slowly add it to the solution. Allow the mixture to settle for complete precipitation of the lavender product. The supernatant should be nearly colorless.

3. **Isolate the Product.** Premeasure the mass (±0.01 g) of a piece of filter paper, fitted for a Büchner funnel. Vacuum filter the product; wash with two 5-mL volumes of cold 95% ethanol. Air-dry the product and filter paper on a watchglass. Determine the mass of the filter paper and product. Calculate the percent yield.

4. **Obtain the Instructor's Approval.** Transfer your product to a *clean, dry* test tube, stopper and (properly) label the test tube, and submit it to your instructor for approval.

F. Synthesis of Hexaamminenickel (II) Chloride

1. **Form the Complex Ion.** In a previously mass-measured 125-mL Erlenmeyer flask, measure and dissolve 6 g (±0.01 g) of nickel(II) chloride hexahydrate, $[Ni(H_2O)_6]Cl_2$, in 10 mL of warm (~50°C) deionized water. Cool the mixture in an ice bath. In a fume hood, slowly add 10 mL ethylenediamine. (**Caution:** *Avoid skin contact and inhalation.*)

2. **Precipitate and Isolate the Product.** Complete as in Parts F.2 and F.3.

G. Synthesis of *Tris*(ethylenediamine)nickel (II) Chloride Dihydrate

3. **Obtain the Instructor's Approval.** Transfer your product to a *clean, dry* test tube, stopper and (properly) label the test tube, and submit it to your instructor for approval.

Disposal: Dispose of the waste solutions from Parts E, F, and G in the "Waste Metal Ion Solutions" container.

CLEANUP: Rinse all glassware twice with tap water and discard in the "Waste Metal Ion Solutions" container. Rinse the glassware twice with deionized water and discard in the sink; follow with a generous supply of tap water.

The Next Step

The formation of coordination compounds is common in nature and the laboratory. The complexing of trace metal ions in soils maintains a stable plant nutrient base. Complexing metal ions in electroplating baths results in the uniform deposition of metal ions at cathodes. The formation of the iron-hemoglobin complex is vital to life. Complexing "hot" metal ions from nuclear reactors assists in cleaning reactor cores. Research the presence of a transition a metal ion complex in nature, and design a plan for its analysis. Research the synthesis and analysis of a transition metal ion complex. Most are quite colorful!

Experiment 36 *Prelaboratory Assignment*

Transition Metal Complexes

Date ________ Lab Sec. ______ Name ______________________ Desk No. ________

1. Consider the coordination compound, $[CoCl(NH_3)_4(H_2O)]SO_4$. Use the definitions in the Introduction to identify the following with the formula and charge (if applicable).
   a. the ligand(s)

   b. the complex

   c. the coordination sphere

   d. the coordination number of cobalt

2. Write the formula of the complex ion formed between
   a. the ligand, Cl^-, and the platinum(II) ion with a coordination number of four.

   b. the ligand, NH_3, and the nickel(II) ion with a coordination number of six.

   c. the ligand, ethylenediamine (abbreviated "en"), and the chromium(III) ion with a coordination number of six.

   d. the ligand, dihydrogen ethylenediaminetetraacetate, abbreviated H_2Y^{2-} and the zinc(II) ion with a coordination number of four.

3. The cyanide ion, CN^-, is a much *stronger* ligand than the chloride ion, Cl^-. Explain.

4. Complex ions with the ligand ethylenediamine are more stable than those with the ligand NH_3, assuming the same metal ion. Explain.

5. When 6 M NaOH is slowly added to a solution containing chromium(III) ion, a precipitate forms. However when an excess of 6 M NaOH is added, the precipitate dissolves forming a complex ion with a coordination number of four.
   a. Write the formula of the precipitate.

   b. Write the formula of the complex ion (for example, see Experiment 15).

6. Experimental Procedure, Parts A–D. Identify the chemicals that are cited as **caution**.

7. Experimental Procedure, Parts A–D. Of the *four* ligands in the study, which is a/are polydentate(s)?

8. Experimental Procedure, Part E. A 6.0-g sample of $[Cu(H_2O)_4]SO_4 \cdot H_2O$ (molar mass $= 249.68$ g/mol) is dissolved in deionized water. If an excess of ammonia is added to the solution, solid $[Cu(NH_3)_4]SO_4 \cdot H_2O$ (molar mass $= 245.61$ g/mol) forms. What is the theoretical yield of product from the reaction?

Transition Metal Complexes

Date _____ Lab Sec. _____ Name _____ Desk No. _____

A. Chloro Complexes of the Copper(II), Nickel(II), and Cobalt(II) Ions

Solution	①Color/H_2O	②Color/HCl	Formula of Complex	③Effect of H_2O
0.1 M $CuSO_4$				
0.1 M $Ni(NO_3)_2$				
0.1 M $CoCl_2$				

For each metal ion state whether the aqua complex or the chloro complex is more stable:

Cu^{2+} _____; Ni^{2+} _____; Co^{2+} _____

B. Complexes of the Copper(II) Ion

Ligand	④Color	Formula of Complex	⑤Effect of OH^-
NH_3			
Ethylenediamine			
SCN^-			
H_2O			
_____			

C. Complexes of the Nickel(II) Ion

Ligand	⑥Color	Formula of Complex	⑦Effect of OH^-
NH_3			
Ethylenediamine			
SCN^-			
H_2O			
_____			

D. Complexes of the Cobalt(II) Ion

Ligand	⑧Color	Formula of Complex	⑨Effect of OH^-
NH_3			
Ethylenediamine			
SCN^-			
H_2O			
_____			

Review of Data

Of the complexes formed with copper(II), nickel(II), and colbalt(II), which metal ion appears to form the most stable complexes. Explain. Also see Laboratory Question 5.

Synthesis of a Coordination Compound

Name and formula of coordination compound   _____

1. Mass of Erlenmeyer flask (g)   _____

2. Mass of Erlenmeyer flask + starting material   _____

3. Mass of starting material (g)   _____

4. Mass of filter paper (g)   _____

5. Mass of filter paper + product (g)   _____

6. Mass of product (g)   _____

7. Instructor's approval   _____

8. Theoretical yield of product* (g)   _____

9. Percent yield* (%)   _____

*Show calculation.

Laboratory Questions

Circle the questions that have been assigned

1. Part A.2. Is the chloride ion or water a stronger ligand? Explain.

2. Part B.1. Is water or ammonia a stronger ligand. Explain.

3. Part B.2. Is ammonia or ethylenediamine a stronger ligand? Explain.

4. Parts A–D. Of the five ligands, Cl^-, NH_3, $H_2NCH_2CH_2NH_2$, SCN^-, and H_2O, studied in this experiment, which ligand appeared to be the strongest ligand? Why? Which ligand appeared to be the weakest? Why?

5. Parts A–D. Along period 4 of the Periodic Table, cobalt, nickel, and copper appear in succession. From your data, does a trend in the stability of complexes that they form seem to exist? Explain.

6. Part E.2. a. A solution of potassium cyanide, KCN, instead of conc NH_3 is added. Write the formula of the expected complex ion.
   b. A solution of potassium chloride, KCl, instead of conc NH_3 is added. Write the formula of the expected complex ion.

7. Part E.2. Identify the precipitate that forms *before* the addition of excess *conc* NH_3.

8. Part E.3. Why is 95% ethanol used to wash the solid [the tetraamminecopper(II) sulfate monohydrate product] on the filter paper instead of deionized water?

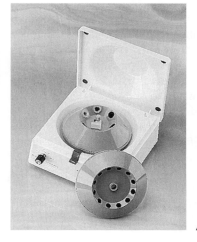

Preface to Qualitative Analysis

A centrifuge compacts a precipitate by centrifugal force.

Some rocks have a reddish tint; others are nearly black. Table salt is white, but not all salts are white. A quick, yet simple, identification of the ions of a salt or salt mixture is often convenient. Gold prospectors were quick to identify the presence of gold and/or silver. It is the characteristic physical and chemical properties of an ion that will allow us in the next series of experiments to identify its presence in a sample. For example, the Ag^+ ion is identified as being present in a solution by its precipitation as the chloride, $AgCl(s)$. Although other cations precipitate as the chloride, silver chloride is the only one that is soluble in an ammoniacal solution.[1]

Many ions have similar chemical properties, but each ion also has unique chemical properties. In order to characteristically identify a particular ion in a mixture, the interferences of ions of similar properties must be eliminated. The chemist must take advantage of the "unique" chemical properties of the ion in question to determine its presence in a mixture. A procedure that follows this pattern of analysis is called **qualitative analysis**, one that will become more familiar in the next few experiments.

Qualitative analysis: a systematic procedure by which the presence (or absence) of a substance (usually a cation or anion) can be determined

With enough knowledge of the chemistry of the various ions, a unique separation and identification procedure for each ion can be developed. Some procedures are quick, one-step, tests; others are more exhausting. The experimental procedure, however, must systematically eliminate all other ions that may interfere with the specific ion test.

The separation and identification of the ions in a mixture require the application of many chemical principles, many of which we will cite as we proceed. An *understanding* of the chemistry of precipitate formation, ionic equilibrium, acids and bases, pH, oxidation and reduction reactions, and complex formation is necessary for their successful separation and identification. To help you understand these principles and test procedures, each experiment presents some pertinent chemical equations, but you are also asked to write equations for other reactions that occur in the separation and identification of the ions.

To complete the procedures for the separation and identification of ions, you will need to practice good laboratory techniques and, in addition, develop several new techniques. The most critical techniques in qualitative analysis are the maintenance of clean glassware and the prevention of contamination of the testing reagents.

---

[1]This was one test procedure that prospectors for silver used in the early prospecting days.

A. Measuring and Mixing Test Solutions

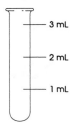

- 3 mL
- 2 mL
- 1 mL

Figure D4.1 Use the 10-mL graduated cylinder or pipet to transfer 1 mL, 2 mL, and 3 mL of water to a small test tube. Mark the test tube at each mark.

Most of the testing for ions will be performed in small (~3 mL) test tubes or centrifuge tubes that fit the centrifuges used in your laboratory. Reagents will be added with dropper bottles or dropping pipets (~15–20 drops/mL; you should do a preliminary check with your dropping pipet to determine the drops/mL). If the procedure dictates the addition of 1 mL, do *not* use a graduated cylinder to transfer the 1 mL, but instead use the dropping pipet or estimate the addition of 1 mL in the (~3-mL) test tube (Figure D4.1). Do *not* mix the different dropping pipets with the various test reagents you will be using and do *not* contaminate a reagent by inserting your pipet or dropping pipet into it. Instead, if the procedure calls for a larger volume, first dispense a small amount of reagent into one of your *small* beakers or test tubes.

When mixing solutions in a test tube, agitate by tapping the side of the test tube, break up a precipitate with a stirring rod, or stopper the test tube and invert, but *never* use your thumb (Figure D4.2)!

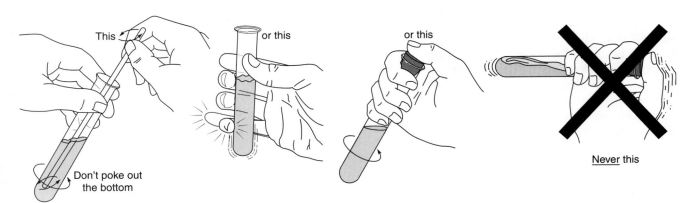

This → Don't poke out the bottom

or this ✓

or this

Never this

Figure D4.2 Technique for mixing solutions in a test tube.

B. Testing for Complete Precipitation

Precipitating reagent: a solution containing an ion(s) that, when added to a second solution, causes a precipitate to form

Oftentimes it is advisable to test the supernatant to determine if complete precipitation of an ion has occurred. After the mixture has been centrifuged, add a drop of the **precipitating reagent** to the supernatant (Figure D4.3). If a precipitate forms, add several more drops, disperse the mixture with a stirring rod or by gentle agitation, and centrifuge. Repeat the test for complete precipitation.

C. Washing a Precipitate

A precipitate must often be washed to remove occluded impurities. Add deionized water or wash liquid to the precipitate, disperse the solid thoroughly with a stirring rod or by gentle agitation, centrifuge, and decant. Usually the wash liquid can be discarded. Two washings are usually satisfactory. Failure to properly wash precipitates often leads to errors in the analysis (and arguments with your laboratory instructor!) because of the presence of occluded contaminating or interfering ions.

As you will be using the centrifuge frequently in the next several experiments, *be sure* to read carefully Technique 11F in the Laboratory Techniques section of this manual.

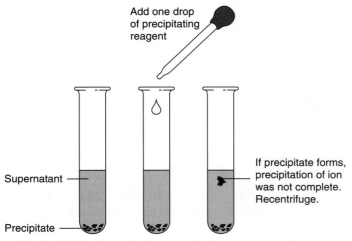

Add one drop
of precipitating
reagent

Supernatant

Precipitate

If precipitate forms,
precipitation of ion
was not complete.
Recentrifuge.

Figure D4.3 A test for the complete precipitation of an ion.

Many procedures call for a solution to be heated. Heating a mixture either accelerates the rate of a chemical reaction or causes the formation of larger crystals of precipitate, allowing its separation to be more complete. ***Never*** *heat the small test tubes directly with a flame.* Heat one or several test tubes in a hot water bath; a 150-mL beaker containing 100 mL of deionized water is satisfactory. The test tube can be placed directly into the bath, supported against the wall of the beaker (Technique 13B). Read the Experimental Procedure before lab; if a hot water bath is needed, start heating the water at the beginning of the laboratory period and keep it warm with a hot plate or **"cool" flame.**

A solution can be cooled by placing the test tube under cold running tap water or by submerging the test tube in a beaker of ice water.

D. Heating and Cooling Solutions

Cool flame: a nonluminous flame supplied with a reduced supply of fuel.

A flow diagram is often used to organize the sequence of test procedures for the separation and identification of the large number of ions in a mixture. A flow diagram uses several standard notations:

E. Constructing Flow Diagrams

- Brackets, [ ], indicate the use of a test reagent written in molecular form.
- A longer single horizontal line, _____, indicates a separation of a precipitate from a solution, most often with a centrifuge.
- A double horizontal line, ══, indicates the soluble ions in the solution.
- Two short vertical lines, ‖, indicate the presence of a precipitate; these lines are drawn to the *left* of the single horizontal line.
- One short vertical line, │, indicates a supernatant and is drawn to the *right* of the single horizontal line.
- Two branching diagonal lines, ∧, indicate a separation of the existing solution into two portions.
- A rectangular box, □, placed around a compound or the result of a test confirms the presence of the ion.

The flow diagram for the anions is presented in Experiment 37. Study it closely and become familiar with the symbols and notations as you read the Introduction and Experimental Procedure. Partially completed flow diagrams for the various groups of cations are presented in the Prelaboratory Assignments of Experiments 38 and 39.

F. How to Effectively Do "Qual"

The following suggestions are offered "before" and "during" the following "qual" experiments:

- Use good laboratory techniques during the analyses. Review the suggested techniques that appear as icons in the Experimental Procedure prior to beginning the analysis.
- *Always* read the Experimental Procedure in detail. Is extra equipment necessary? Is a hot water bath needed? Maintain a water bath during the laboratory period if one is needed. What cautions are to be taken?
- Understand the principles of the separation and identification of the ions. Is this an acid–base separation, redox reaction, or complex formation? Why is this reagent added at this time?
- Closely follow, simultaneously, the principles used in each test, the flow diagram, the Experimental Procedure, and the Report Sheet during the analysis.
- Mark with a "magic marker" 1-, 2-, and 3-mL intervals on the small test tube used for testing your sample to quickly estimate volumes (see Figure D4.1).

- Keep a number of *clean* dropping pipets, stirring rods, and small test tubes available; always rinse each test tube several times with deionized water immediately after use.[2]
- Estimate the drops/mL of one or more of your dropping pipets.
- Keep a wash bottle filled with deionized water available at all times.
- Maintain a "file" of confirmatory tests of the ions in the test tubes that result from the analysis on your *reference* solution; in that way, observations and comparisons of the *test* solution can be quick.

Caution: *In the next several experiments you will be handling a large number of chemicals (acids, bases, oxidizing and reducing agents, and, perhaps, even some toxic chemicals), some of which are more concentrated than others and must be handled with care and respect!*
Re-read the Laboratory Safety section, pages 1–4, of this manual.

Carefully, handle all chemicals. **Read the label!** *Do not intentionally inhale the vapors of any chemical unless you are specifically told to do so; avoid skin contact with any chemicals—wash the skin immediately in the laboratory sink, eye wash fountain, or safety shower; clean up any spilled chemical—if you are uncertain of the proper cleanup procedure, flood with water, and consult your laboratory instructor; be aware of the techniques and procedures of neighboring chemists—discuss potential hazards with them.*

*And finally, **dispose of the waste chemicals** in the appropriately labeled waste containers. Consult your laboratory instructor to ensure proper disposal.*

---

[2]Failure to maintain clean glassware during the analysis causes more spurious data and reported errors in interpretation than any other single factor in qualitative analysis.

Experiment **37**

Qual: Common Anions

Calcium ion and carbonate ion combine to form a calcium carbonate precipitate, a preliminary test for the presence of carbonate ion in a solution.

OBJECTIVES

- To observe and utilize some of the chemical and physical properties of anions
- To separate and identify the presence of a particular anion in a solution containing a mixture of anions

TECHNIQUES

The following techniques are used in the Experimental Procedure

INTRODUCTION

Common anions in aqueous solution are either single atom anions (Cl^-, Br^-, I^-) or polyatomic anions usually containing oxygen (OH^-, SO_4^{2-}, CO_3^{2-}, PO_4^{3-}). In nature the most common anions are chloride, silicate, carbonate, phosphate, sulfate, sulfide, nitrate, aluminate, and combinations thereof.

   Specific anion tests are subject to interference from other anions and cations. Therefore, to characteristically identify an anion in a mixture, preliminary elimination of the interferences is necessary.[1]

   Only six of the many known inorganic anions will be identified in this experiment: phosphate, PO_4^{3-}, carbonate, CO_3^{2-}, chloride, Cl^-, iodide, I^-, sulfide, S^{2-}, and nitrate, NO_3^-. The chemical properties of several of these anions have been seen in previous experiments in this manual (for example, see Experiments 3 and 24). Many anions can be detected directly in the sample solution by the addition of a single test reagent. However, some anion detection procedures require a systematic removal of the interferences before the use of the test reagent. For example, a test for the presence of PO_4^{3-} requires the prior removal of AsO_4^{3-}; a test for CO_3^{2-} requires the prior removal of SO_3^{2-}.

   The separation and identification of the anions are outlined in the **flow diagram** on the next page (see D4.E). Follow the diagram as you read through the Introduction and follow the Experimental Procedure.

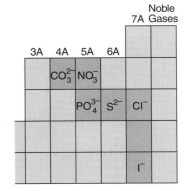

Common anions detected in this experiment.

Flow diagram: a diagram that summarizes a procedure for following a rigid sequence of steps

---

[1] For more information on anion qualitative analysis, go to http://www.chemlin.net/chemistry.

Flow Diagram for Anion "Qual" Scheme

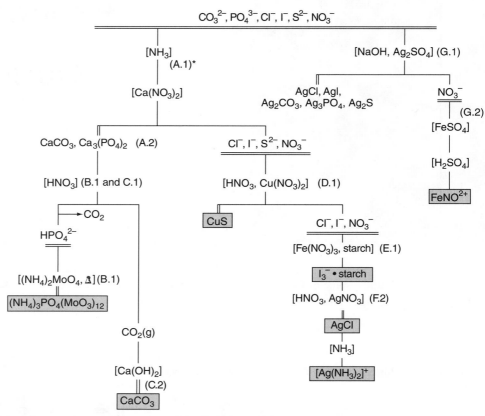

* Numbers in parentheses refer to parts of the Experimental Procedure

---

Phosphate Ion

The phosphate ion, PO_4^{3-}, is a strong Brønsted base (proton acceptor)

All phosphate salts are *insoluble* except those of the Group 1A cations and ammonium ion (Appendix G). The phosphate salt of calcium forms a *white precipitate* in a basic solution but subsequently dissolves in an acidic solution.

$$3\ Ca^{2+}(aq) + 2\ PO_4^{3-}(aq) \rightarrow Ca_3(PO_4)_2(s) \qquad (37.1)$$
$$\downarrow 2\ \boldsymbol{H^+(aq)}$$
$$3\ Ca^{2+}(aq) + 2\ HPO_4^{2-}(aq)$$

Ammonium molybdate added to an acidified solution of the hydrogen phosphate ion precipitates *yellow* ammonium phosphomolybdate, confirming the presence of phosphate ion in the test solution.

$$HPO_4^{2-}(aq) + 12\ (NH_4)_2MoO_4(aq) + 23\ H^+(aq) \rightarrow$$
$$\boxed{(NH_4)_3PO_4(MoO_3)_{12}(s)} + 21\ NH_4^+(aq) + 12\ H_2O(l) \quad (37.2)$$

The ⬜ is used to note the confirmation of the presence of an ion in the test solution

The rate of precipitate formation depends on the concentration of the phosphate ion in solution.

The reactions of the arsenate ion (AsO_4^{3-}) are identical to those of the phosphate ion and therefore would, if present, interfere with the test.

Carbonate Ion

All carbonate salts are *insoluble* except those of the Group 1A cations and ammonium ion (Appendix G). Acidification of a solution containing carbonate ion produces carbon dioxide gas (Figure 37.1).

$$CO_3^{2-}(aq) + 2\ H^+(aq) \rightarrow H_2O(l) + CO_2(g) \qquad (37.3)$$

When the evolved CO_2 (an acidic anhydride) comes into contact with a basic solution containing calcium ion, the carbonate ion re-forms and reacts with the calcium ion, forming a *white precipitate* of calcium carbonate:

$$CO_2(g) + 2\,OH^-(aq) \rightarrow CO_3^{2-}(aq) + H_2O(l) \qquad (37.4)$$

$$\begin{array}{c} \big| \; Ca^{2+}(aq) \\ \underline{\quad\quad\quad} \rightarrow \boxed{CaCO_3(s)} \qquad (37.5) \end{array}$$

The precipitate confirms the presence of carbonate ion in the test solution. The sulfite ion, SO_3^{2-}, if present, would interfere with the test; under similar conditions, it produces sulfur dioxide gas and insoluble calcium sulfite.

Figure 37.1 Acidifying a solution containing carbonate ion produces carbon dioxide gas.

Sulfide Ion

Most sulfide salts are insoluble, including CuS. When Cu^{2+} is added to a solution containing sulfide ion, a *black precipitate* of copper(II) sulfide, CuS, forms, confirming the presence of sulfide ion in the test solution:

$$Cu^{2+}(aq) + S^{2-}(aq) \rightarrow \boxed{CuS(s)} \qquad (37.6)$$

Chloride and Iodide Ions

The salts of the chloride and iodide ions are soluble with the exception of the Ag^+, Pb^{2+}, and Hg_2^{2+} halides. A simple reaction with silver ion would cause a mixture of the silver halides to precipitate, and therefore no separation or identification could be made. Instead, differences in the ease of oxidation of the chloride and iodide ions are used for their identification (see Experiment 11, Part D). The iodide ion is most easily oxidized. A weak oxidizing agent oxidizes only the iodide ion. In this experiment iron(III) ion oxidizes iodide ion to the yellow-brown triiodide complex, I_3^-

$$2\,Fe^{3+}(aq) + 3\,I^-(aq) \rightarrow 2\,Fe^{2+}(aq) + I_3^-(aq) \qquad (37.7)$$

The I_3^- then reacts with starch to form a *deep-blue complex*, $I_3^- \cdot$starch, confirming the presence of iodide ion in the sample.

$$I_3^-\,(aq) + starch(aq) \rightarrow I_3^-\cdot starch\ (aq,\ deep\text{-}blue) \qquad (37.8)$$

The chloride ion is then precipitated as a *white precipitate* of silver chloride.[2]

$$Cl^-(aq) + Ag^+(aq) \rightarrow \boxed{AgCl(s)} \qquad (37.9)$$

To further confirm the presence of chloride ion in the test solution, aqueous ammonia is added to dissolve the silver chloride which again precipitates with the addition of nitric acid:

$$AgCl(s) + 2\,NH_3(aq) \rightleftharpoons [Ag(NH_3)_2]^+(aq) + Cl^-(aq) \qquad (37.10)$$

$$[Ag(NH_3)_2]^+(aq) + Cl^-(aq) + 2\,H^+(aq) \rightarrow AgCl(s) + 2\,NH_4^+(aq) \qquad (37.11)$$

Nitrate Ion

As all nitrate salts are soluble, no precipitate can be used for identification of the nitrate ion. The nitrate ion is identified by the "brown ring" test. The nitrate ion is reduced to nitric oxide by iron(II) ions in the presence of concentrated sulfuric acid:

$$NO_3^-(aq) + 3\,Fe^{2+}(aq) + 4\,H^+(aq) \xrightarrow{conc\ H_2SO_4}$$
$$3\,Fe^{3+}(aq) + NO(aq) + 2\,H_2O(l) \qquad (37.12)$$

---

[2]A faint cloudiness with the addition of Ag^+ is inconclusive as the chloride ion is one of those "universal impurities" in aqueous solutions.

The nitric oxide, NO, combines with excess iron(II) ions, forming the *brown* $FeNO^{2+}$ ion at the interface of the aqueous layer and a concentrated sulfuric acid layer (where acidity is high) that underlies the aqueous layer.

$$Fe^{2+}(aq) + NO(aq) \rightarrow \boxed{FeNO^{2+}(aq)} \qquad (37.13)$$

$FeNO^{2+}$ is more stable at low temperatures. This test has many sources of interference: (1) sulfuric acid oxidizes bromide and iodide ions to bromine and iodine, and (2) sulfites, sulfides, and other reducing agents interfere with the reduction of NO_3^- to NO. A preparatory step of adding sodium hydroxide and silver sulfate removes these interfering anions, leaving only the nitrate ion in solution.

EXPERIMENTAL PROCEDURE

Procedure Overview: Two solutions are tested with various reagents in this analysis: (1) a reference solution containing all six of the anions for this analysis and (2) a test solution containing any number of the anions. Separations and observations are made and recorded. Equations that describe the observations are also recorded. Comparative observations of the two solutions result in the identification of the anions in the test solution. All tests are qualitative; only identification of the anion(s) is required.

To become familiar with the identification of these anions, take a sample that contains all of the anions (the reference solution) and analyze it according to the procedure. At each circled, superscript (e.g., ①), *stop* and record on the Report Sheet. After each anion is confirmed, *save* it in the test tube so that its appearance can be compared to that of your test solution.

To analyze for anions in your test solution, place the test solution alongside the reference solution during the analysis. As you progress through the procedure, perform the same test on both solutions and make comparative observations. Check (√) the findings on the Report Sheet. Do not discard any solutions (but keep all solutions labeled) until the experiment is complete.

The test solution may be a water sample from some location in the environment, e.g., a lake, a stream, or a drinking water supply. Ask your instructor about this option.

Before proceeding, review the techniques outlined in Dry Lab 4, Parts A–D (D4.A–D). The review of these procedures may expedite your analysis with less frustration.

Contamination by trace amounts of anions in test tubes and other glassware leads to "unexplainable" results in qualitative analysis. Thoroughly clean all glassware with soap and tap water; rinse twice with tap water and twice with deionized water before use (see D4.F).

> *Disposal:* Dispose of all test solutions and precipitates in the appropriate waste container.

Caution: *A number of acids and bases are used in the analysis of these anions. Handle each of these solutions with care. Read the Laboratory Safety section for instructions in handling acids and bases.*

The expression "small test tube" that is mentioned throughout the Experimental Procedure refers to a 75-mm test tube (~3 mL volume) *or* a centrifuge tube of the size that fits into your laboratory centrifuge. Consult with your laboratory instructor.

A. Separation of Carbonate and Phosphate Anions

The Experimental Procedure is written for a single solution. If you are simultaneously identifying anions in *both* a reference solution *and* a test solution, adjust the procedure accordingly. If the test solution is a sample with an environmental origin, gravity filter 10–15 mL before beginning the Experimental Procedure.

1. **Precipitate the CO_3^{2-} and PO_4^{3-}.** Place about 1.5 mL of the reference solution in a small test tube (see D4.A). Test the solution with pH paper. If acidic, add drops of 3 M NH_3 until the solution is basic; then add 3–4 more drops. Add 10–12 drops of 0.1 M $Ca(NO_3)_2$ until the precipitation of the anions is complete (see D4.B).①

2. **Separate the Solution from Precipitate.** Centrifuge the solution. Decant the supernatant② into a small test tube and save for Part D. Wash the precipitate *twice* with about 1 mL of deionized water (see D4.C). Discard the washings as directed by your instructor. Save the precipitate for Part B.

---

1. **Confirmatory Test.** Dissolve the precipitate from Part A.2 with 6 M HNO_3 (**Caution!**). Add 1 mL of 0.5 M $(NH_4)_2MoO_4$. Shake and warm slightly in a warm water (~60°C) bath and let stand for 10–15 minutes (see D4.D). A *slow* formation of a *yellow precipitate* confirms the presence of the phosphate ion③ in the solution.[3]

B. Test for Phosphate Ion

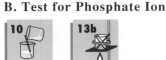

---

1. **Precipitate the CO_3^{2-}.** Repeat Part A. Centrifuge the mixture; save the precipitate, but discard the supernatant or save for Part D. Dip a glass rod into a saturated $Ca(OH)_2$ solution.

C. Test for Carbonate Ion

2. **Confirmatory Test.** Add 5–10 drops of 6 M HNO_3 to the precipitate and immediately insert the glass rod into the test tube (Figure 37.2). *Do not* let the glass rod touch the test tube wall or the solution. The evolution of the CO_2 gas causes the formation of a *milky solution* on the glass rod, confirming the presence of carbonate ion④ in the solution.

---

1. **Confirmatory Test.** To the supernatant from Part A.2 and/or C.1, add 2–4 drops of 6 M HNO_3 until the solution is acid to pH paper and then drops of 1 M $Cu(NO_3)_2$ until precipitation is complete. Be patient, allow ~2 minutes to form.⑤ Centrifuge; save the supernatant for Part E. The *black precipitate* confirms the presence of sulfide ion in the solution.

D. Test for Sulfide Ion

---

1. **Confirmatory Test.** To about 1 mL of the supernatant from Part D.1 add 5 drops of 0.2 M $Fe(NO_3)_3$. Agitate the solution. The formation of I_3^- is slow—allow 2–3 minutes. Add 2 drops of 1% starch solution. The *deep-blue* I_3^-•starch complex confirms the presence of iodide ion in the sample.⑥

E. Test for Iodide Ion

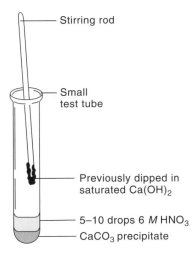

Stirring rod

Small test tube

Previously dipped in saturated $Ca(OH)_2$

5–10 drops 6 M HNO_3

$CaCO_3$ precipitate

Figure 37.2 Position a stirring rod dipped into a saturated $Ca(OH)_2$ solution just above the solid/HNO_3 mixture.

---

[3]A *white* precipitate may form if the solution is heated too long or if the solution is not acidic enough. The precipitate is MoO_3, *not* a pale form of the phosphomolybdate precipitate.

F. Test for Chloride Ion

1. Add 2–3 drops of 6 M HNO$_3$ to the solution from Part E.1.

2. **Confirmatory Test.** Add drops of 0.01 M AgNO$_3$ to the aqueous solution (sample) and centrifuge. A *white precipitate* indicates the likely presence of Cl$^-$.[7] Discard the supernatant. The addition of 6 M NH$_3$ quickly dissolves the precipitate if Cl$^-$ is present.[8] Reacidification of the solution with 6 M HNO$_3$ re-forms the silver chloride precipitate.

G. Test for Nitrate Ion

1. **Precipitate the "Other" Anions.** Place $1\frac{1}{2}$ mL of the reference solution into a small test tube. Add drops of 3 M NaOH until the solution is basic to pH paper. Add drops of a saturated (0.04 M) Ag$_2$SO$_4$ solution until precipitation appears complete. Centrifuge and save the supernatant for Part G.2.[9] Test for complete precipitation in the supernatant (see D4.B) and, if necessary, centrifuge again.

2. **Confirmatory Test.** Decant 0.5 mL of the supernatant into a small test tube and acidify (to pH paper) with 3 M H$_2$SO$_4$. Add ~0.5 mL (see D4.A) of a saturated iron(II) sulfate, FeSO$_4$, solution and agitate. Cool the solution in an ice bath. Holding the test tube at a 45° angle (Figure 37.3), add, with a dropping pipet, *slowly and cautiously,* down its side, about 0.5 mL of conc H$_2$SO$_4$. (**Caution:** *Concentrated H$_2$SO$_4$ causes severe skin burns.*)[4] Do *not* draw conc H$_2$SO$_4$ into the bulb of the dropping pipet. Do *not* agitate the solution. The more dense conc H$_2$SO$_4$ underlies the aqueous layer. Use extreme care to avoid mixing the conc H$_2$SO$_4$ with the solution. Allow the mixture to stand for several minutes. A *brown ring* at the interface between the solution and the conc H$_2$SO$_4$ confirms the presence of the nitrate ion[10] in the test solution.

> *Disposal:* Dispose of the concentrated acid solution in the "Waste Acids" container.

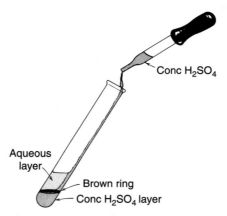

Aqueous layer

Conc H$_2$SO$_4$

Brown ring

Conc H$_2$SO$_4$ layer

Figure 37.3 Test for the nitrate ion.

---

[4]The conc H$_2$SO$_4$ has a greater density than water and therefore underlies the aqueous layer.

Qual: Common Anions

Date _____ Lab Sec. _____ Name _____ Desk No. _____

1. A review of the Dry Lab 4 will make this experiment proceed more smoothly. Complete the following:
   a. The approximate volume of a standard 75-mm ("small") test tube is _____ mL.

   b. The clear solution above a precipitate is called the _____.

   c. A _____ is an instrument used to separate and compact a precipitate in a test tube.

   d. The number of drops of water equivalent to 1 mL is about _____.

2. The following references are made to Dry Lab 4, Preface to Qualitative Analysis, in this experiment. Identify what each reference provides for an effective separation and analysis.

Experimental Procedure	Dry Lab 4 Reference	Information Provided
Part A.1	D4.A	
Part A.1	D4.B	
Part A.2	D4.C	
Part B.1	D4.D	

3. a. Experimental Procedure, Part A.1. Describe the technique for mixing solutions in a "small" test tube.

   b. Experimental Procedure, Part A.1. Describe the technique for washing a precipitate.

4. Refer to Dry Lab 4.E. On a flow diagram, what is the meaning of . . .
   a. a single horizontal line, —?

   b. a pair of short vertical lines, ‖ ?

   c. a pair of horizontal lines, ═?

   d. the brackets, [ ], around a reagent?

5. Three anions in this experiment are identified by the complexes they form. Which anions are so identified?

6. Four anions are confirmed present by the formation of a precipitate. Which anions are so confirmed? Write the formula and indicate the color of the precipitates.

7. Identify a single reagent that distinguishes the carbonate ion from the chloride ion. Write the balanced equation(s) that makes the distinction.

Qual: Common Anions

Date _____ Lab Sec. _____ Name _____ Desk No. _____

Procedure Number and Ion	Test Reagent or Technique	Evidence of Chemical Change	Chemical(s) Responsible for Observation	Equation(s) for Observed Reaction	Check (√) if Observed in Unknown
①	_____	_____	_____	_____	_____
②	_____	_____	_____	_____	_____
③ PO_4^{3-}	_____	_____	_____	_____	☐
④ CO_3^{2-}	_____	_____	_____	_____	☐
⑤ S^{2-}	_____	_____	_____	_____	☐
⑥ I^-	_____	_____	_____	_____	☐
⑦ Cl^-	_____	_____	_____	_____	☐
⑧	_____	_____	_____	_____	_____
⑨	_____	_____	_____	_____	_____
⑩ NO_3^-	_____	_____	_____	_____	☐

Anions present in unknown test solution no. _____: _____

Instructor's approval: _____

Laboratory Questions

Circle the questions that have been assigned.

1. Part A.1. The *test* solution is made basic and drops of 0.1 M $Ca(NO_3)_2$ are added but no precipitate forms. To what part of the Experimental Procedure do you proceed? Explain.

2. Part A.1. The *reference* solution is made acidic instead of basic. How would this change the composition of the precipitate and the test in Part C.1? Explain.

3. Part D.1. 6 M HNO_3 could not be found on the reagent shelf. Instead 6 M HCl is added to the *test* solution.
   a. What affect does this have on the test for sulfide ion? Explain.
   b. What affect does this have on subsequent tests of the supernatant from Part D.1? Explain.

4. Part F.2. 6 M NH_3, a basic solution, cannot be found on the reagent shelf, but 6 M NaOH, also a base, is available. What would be observed if the 6 M NaOH is substituted for the 6 M NH_3 in testing the *reference* solution? Explain.

5. Part F.2. Both iodide ion and chloride ion are in the *test* solution. Describe the appearance of the contents of the test tube after drops of 0.01 M $AgNO_3$ have been added and the solution has been centrifuged.

6. Part G.2. There is no other "brown ring" test commonly known in chemistry. What is the substance producing the brown ring?

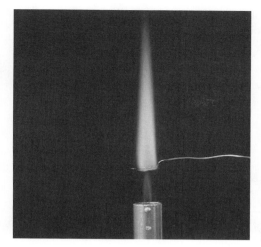

Experiment 38

Qual I. Na$^+$, K$^+$, NH$_4^+$, Mg^{2+}, Ca^{2+}, Cu^{2+}

Flame tests provide a positive identification for many metal ions.

OBJECTIVES

- To observe and utilize the chemical and physical properties of Na$^+$, K$^+$, NH$_4^+$, Mg^{2+}, Ca^{2+}, and Cu^{2+}
- To separate and identify the presence of one or more of the cations, Na$^+$, K$^+$, NH$_4^+$, Mg^{2+}, Ca^{2+}, Cu^{2+} in an aqueous solution

TECHNIQUES

The following techniques are used in the Experimental Procedure

INTRODUCTION

In identifying a particular cation in a mixture of cations, it would be ideal if we could detect each cation in the mixture by merely adding a specific reagent that would produce a characteristic color or precipitate; however, no such array of reagents has been developed. Instead, the cations must first be separated into groups having similar chemical properties. From there, the cations in each group respond to a specific reagent to produce the characteristic color or precipitate for identification. The qualitative analysis of a mixture of cations requires such a plan of investigation.[1]

The analysis of Qual I cations requires good laboratory technique for their separation and confirmation, as well as careful preparation and understanding of the procedure. If in following the prescribed steps for separation and identification, you only rely on a "cookbook" procedure, expect unexplainable results. Please read carefully the remainder of this Introduction and the Experimental Procedure, review your laboratory techniques, and complete the Prelaboratory Assignment before beginning the analysis; it will save you time and minimize frustration.

This experiment is the first of two in which a set of reagents and techniques are used to identify the presence of a particular cation among a larger selection of cations.[2] We will approach this study as an experimental chemist: we will conduct some tests, write down our observations, and then write balanced equations that agree with our data.

---

[1]For a more detailed grouping of cations for qualitative analysis, go to *www.chemistry.about.com* and search qualitative analysis.

[2]The ions Hg$_2^{2+}$, Pb^{2+}, Ag$^+$, Bi^{3+} are not included in the cation analysis because of the disposal problems associated with their potential environmental effects.

As an introduction to the chemistry of the cations in this experiment, keep in mind that most Na^+, K^+, and NH_4^+ salts are soluble. Thus, these ions can be separated from a large number of other cations.[3]

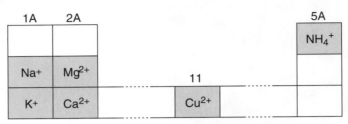

Cations of the Qual I group.

Sodium and potassium ions form a very limited number of salts with low solubilities and only carefully controlled conditions provide confirmatory tests; however, both ions provide very sensitive, distinctive flame tests.

All ammonium salts are water soluble. The ammonium ion is a weak acid; its conjugate base is ammonia, which is a gas. When the pH of an aqueous solution of ammonium ion is increased, ammonia gas is evolved and is easily detected with red litmus.

The Mg^{2+}, Ca^{2+}, and Cu^{2+} ions are separated and identified by taking advantage of their chemical properties: Mg^{2+} and Ca^{2+} are confirmed present in a mixture by their HPO_4^{2-} and $C_2O_4^{2-}$ precipitates, respectively. The Cu^{2+} is reduced to copper metal and then subsequently oxidized to form its characteristic deep-blue color in the presence of ammonia.

Sodium Ion

A flame test to detect the sodium ion produces a characteristic, fluffy yellow flame. The color is so intense that it masks the color produced by other ions and is comparatively persistent.[4]

Potassium Ion

A flame test is also used to detect the presence of potassium ion in a sample. The flame is viewed through cobalt blue glass. The glass absorbs the intense yellow color of sodium, but transmits the lavender color of the potassium flame. The lavender flame is of short duration. A comparative test of a solution known to contain only potassium ion is often necessary for the determination.

Ammonium Ion

To test for ammonium ion, advantage is taken of the following equilibrium:

$$NH_3(aq) + H_2O(l) \rightleftharpoons NH_4^+(aq) + OH^-(aq) \qquad (38.1)$$

When hydroxide ion (a common ion in the equilibrium) is added to a solution containing ammonium ion, this equilibrium shifts *left*. Heating drives *ammonia gas* from the system; its presence is detected using moist litmus paper (its odor is also frequently detected).

The ☐ is used to note the confirmation of the presence of an ion in the test solution

$$\boxed{NH_3(g)} + H_2O(l) \xleftarrow{} \rightleftharpoons NH_4^+(aq) + OH^-(aq) \qquad (38.2)$$
$$\uparrow OH^-(aq)$$

---

[3]Review the solubility rules for salts in Appendix G or in your text.
[4]The sodium flame test is so sensitive that the sodium ion washed (with several drops of deionized water) from a fingerprint can be detected.

The copper(II) ion must be separated from the Mg^{2+} and Ca^{2+} to avoid any interference with their confirmatory tests. Copper(II) ion is readily reduced to copper metal by zinc metal.

Copper Ion

$$Cu^{2+}(aq) + Zn(s) \rightarrow Cu(s) + Zn^{2+}(aq) \qquad (38.3)$$

The $Zn^{2+}(aq)$ does not interfere with the tests for Mg^{2+} and Ca^{2+}.

The copper metal is then oxidized back into an aqueous solution with nitric acid.

$$3\,Cu(s) + 8\,H^+(aq) + 2\,NO_3^-(aq) \rightarrow 3\,Cu^{2+}(aq) + 2\,NO(g) + 4\,H_2O(l) \quad (38.4)$$

The slow addition of aqueous NH_3 then complexes the Cu^{2+} as a soluble deep-blue $[Cu(NH_3)_4]^{2+}$ complex. This confirms the presence of Cu^{2+} in the test solution.

$$Cu^{2+}(aq) + 4\,NH_3(aq) \rightarrow \boxed{[Cu(NH_3)_4]^{2+}(aq)} \qquad (38.5)$$

Structure of $[Cu(NH_3)_4]^{2+}$

A second confirmatory test for the presence of Cu^{2+} is the addition of potassium hexacyanoferrate(II), $K_4[Fe(CN)_6]$, which produces a *red-brown precipitate* of $Cu_2[Fe(CN)_6]$.

$$2\,[Cu(NH_3)_4]^{2+}(aq) + [Fe(CN)_6]^{4-}(aq) \rightarrow \boxed{Cu_2[Fe(CN)_6](s)} + 8\,NH_3(aq) \quad (38.6)$$

Structure of $[Fe(CN)_6]^{4-}$

The Cu^{2+} ion in solution (right) has a less intense color than does the deep-blue $[Cu(NH_3)_4]^{2+}$ complex.

A *white precipitate* of calcium oxalate forms in an ammoniacal solution, confirming the presence of calcium ion in the reference solution.

Calcium Ion

$$Ca^{2+}(aq) + C_2O_4^{2-}(aq) \rightarrow \boxed{CaC_2O_4(s)} \qquad (38.7)$$

The dissolution of calcium oxalate with hydrochloric acid, followed by a flame test, produces a *yellow-red flame,* characteristic of the calcium ion.

Zinc oxalate, also a white insoluble salt, may interfere with the calcium oxalate confirmatory test, but the flame test confirms the calcium ion. Zinc ion does not produce a color for its flame test.

The addition of monohydrogen phosphate ion, HPO_4^{2-}, to an ammoniacal solution containing magnesium ion causes, after heating, the formation of a *white precipitate*.

Magnesium Ion

$$Mg^{2+}(aq) + HPO_4^{2-}(aq) + NH_3(aq) \xrightarrow{\Delta} \boxed{MgNH_4PO_4(s)} \qquad (38.8)$$

Magnesium ion does not produce a characteristic flame test.

Procedure Overview: Two solutions are tested with various reagents in this analysis: (1) a reference solution containing all six of the cations of Qual I and (2) a test solution containing any number of Qual I cations. Separations and observations are made and recorded. Equations that describe the observations are also recorded. Comparative

EXPERIMENTAL PROCEDURE

observations of the two solutions result in the identification of the cations in the test solution. All tests are qualitative; only identification of the cation(s) is required.

To become familiar with the separation and identification of Qual I cations, take a sample that contains the six cations (reference solution) and analyze it according to the Experimental Procedure. At each circled, superscript (e.g., ①), *stop* and record data on the Report Sheet. After the presence of a cation is confirmed, *save* the characteristic appearance of the cation in the test tube so that it can be compared with observations made in the analysis of your test solution.

To analyze for cations in your test solution, place the test solution alongside the reference solution during the analysis. As you progress through the procedure, perform the same test on both solutions and make comparative observations. Check (√) the findings on the Report Sheet. Do not discard any solutions (but keep all solutions labeled) until the experiment is complete.

The test solution may be a water sample from some location in the environment, e.g., a lake, a stream, or a drinking water supply. Ask your instructor about this option.

Before proceeding, review the techniques outlined in Dry Lab 4, Parts A–D (D4.A–D). The review of these procedures may expedite your analysis with less frustration.

Contamination by trace amounts of metal ions in test tubes and other glassware leads to "unexplainable" results in qualitative analysis. Thoroughly clean all glassware with soap and tap water; rinse twice with tap water and twice with deionized water before use (see D4.F).

Caution: *A number of acids and bases are used in the analysis of these cations. Handle each of these solutions with care. Read the Laboratory Safety section for instructions in handling acids and bases.*

The expression "small test tube" that is mentioned throughout the Experimental Procedure refers to a 75-mm test tube (~3 mL volume) *or* a centrifuge tube of the size that fits into your laboratory centrifuge. Consult with your laboratory instructor.

The Experimental Procedure may begin with either Part A or Part D. Consult with your instructor.

A. Test for Sodium Ion

The Experimental Procedure is written for a single reference solution. If you are simultaneously identifying cations in *both* a reference solution *and* a test solution, adjust the procedure accordingly. If the test solution is a sample with an environmental origin, gravity filter 10–15 mL before beginning the Experimental Procedure.

1. **Remove the Interfering Ions.** Place no more than 2 mL of the reference solution in an evaporating dish. Add solid $(NH_4)_2C_2O_4$, (**Caution:** *avoid skin contact*) with stirring, until the solution is basic to pH paper; add a slight excess of the solid and then a pinch of solid $(NH_4)_2CO_3$. Heat the solution *slowly* in a fume hood (NH_3 fumes may be evolved) to a moist residue (Figure 38.1), *not* to dryness! Allow the evaporating dish to cool. Add up to 1 mL (see D4.A) of deionized water, stir, and decant into a small beaker.

2. **Confirmatory Test.** The flame test for sodium ion is reliable but also requires some technique. Clean the flame test wire by dipping it in 6 *M* HCl (**Caution!**) and heating it in the hottest part of a Bunsen flame until the flame is colorless (5 steps in Figure 38.2). Repeat as necessary. Dip the flame test wire into the solution in the beaker and place it in the flame (Figure 38.3). A *brilliant yellow* persistent flame indicates the presence of sodium.①Conduct the sodium flame test on a 0.2 *M* NaCl solution for comparison.

B. Test for Potassium Ion

1. **Confirmatory Test.** Repeat Part A.2. A fleeting lavender flame confirms the presence of potassium. If sodium *is* present, view the flame through cobalt blue glass. Several trials are necessary as the test is judgmental. Conduct the potassium flame test on a 0.2 *M* KCl solution for comparison.②

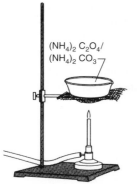

Figure 38.1 Slowly heat the test solution to a moist residue.

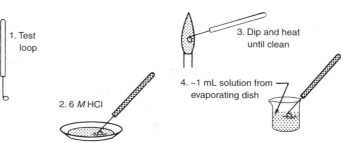

Figure 38.2 Flame test procedure.

C. Test for Ammonium Ion

1. **Prepare the Sample.** Transfer 5 mL of the *original* reference solution to a 100-mL beaker, support it on a wire gauze, and heat until a moist residue forms (do not evaporate to dryness!). Moisten the residue with 1–2 mL of deionized water. Moisten a piece of red litmus paper with water.

2. **Confirmatory Test.** Add 1–2 mL of 6 *M* NaOH to the reference solution, suspend the litmus *above* the solution (Figure 38.4), and very gently warm the mixture—*do not boil*. (**Caution:** *Be careful not to let the* NaOH *contact the litmus paper*.) A change in litmus from *red* to *blue* confirms ammonia.[3] The nose is also a good detector, but it is not always as sensitive as the litmus test.

D. Test for Copper Ion

1. **Reduction of Copper(II) Ion.** Begin with 5 mL of the *original* reference solution in a 150-mm test tube. Polish a 1-cm Zn strip and place it into the solution and let stand for 10–20 minutes.[4] Decant the solution and save for Part E.1. Wash the solid (now copper metal and excess zinc metal) with *at least* three portions (see D4.C) of deionized water and discard the washings.

Figure 38.3 Flame test for the presence of sodium ion.

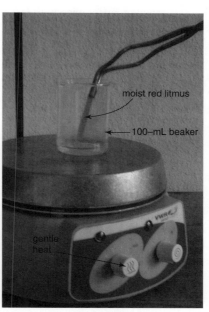

Figure 38.4 Absorbing NH$_3$ vapors on red litmus.

2. **Oxidation of Copper Metal.** Transfer the test tube containing the copper metal to a fume hood. Add drops of 6 M HNO_3 (**Caution:** *6 M HNO_3 causes severe skin burns and the evolved NO gas is toxic*) until the copper metal dissolves[5]... be patient, some heating in a water bath may be necessary.

3. **Confirmatory Test.** *Slowly* add drops of conc NH_3 (**Caution:** *avoid inhalation or skin contact!*) to the solution from Part D.2. The *deep-blue* color confirms the presence of Cu^{2+} in the solution.[6]

4. **A Second Confirmatory Test.** Acidify the solution from Part D.3 to pH paper with 6 M CH_3COOH (**Caution!**). Add 3 drops of 0.2 M $K_4[Fe(CN)_6]$. A *red-brown* precipitate reconfirms the presence of Cu^{2+} ion.[7]

E. Test for Calcium Ion

1. **Sample Preparation.** The supernatant from Part D.1 contains Ca^{2+}, Mg^{2+}, and Zn^{2+} (from the reduction of Cu^{2+}). Add drops of 6 M NH_3 until the solution is just basic to pH paper. Add 2–3 drops of 1 M $K_2C_2O_4$ (see D4.A). A white precipitate[8] confirms the presence of either Ca^{2+} and/or Zn^{2+} as both CaC_2O_4 and ZnC_2O_4 have marginal solubility. If no precipitate forms immediately, warm the solution in a water bath (see D4.D), cool, and let stand. Centrifuge and save the supernatant for Part F.

2. **Confirmatory Test.** Wet the precipitate to a moist paste with a drop of 6 M HCl and perform a flame test. A fleeting *yellow-red* flame is characteristic of calcium ion and confirms its presence.[9]

F. Test for Magnesium Ion

1. **Confirmatory Test.** Add 1–2 drops of 6 M NH_3 to the supernatant from Part E.1. Add 2–3 drops of 1 M Na_2HPO_4, heat in a hot water (~90°C) bath, and allow to stand. The precipitate[10] may be slow in forming; be patient. Observing the white precipitate confirms the presence of Mg^{2+} ion.

> *Disposal:* Dispose of all test solutions and precipitates in the "Waste Metal Salts" container.

CLEANUP: Rinse each test tube twice with tap water. Discard each rinse in the "Waste Metal Salts" container. Thoroughly clean each test tube with soap and tap water; rinse twice with tap water and twice with deionized water.

Qual I. Na$^+$, K$^+$, NH$_4$$^+$, Mg^{2+}, Ca^{2+}, Cu^{2+}

Date _____ Lab Sec. _____ Name _____ Desk No. _____

1. Identify the Qual I cation(s) that is (are) confirmed present in a test solution by
   a. the formation of a precipitate

   b. the color of a soluble complex ion

   c. the characteristic color of a flame test

   d. the evolution of a gas

2. Identify the Qual I cation(s) that is (are) confirmed present in a test solution as a result of a Brønsted acid–base reaction.

3. The following references are made to Dry Lab 4, Preface to Qualitative Analysis, in this experiment. Identify what each reference provides for an effective separation and analysis.

Experimental Procedure	**Dry Lab 4 Reference**	**Information Provided**
Part D.1	D4.C	
Part E.1	D4.A	
Part E.1	D4.D	

4. a. When operating a centrifuge, what is meant by the expression, "balance the centrifuge"?

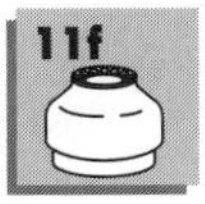

   11f

   b. How full should a test tube (or centrifuge tube) be when placed into a centrifuge?

5. Refer to Dry Lab 4.E to address the following questions.

   a. On a flow diagram, ════════ , indicates _____.

   b. On a flow diagram, ⬜⬜⬜⬜ means _____.

   c. On a flow diagram, ‖ means _____.

6. Identify the reagent that separates
   a. Mg^{2+} from Cu^{2+}. Explain the chemistry of the separation.

   b. Ca^{2+} from Na^+. Explain the chemistry of the separation.

7. Write the balanced redox equation for the separation of Cu^{2+} from the other Qual I cations.

8. Complete the following flow diagram for the Qual I cations. Refer to Dry Lab 4.E for a review of the symbolism on a flow diagram.

*Numbers in parentheses refer to parts of the Experimental Procedure.

Qual I. Na^+, K^+, NH_4^+, Mg^{2+}, Ca^{2+}, Cu^{2+}

Date _____ Lab Sec. _____ Name _____ Desk No. _____

Procedure Number and Ion	Test Reagent or Technique	Evidence of Chemical Change	Chemical(s) Responsible for Observation	Equation(s) for Observed Reaction	Check (√) if Observed in Unknown
① Na^+	flame	_____	_____	_____	☐
② K^+	flame	_____	_____	_____	☐
③ NH_4^+		_____	_____	_____	☐
④ Cu^{2+}		_____	_____	_____	_____
⑤		_____	_____	_____	_____
⑥		_____	_____	_____	☐
⑦		_____	_____	_____	☐
⑧ Ca^{2+}		_____	_____	_____	☐
⑨	flame	_____	_____	_____	☐
⑩ Mg^{2+}		_____	_____	_____	☐

Cations present in unknown test solution no. _____ : _____

Instructor's approval: _____

Laboratory Questions

Circle the questions that have been assigned.

1. Explain why a positive flame test for sodium is *not* an absolute confirmation of sodium ion in a test sample.

2. Part A.1. The $(NH_4)_2C_2O_4$ addition is omitted in the procedure. How does this affect the appearance of the flame test in Part A.2?

3. Part C.2. Instead of 6 M NaOH being added to the solution, 6 M HCl is added. How will this affect the test for the presence of ammonium ion in the solution? Explain.

4. Part D.1. What is the fate of Zn^{2+} in the experiment? Explain.

5. Part D.2. Instead of 6 M HNO_3 being added to the solution, 6 M HCl is added (both are strong acids). How will this affect the test for the presence of copper(II) ion in the solution? Explain.

6. Part D.3. Instead of conc NH_3 being added to the solution, 6 M NaOH is added (both are bases). How will this affect the test for the identification of copper(II) ion in the solution? Explain.

7. Part E.1. Instead of 6 M NH_3 being added to the solution, 6 M NaOH is added (both are bases) before the addition of the $K_2C_2O_4$. What would be the appearance of the solution? Explain.

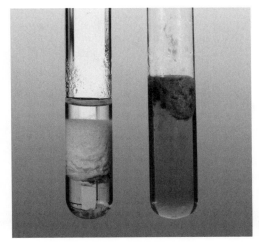

Qual II. Ni²⁺, Fe³⁺, Al³⁺, Zn²⁺

Nickel(II) ions (left) and iron(III) ions (right) readily form hydroxide precipitates.

- To observe and utilize the chemical and physical properties of Ni^{2+}, Fe^{3+}, Al^{3+}, Zn^{2+}
- To separate and identify the presence of one or more of the cations, Ni^{2+}, Fe^{3+}, Al^{3+}, and Zn^{2+} in an aqueous solution

TECHNIQUES

The following techniques are used in the Experimental Procedure

INTRODUCTION

The Qual II cations are perhaps more relevant to our industrial society than are the Qual I cations. The use of iron as the major component of steel, of aluminum for light-weight construction materials, and of zinc for coinage alloys and galvanizing steel are all familiar in our everyday lives. Aqueous solutions containing these cations tend to be quite colorful as well . . . the rust color of Fe^{3+} and the green color of Ni^{2+} . . . as well as the compounds that confirm the presence of Fe^{3+} and Ni^{2+}. However, solutions containing Al^{3+} and Zn^{2+} tend to be colorless, but the compounds that confirm their presence do have color. Only Cu^{2+} of the Qual I cations has color (sky blue) as does its confirmation (deep-dark blue).

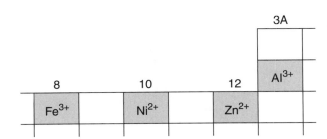

Cations of the Qual II Group

Figure 39.1 Formation of the aluminum hydroxide precipitate.

The separation of Al^{3+} and Zn^{2+} from the Ni^{2+} and Fe^{3+} cations is accomplished by the addition of a highly concentrated NaOH solution (Figure 39.1). Thereafter, the Al^{3+} and the Zn^{2+} are further separated and characteristically identified through pH control. Once separated from the Al^{3+} and Zn^{2+}, the Fe^{3+} and Ni^{2+} are independently identified with specific test reagents.

Separation of Ni²⁺, and Fe³⁺ from Zn²⁺ and Al³⁺

Gelatinous: jellylike due to adsorbed and occluded water molecules

Amphoteric (also amphiprotic): the chemical property of a substance as having both acidlike and baselike properties

Structures of $[Al(OH)_4]^-$ and $[Zn(OH)_4]^{2-}$

A strong base, OH^-, added to an aqueous solution of the four cations precipitates Ni^{2+} and Fe^{3+} as **gelatinous** hydroxides (see opening photo), but the Zn^{2+} and Al^{3+} hydroxides, being **amphoteric**, redissolve in excess base, forming the aluminate, $[Al(OH)_4]^-$, and zincate, $[Zn(OH)_4]^{2-}$, ions.

$$Al^{3+}(aq) + 4\,OH^-(aq) \rightleftharpoons [Al(OH)_4]^-(aq) \tag{39.1}$$

$$Zn^{2+}(aq) + 4\,OH^-(aq) \rightleftharpoons [Zn(OH)_4]^{2-}(aq) \tag{39.2}$$

The gelatinous hydroxides of Ni^{2+} and Fe^{3+} dissolve with the addition of nitric acid, after the soluble aluminate and zincate ions are separated.

Iron(III) Ion

Structure of $[Ni(NH_3)_6]^{2+}$

When a reference solution containing the Ni^{2+} and Fe^{3+} ions is treated with an excess of ammonia, NH_3, *brown* $Fe(OH)_3$ precipitates and the soluble *blue* hexaammine complex, $[Ni(NH_3)_6]^{2+}$, forms in solution.

$$Fe^{3+}(aq) + 3\,NH_3(aq) + 3\,H_2O(l) \rightarrow Fe(OH)_3(s) + 3\,NH_4^+(aq) \tag{39.3}$$

$$Ni^{2+}(aq) + 6\,NH_3(aq) \rightleftharpoons [Ni(NH_3)_6]^{2+}(aq) \tag{39.4}$$

Acid dissolves the $Fe(OH)_3$ precipitate; addition of thiocyanate ion, SCN^-, forms the *blood-red* thiocyanatoiron(III) complex, $[FeNCS]^{2+}$, a confirmation of the presence of Fe^{3+} in the test solution (see Experiment 34).

$$Fe^{3+}(aq) + SCN^-(aq) \rightarrow \boxed{[FeNCS]^{2+}(aq)} \tag{39.5}$$

Other forms of the Fe^{3+}–SCN^- complex are $[Fe(NCS)_2]^+$, $[Fe(NCS)_4]^-$, and $[Fe(NCS)_6]^{3-}$ depending on the SCN^- concentration.

Nickel Ion

Structure of $Ni(HDMG)_2(s)$

A supernatant from the test for iron(III) ion is used to test for the presence of nickel ion. The confirmation of Ni^{2+} ion in the reference solution is the appearance of a bright *pink* (or brick-red) *precipitate* formed with the addition of dimethylglyoxime, H_2DMG,[1] to a solution of the hexaamminenickel(II) complex ion.

$$[Ni(NH_3)_6]^{2+}(aq) + 2\,H_2DMG(aq) \rightarrow$$
$$\boxed{Ni(HDMG)_2(s)} + 2\,NH_4^+(aq) + 4\,NH_3(aq) \tag{39.6}$$

Aluminum Ion

A solution containing $[Al(OH)_4]^-$ and $[Zn(OH)_4]^{2-}$ ions, acidified with HNO_3, re-forms the Al^{3+} and Zn^{2+} ions. The subsequent addition of ammonia reprecipitates Al^{3+} as the gelatinous hydroxide, but Zn^{2+} forms the soluble tetraammine complex, $[Zn(NH_3)_4]^{2+}$:

$$[Al(OH)_4]^-(aq) + 4\,H^+(aq) \rightarrow Al^{3+}(aq) + 4\,H_2O(l) \tag{39.7}$$
$$\downarrow 3\,NH_3(aq)$$
$$Al(OH)_3(s) + 3\,NH_4^+(aq) + H_2O(l) \tag{39.8}$$

$$[Zn(OH)_4]^{2-}(aq) + 4\,H^+(aq) \rightarrow Zn^{2+}(aq) + 4\,H_2O(l) \tag{39.9}$$
$$\downarrow 4\,NH_3(aq)$$
$$[Zn(NH_3)_4]^{2+}(aq) \tag{39.10}$$

$Al(OH)_3$, an opaque, blue-white, gelatinous precipitate (Figure 39.1), is not easy to detect. To confirm the presence of Al^{3+} ion, the $Al(OH)_3$ precipitate is subsequently

---

[1]Dimethylglyoxime, abbreviated as H_2DMG for convenience in this experiment, is an organic chelating agent (see Experiment 36), specific for the precipitation of the nickel ion.

dissolved with HNO_3, aluminon reagent[2] is added, and the $Al(OH)_3$ is reprecipitated with the addition of NH_3. In the process of reprecipitating, the aluminon reagent, a red dye, adsorbs onto the surface of the gelatinous $Al(OH)_3$ precipitate, giving it a *pink/red* appearance, confirming the presence of Al^{3+} in the test solution.

$$Al^{3+}(aq) + 3\ NH_3(aq) + 3\ H_2O(l) + \text{aluminon}(aq) \rightarrow$$
$$\boxed{Al(OH)_3 \cdot \text{aluminon}(s)} + 3\ NH_4^+(aq) \quad (39.11)$$

When potassium hexacyanoferrate(II), $K_4[Fe(CN)_6]$, is added to an acidified solution of the $[Zn(NH_3)_4]^{2+}$ ion, a *light green precipitate* of $K_2Zn_3[Fe(CN)_6]_2$ forms, confirming the presence of Zn^{2+} in the test solution.

Zinc Ion

$$3\ [Zn(NH_3)_4]^{2+}(aq) + 4\ H^+(aq) \rightarrow 3\ Zn^{2+}(aq) + 4\ NH_4^+(aq) \quad (39.12)$$
$$\downarrow 2\ K_4[Fe(CN)_6](aq)$$
$$\boxed{K_2Zn_3[Fe(CN)_6]_2(s)} + 6\ K^+(aq) \quad (39.13)$$

As a guide to an understanding of the separation and identification of these four cations, carefully read the Experimental Procedure and then complete the flow diagram on the Prelaboratory Assignment. Review Dry Lab 4 for an understanding of the symbolism in the flow diagram.

Procedure Overview: Two solutions are tested with various reagents in this analysis: (1) a reference solution containing the Ni^{2+}, Fe^{3+}, Al^{3+}, and Zn^{2+} ions of Qual II and (2) a test solution containing any number of Qual II cations. Separations and observations are made and recorded. Equations that describe the observations are also recorded. Comparative observations of the two solutions result in the identification of the cations in the test solution. All tests are qualitative; only identification of the cation(s) is required.

EXPERIMENTAL PROCEDURE

To become familiar with the identification of these cations, take a sample that contains the four cations (reference solution) and analyze it according to the procedure. At each circled, superscript (e.g., ①), *stop* and record on the Report Sheet. After each cation is confirmed, *save* it in the test tube so that its appearance can be compared with that of your test solution.

To analyze for cations in your test solution, place the test solution alongside the reference solution during the analysis. As you progress through the procedure, perform the same test on both solutions and make comparative observations. Check ($\sqrt{}$) the findings on the Report Sheet. Do not discard any solutions (but keep all solutions labeled) until the experiment is complete.

The test solution may be a water sample from some location in the environment, e.g., a lake, a stream, or a drinking water supply. Ask your instructor about this option.

Before proceeding, review the techniques outlined in Dry Lab 4, Parts A–D (D4.A–D). The review of these procedures may expedite your analysis with less frustration.

Contamination by trace amounts of metal ions in test tubes and other glassware leads to "unexplainable" results in qualitative analysis. Thoroughly clean all glassware with soap and tap water; rinse twice with tap water and twice with deionized water before use (see D4.F).

Caution: *A number of acids and bases are used in the analysis of these cations. Handle each of these solutions with care. Read the Laboratory Safety section for instructions in handling acids and bases.*

The expression "small test tube" that is mentioned throughout the Experimental Procedure refers to a 75-mm test tube ($\sim$3 mL volume) *or* a centrifuge tube of the size that fits into your laboratory centrifuge. Consult with your laboratory instructor.

---

[2]The aluminon reagent is the ammonium salt of aurin tricarboxylic acid, a red dye.

A. Separation of Ni²⁺ and Fe³⁺ from Zn²⁺ and Al³⁺

The Experimental Procedure is written for a single reference solution. If you are simultaneously identifying cations in *both* a reference solution *and* a test solution, adjust the procedure accordingly. If the test solution is a sample with an environmental origin, gravity filter 10–15 mL before beginning the Experimental Procedure.

1. **Separate the Hydroxide Precipitates from the Amphoteric Hydroxides.** To 2 mL of the reference solution (in a small test tube) add 10 drops of 6 *M* NaOH (**Caution!**) (see D4.A). Centrifuge and save the precipitate.① Test for complete precipitation (see D.4B) by adding several drops of 6 *M* NaOH to the supernatant. Decant the supernatant② into a small test tube and save for Part D.

2. **Dissolve the Hydroxide Precipitates.** Dissolve the precipitate③ with a minimum number of drops of conc HNO_3. (**Caution:** *Be careful!!*) If necessary, heat the solution in the hot water bath for several minutes (see D4.D).

B. Test for Iron Ion (III)

1. **Separate Fe³⁺ from Ni²⁺ Ions.** To the solution from Part A.2, add 5 drops of 4 *M* NH_4Cl and then drops of conc NH_3 (**Caution:** *do not inhale—use a fume hood if available*) until the solution is basic to pH paper; add an additional 2 drops of conc NH_3 to ensure the complexing of the Ni²⁺. Centrifuge, save the precipitate,④ and transfer the supernatant⑤ to a small test tube for testing in Part C.

2. **Confirmatory Test.** Dissolve the precipitate with 6 *M* HCl and add 5 drops of 0.1 *M* NH₄SCN.⑥ The *blood-red* solution due to the thiocyanatoiron(III) complex confirms the presence of iron(III) ion in the test solution.

C. Test for Nickel Ion

1. **Confirmatory Test.** To the supernatant solution from Part B.1, add 3 drops of dimethylglyoxime solution.⑦ Appearance of a *pink* (brick-red) *precipitate* confirms the presence of nickel ion in the test solution.

D. Test for Aluminum Ion

To digest the precipitate: to make the precipitate more compact

1. **Separate Al³⁺ from Zn²⁺.** Acidify the supernatant from Part A.1 to pH paper with 6 *M* HNO_3. Add drops of 6 *M* NH_3 until the solution is now basic to pH paper; then add 5 more drops. Heat the solution in the hot water bath for several minutes to **digest** the gelatinous **precipitate.**⑧ Centrifuge and decant the supernatant⑨ into a small test tube and save for the Zn²⁺ analysis in Part E.

2. **Confirmatory Test.** Wash the precipitate (see D.4C) *twice* with 1 mL of hot, deionized water and discard each washing. Centrifugation is necessary after each washing. Add 6 *M* HNO_3 until the precipitate *just* dissolves. Add 2 drops of the aluminon reagent, stir, and add drops of 6 *M* NH_3 until the solution is again basic and a precipitate re-forms. Centrifuge the solution; if the $Al(OH)_3$ *precipitate* is now *pink or red* and the solution is colorless, Al³⁺ is present in the sample.⑩

E. Test for Zinc Ion

1. **Confirmatory Test.** To the supernatant from Part D.1, add 6 *M* HCl until the solution is acid to pH paper; then add 3 drops of 0.2 *M* $K_4[Fe(CN)_6]$ and stir. A very light green precipitate⑪ confirms the presence of Zn²⁺ in the sample. The precipitate is slow to form and difficult to see. Centrifugation may be necessary.

Disposal: Dispose of all test solutions and precipitates in the "Waste Metal Salts" container.

CLEANUP: Rinse each test tube twice with tap water. Discard each rinse in the "Waste Metal Salts" container. Thoroughly clean each test tube with soap and tap water; rinse twice with tap water and twice with deionized water.

The Next Step

The qualitative analysis of inorganic cations and anions is a study in itself. If further interest in the separation and identification of ions seems intriguing, research various qualitative analysis schemes online. Complete textbooks are also written on the subject.

Qual II. Ni^{2+}, Fe^{3+}, Al^{3+}, Zn^{2+}

Date _____ Lab Sec. _____ Name _____ Desk No. _____

1. Confirmatory tests for the various ions are often colorful. Identify the confirmatory compound/ion and the color for each of the following ions:

Ion	Confirmatory Compound/Ion	Color
Fe^{3+}	_____	_____
Ni^{2+}	_____	_____
Al^{3+}	_____	_____
Zn^{2+}	_____	_____

2. Identify the Qual II cation that is confirmed present in a test solution as a result of the formation of a soluble complex ion.

3. Identify the reagent that separates

   a. Fe^{3+} from Al^{3+}. Explain the chemistry of the separation.

   b. Fe^{3+} from Ni^{2+}. Explain the chemistry of the separation.

   c. Al^{3+} from Zn^{2+}. Explain the chemistry of the separation.

4. The following references are made to Dry Lab 4, Preface to Qualitative Analysis, in this experiment. Identify what each reference provides for an effective separation and analysis. These techniques are used at other points in this experiment as well.

Experimental Procedure	Dry Lab 4 Reference	Information Provided
Part A.1	D4.A	
Part A.1	D4.B	
Part A.2	D4.D	
Part D.2	D4.C	

5. Aluminum hydroxide is amphoteric but ferric hydroxide is not. Write equations to show the difference in this chemical property of the two ions.

6. Refer to Dry Lab 4.B. Describe the technique used to test for the completeness of the precipitation of an ion.

7. Complete the Qual III flow diagram (see D4.E).

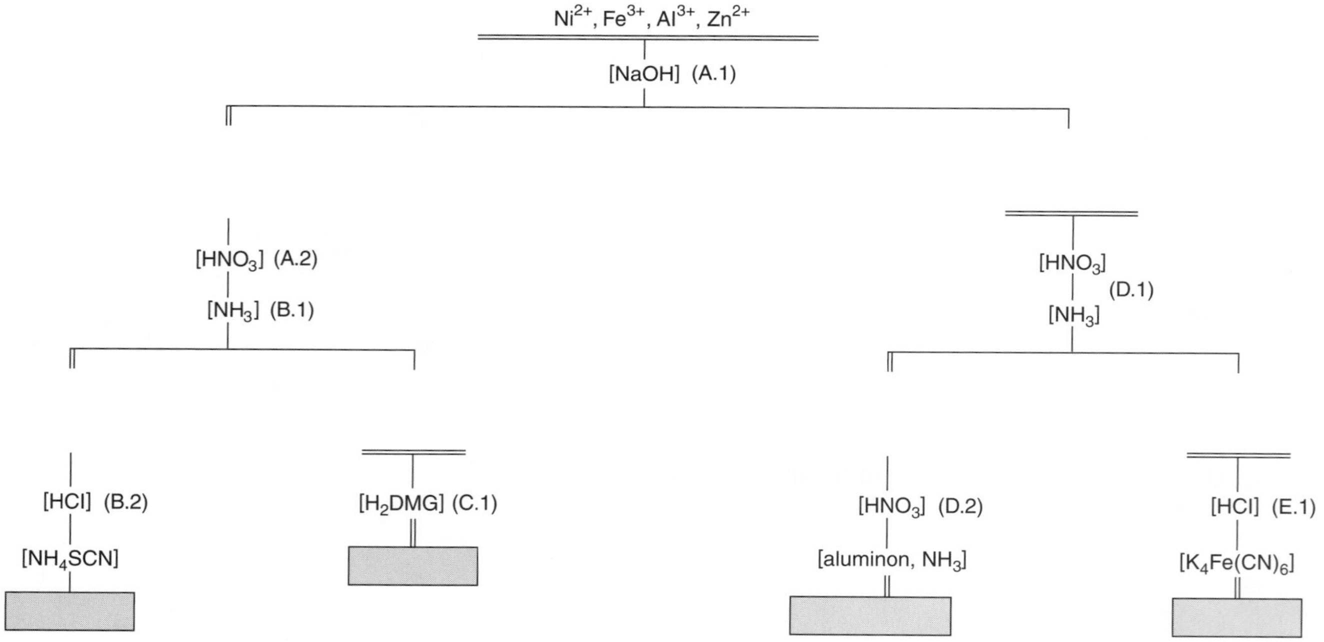

$$Ni^{2+}, Fe^{3+}, Al^{3+}, Zn^{2+}$$

[NaOH] (A.1)

[HNO$_3$] (A.2)

[NH$_3$] (B.1)

[HCl] (B.2)

[NH$_4$SCN]

[H$_2$DMG] (C.1)

[HNO$_3$] (D.1)

[NH$_3$]

[HNO$_3$] (D.2)

[aluminon, NH$_3$]

[HCl] (E.1)

[K$_4$Fe(CN)$_6$]

*Numbers in parentheses refer to parts of the Experimental Procedure.

Qual II. Ni^{2+}, Fe^{3+}, Al^{3+}, Zn^{2+}

Date _____ Lab Sec. _____ Name _____ Desk No. _____

Procedure Number and Ion	Test Reagent or Technique	Evidence of Chemical Change	Chemical(s) Responsible for Observation	Equation(s) for Observed Reaction	Check (√) if Observed in Unknown
①	_____	_____	_____	_____	_____
②	_____	_____	_____	_____	_____
③	_____	_____	_____	_____	_____
④ Fe^{3+}	_____	_____	_____	_____	_____
⑤	_____	_____	_____	_____	_____
⑥	_____	_____	_____	_____	☐
⑦ Ni^{2+}	_____	_____	_____	_____	☐
⑧ Al^{3+}	_____	_____	_____	_____	_____
⑨	_____	_____	_____	_____	_____
⑩	_____	_____	_____	_____	☐
⑪ Zn^{2+}	_____	_____	_____	_____	☐

Cations present in unknown test solution no. _____: _____

Instructor's approval: _____

Laboratory Questions

Circle the questions that have been assigned.

1. Part A.1. Instead of 6 M NaOH being added to the test solution, 6 M NH_3 is added (both are bases). How will this affect the separation of the ions in the test solution? Explain.

2. Part B.1. After adding NH_3 to the solution, a red-brown precipitate was the only evidence of color change. What can be concluded from this observation?

3. Part B.1. Instead of conc NH_3 being added to the test solution, 6 M NaOH is added (both are bases). How will this affect the separation of the Fe^{3+} from the Ni^{2+} ions in the test solution? Explain.

4. Part D.1. After adding the NH_3, no precipitate ever formed. What can you conclude from this observation? What is the next step in the analysis?

5. Part D.2. Aluminon is a dye. Explain "how" aluminon is used in the detection of the aluminum ion in the test solution.

6. Part E.1. Why is the test solution acidified with 6 M HCl before the addition of $K_4[Fe(CN)_6]$? Explain.

Conversion Factors[1]

The magnitude of a measurement must be familiar to a chemist.

Length
1 meter (m) = 39.37 in. = 3.281 ft = distance light travels in 1/299,792,548th of a second
1 inch (in.) = 2.54 cm (exactly) = 0.0254 m
1 kilometer (km) = 0.6214 (statute) mile
1 angstrom (Å) = 1×10^{-10} m = 0.1 nm
1 micron or micrometer (μm) 1×10^{-6} m

Mass
1 gram (g) = 0.03527 oz = 15.43 grains
1 kilogram (kg) = 2.205 lb = 35.27 oz
1 metric ton = 1×10^{6} g = 1.102 short ton
1 pound (lb) = 453.6 g = 7000 grains
1 ounce (oz) = 28.35 g = 437.5 grains

Temperature
°F = 1.8°C + 32
K = °C + 273.15

Volume
1 liter (L) = 1 dm^3 = 1.057 fl qt = 1×10^3 mL = 1×10^3 cm^3 = 61.02 in.^{3} = 0.2642 gal
1 fluid quart (fl qt) = 946.4 mL = 0.250 gal = 0.00595 bbl (oil)
1 fluid ounce (fl oz) = 29.57 mL
1 cubic foot (ft^3) = 28.32 L = 0.02832 m^3

Pressure
1 atmosphere (atm) = 760 torr (exactly) = 760 mm Hg = 29.92 in. Hg = 14.696 lb/in.^{2} = 1.013 bar = 101.325 kPa
1 pascal (Pa) = 1 kg/(m $\bullet$ s^2) = 1 N/m^2
1 torr = 1 mm Hg = 133.3 N/m^2

Energy
1 joule (J) = 1 kg $\bullet$ m^2/s^2 = 0.2390 cal = 9.48×10^{-4} Btu 1 = 1×10^7 ergs
1 calorie (cal) = 4.184 J = 3.087 ft$\bullet$lb
1 British thermal unit (Btu) = 252.0 cal = 1054 J = 3.93×10^{-4} hp $\bullet$ hr = 2.93×10^{-4} kW $\bullet$ hr
1 liter atmosphere (L $\bullet$ atm) = 24.2 cal = 101.3 J
1 electron volt (eV) = 1.602×10^{-19} J
1 kW $\bullet$ hr (kWh) = 3413 Btu = 8.606×10^5 cal = 3.601×10^6 J = 1.341 hp$\bullet$hr

Constants and Other Conversion Data
velocity of light (c) = 2.9979×10^8 m/s = 186,272 mi/s
gas constant (R) = 0.08206 L $\bullet$ atm/(mol $\bullet$ K) = 8.314 J/(mol$\bullet$K) = 1.986 cal/(mol $\bullet$ K) = 62.37 L $\bullet$ torr/(mol $\bullet$ K)
Avogadro's number (N_o) = 6.0221×10^{23}/mol
Planck's constant (h) = 6.6261×10^{-34} J $\bullet$ s/photon
Faraday's constant ($\widetilde{\mathscr{F}}$) = 96,485 C/mol e^-

---

[1]For additional conversions, go to http://www.onlineconversion.com.

Appendix B

Treatment of Data

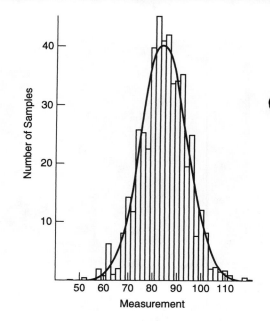

Bar graph and standard error curve for the multiple analysis of a sample.

Confidence in a scientific theory depends on the reliability of the experimental data on which the theory is based. For this reason, a scientist must be concerned about the quality of the data he or she collects. Of prime importance is the **accuracy** of the data—how closely the measured values lie to the true values.

To obtain accurate data we must use instruments that are carefully calibrated for a properly designed experimental procedure. Miscalibrated equipment, such as a balance or buret, may result in reproducible, but erred data. Flawed instruments or experimental procedures result in **systematic errors**—errors that can be detected and corrected. As a result of systematic errors, the data may have good **precision**, but not necessarily have good **accuracy**. To have good accuracy of data, the systematic errors must be minimized.

Because scientists collect data, **random errors** may also occur in measurements. Random errors are a result of reading/interpreting the value from the measuring instrument. For example, reading the volume of a liquid in a graduated cylinder to the nearest milliliter depends on the "best view" of the bottom of the meniscus, the judgment of the bottom of the meniscus relative to the volume scale, and even the temperature of the liquid. A volume reading of 10.2 mL may be read as 10.1 or 10.3, depending upon the chemist and the laboratory conditions. When the random errors are small, all measurements are close to one another and we say the data are of high precision. When the random errors are large, the values cover a much broader range and the data are of low precision. *Generally,* data of high precision are also of high accuracy, especially if the measuring device is properly calibrated.

Remember that all measurements are to be expressed with the correct number of significant figures, the number being reflective of the measuring instrument.

Systematic errors: Determinate errors that arise from flawed equipment or experimental design

Precision: Data with small deviations from an average value have high precision

Accuracy: Data with small deviations from an accepted or accurate value have good accuracy

Random errors: Indeterminate errors that arise from the bias of a chemist in observing and recording measurements

Average (or Mean) Value, $\bar{x}$

Methods of analyzing experimental data, based on statistics, provide information on the degree of precision of the measured values. Applying the methods is simple, as you will see, but to understand their significance, examine briefly the **standard error curve** (Figure B.1a).

If we make a large number of measurements of a quantity, the values would fluctuate about the average value (also called the mean value). The **average, or mean, value** is obtained by dividing the sum of all the measured values by the total number of

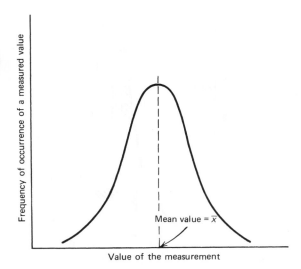

Figure B.1a The standard error curve.

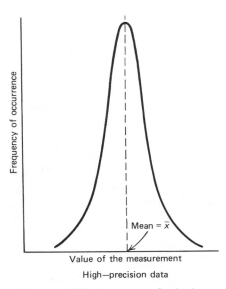

Figure B.1b Error curve for high-precision data.

values. If x_1, x_2, x_3, etc., are measured values, and there are n of them, then the average value, $\bar{x}$, is computed as

$$\text{average (or mean) value, } \bar{x} = \frac{x_1 + x_2 + x_3 + \cdots + x_n}{n} \qquad \text{(B.1)}$$

Most values lie close to the average, but some lie further away. If we plot the frequency with which a measured value occurs versus the value of the measurement, we obtain the curve in Figure B.1a. When the random errors are small (high-precision data, Figure B.1b), the curve is very narrow and the peak is sharp. When the random errors are large (low-precision data, Figure B.1c), the data are more "spread out," and the error curve is broader and less sharp.

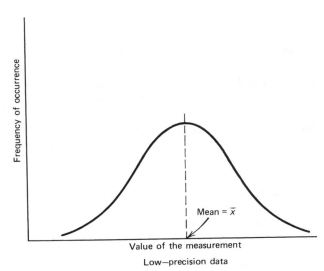

Figure B.1c Error curve for low-precision data.

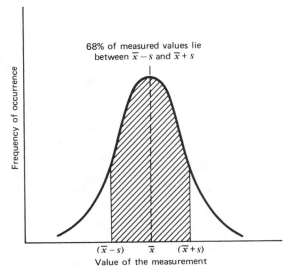

Figure B.2a Relationship of the standard deviation to the error curve.

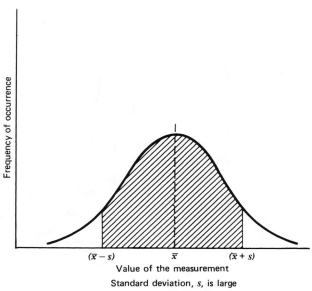

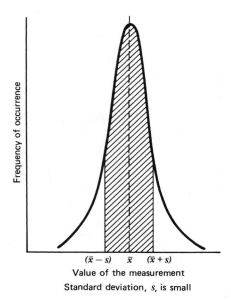

Figure B.2b Data set with a large standard deviation.

Figure B.2c Data set with a small standard deviation.

Standard Deviation, s

Statistics gives us methods for computing quantities that tell us about the width of the error curve for our data and, therefore, about the precision of the data, even when the amount of data is relatively small. One of the most important statistical measures of precision is the **standard deviation**, s. To calculate the standard deviation, we first compute the average value, $\bar{x}$. The next step is to compute the *deviation*, d, from the average value for *each* measurement—the difference between the average and each measured value:

$$\text{deviation, } d_i = \bar{x} - x_i \qquad \text{(B.2)}$$

d_i is the deviation for the measured value, x_i. The standard deviation is obtained by squaring the deviations of all measurements, adding the squared values together, dividing this sum by $n - 1$ (where n is the number of measurements), and then taking the square root.

$$\text{standard deviation } s = \sqrt{\frac{d_1^2 + d_2^2 + \cdots + d_n^2}{(n - 1)}} \qquad \text{(B.3)}$$

The standard deviation means that if we make yet another measurement, the probability that its value will lie within $\pm s$ of the average value is 0.68. In other words, 68% of the measurements lie within $\pm s$ of the average value (i.e., within the range $\bar{x} - s$ to $\bar{x} + s$). On the error curve in Figure B.2a, this represents the measurements falling within the shaded area. If we obtain a large calculated s from a set of measured values, it means that the error curve for our data is broad and the precision of the data is low (Figure B.2b); a small value of s for a set of data means that the error curve is narrow and the precision of the data is high (see Figure B.2c). Thus, s is a statistical measure of the precision of the data.

For most scientific data, three results is the *absolute minimum* number for determining the standard deviation of the data. Chemists tend to require four or more results for a meaningful interpretation of the standard deviation value of the data.

Relative Standard Deviation

The ratio of the standard deviation to the average value of the data often gives a better appreciation for the precision of the data. The ratio, called the relative standard deviation (RSD), is either expressed in parts per thousand (ppt) or parts per hundred (pph or %).

When expressed as a percentage, the RSD is referred to as %RSD or the coefficient of variation (CV) of the data.

$$\text{RSD} = \frac{s}{\bar{x}} \times 1000 \text{ ppt} \qquad (B.4)$$

$$\text{\%RSD (or CV)} = \frac{s}{\bar{x}} \times 100\% \qquad (B.5)$$

The RSD or CV expresses precision of the data—the smaller the RSD or CV, the greater is the precision for the average value of the data.

As an example that illustrates how these statistical methods are applied, suppose that four analyses of an iron ore sample give the following data with four significant figures:

Trial	Mass of Iron per kg Ore Sample
1	39.74 g/kg
2	40.06 g/kg
3	39.06 g/kg
4	40.92 g/kg

$$\text{average (or mean) value, } \bar{x} = \frac{39.74 + 40.06 + 39.06 + 40.92}{4} = 39.94 \text{ g/kg}$$

To calculate the standard deviation and percent relative standard deviation (or coefficient of variation), compute the deviations and their squares. Let's set up a table.

Trial	Measured Values	$d_i = \bar{x} - x_i$	d_i^2
1	39.74 g	0.20	0.040
2	40.06 g	−0.12	0.014
3	39.06 g	0.88	0.77
4	40.92 g	−0.98	0.96
	$\bar{x} = 39.94$ g		Sum = 1.78

$$\text{standard deviation, } s = \sqrt{\frac{1.78}{4 - 1}} = 0.77$$

$$\text{\%RSD (or CV)} = \frac{0.77}{39.94} \times 100 = 1.93\%$$

The precision of our analysis is expressed in terms of a standard deviation; the amount of iron in the sample is reported as 39.94 ± 0.77 g Fe/kg of sample, meaning that 68% of subsequent analyses should be in the range of 39.94 ± 0.77 g Fe/kg of sample. The percent relative standard deviation, %RSD (or coefficient of variation, CV), of the precision of the data is 1.93%.

Relative Error

Scientists check the *accuracy* of their measurements by comparing their results with values that are well established and are considered "accepted values." Many reference books, such as the Chemical Rubber Company's (CRC) *Handbook of Chemistry and Physics*, are used to check a result against an accepted value. To report the relative error in *your* result, take the absolute value of the difference between your value and the accepted value, divide this difference by the accepted value. Taking x to be your measured value and y to be the accepted value,

$$\text{relative error} = \frac{|x - y|}{y} \qquad (B.6)$$

Relative error may be expressed as percent or parts per thousand, multiplying the relative error by 100 or 1000.

Appendix C

Graphing Data

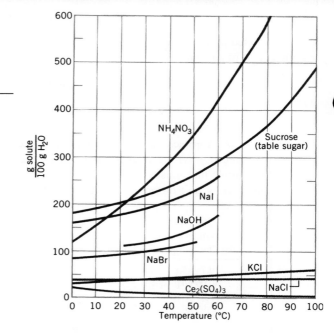

Plotted data show how the solubilities of salts vary with temperature.

A well-designed graph of experimental data is a very effective organization of the data for observing trends, discovering relationships, or predicting information. It is therefore worthwhile to learn how to effectively construct and present a graph and how to extract information from it.

Graph Construction

In general, a graph is constructed on a set of perpendicular axes (Figure C.1); the vertical axis (the y axis) is the **ordinate**, and the horizontal axis (the x axis) is the **abscissa**.

Constructing a graph involves the following five steps whether the graph is constructed manually or with the appropriate software, such as Excel.

1. **Select the Axes.** First choose which variable corresponds to the ordinate and which one corresponds to the abscissa. Usually the dependent variable is plotted

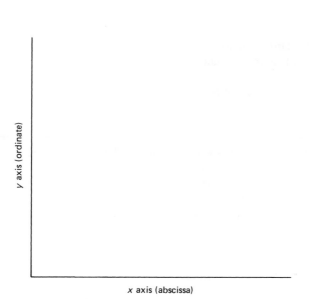

Figure C.1 A graph is usually constructed on a set of perpendicular axes.

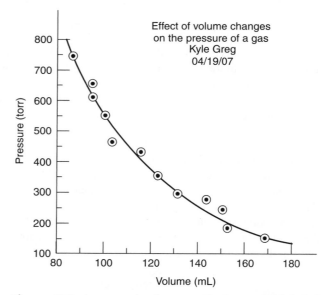

Figure C.2 An example of a properly drawn and labeled graph showing how the pressure of a gas depends on the volume of that gas.

along the ordinate and values of the independent variable along the abscissa. For example, if we observe how the pressure of a gas responds to a change in volume, pressure is the dependent variable. We therefore assign pressure to the vertical axis and volume to the horizontal axis; we say we are plotting pressure versus volume. Be sure to "label each axis" by indicating the units that correspond to the variables being plotted (Figure C.2).

2. **Set the Scales for the Axes.** Construct the graph so that the data fill as much of the space of the graph paper as possible. Therefore, choose scales for the x and y axes that cover the range of the experimental data. For example, if the measured pressure range is 150 to 740 torr, choose a pressure scale that ranges from 100 to 800 torr. This covers the entire data range and allows us to mark the major divisions at intervals of 100 torr (Figure C.2). When choosing the scale, always choose values for the major divisions that make the smaller subdivisions easy to interpret. With major divisions at every 100 torr, minor divisions occur at every 50 torr. This makes plotting values such as 525 torr very simple.

   Construct the scale for the x-axis in the same manner. In Figure C.2, the volumes range from 170 mL at a pressure of 150 torr to 85 mL at a pressure of 750 torr. The scale on the x-axis ranges from 80 to 180 mL and is marked off in 20-mL intervals. Label each axis with the appropriate units.

   There are a few additional points to note about marking the scales of a graph:

   - The values plotted along the axes do not have to begin at zero at the origin; in fact, they seldom do.
   - The size of the minor subdivisions should permit estimation of all the significant figures used in obtaining the data (if pressure measurements are made to the nearest torr, then the pressure scale should be interpreted to read to the nearest torr).
   - If the graph is used for extrapolation, be sure that the range of scales covers the range of the extrapolation.

3. **Plot the Data.** Place a dot for each data point at the appropriate place on the graph. Draw a small circle around the dot. *Ideally*, the size of the circle should approximate the estimated error in the measurement. For most software graphing programs, error bars can be added to the data points to better represent the precision of the data. If you plot two or more different data sets on the same graph, use different-shaped symbols (triangle, square, diamond, etc.) around the data points to distinguish one set of data from another.

4. **Draw a Curve for the Best Fit.** Draw a *smooth* curve that best fits your data. This line does not have to pass through the centers of all the data points, or even through any of them, but it should pass as closely as possible to all of them. Most software has the option of adding a trendline to the plotted data. Generally several options as to the type of trendline are offered—select the one that best fits your data.

   Note that the line in Figure C.2 is not drawn through the circles. It stops at the edge of the circle, passes "undrawn" through it, and then emerges from the other side.

5. **Title Your Graph.** Place a descriptive title in the upper portion of the graph, well away from the data points and the smooth curve. Include your name and date under the title.

Straight-Line Graphs

Often, the graphical relationship between measured quantities produces a straight line. This is the case, for example, when we plot pressure versus temperature for a fixed volume of gas. Such linear relationships are useful because the line corresponding to the best fit of the data points can be drawn with a straight edge, and because quantitative (extrapolated) information about the relationship is easily obtained directly from the graph.

   Algebraically, a straight line is described by the equation

$$y = mx + b \tag{C.1}$$

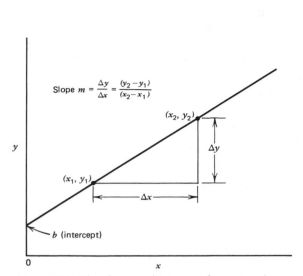

Figure C.3 The slope and intercept for a straight line, $y = mx + b$.

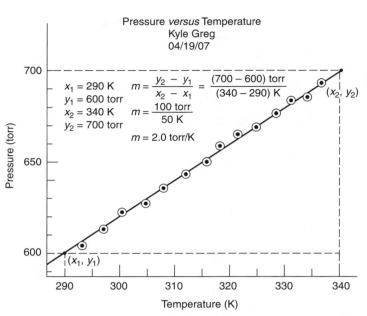

Figure C.4 Determination of the slope of a straight line drawn for a plot of pressure vs. temperature for a gas.

m is the slope of the straight line and b is the point of intersection of the line with the y axis when $x = 0$ (Figure C.3). The slope of the line, which is usually of greatest interest, is determined from the relationship

$$m = \frac{y_2 - y_1}{x_2 - x_1} = \frac{\Delta y}{\Delta x} \qquad (C.2)$$

Figure C.4 illustrates the determination of the slope for a typical plot of pressure versus temperature. First plot the data and then draw the *best straight line*. Next, choose points *on the drawn line* corresponding to the easily readable values along the x axis. Read corresponding y values along the y axis, and then compute the slope.

Using appropriate software, if a straight line is the selected trendline, the equation for the straight line is generally given, from which the slope and y-intercept can be obtained. See Figure C.5.

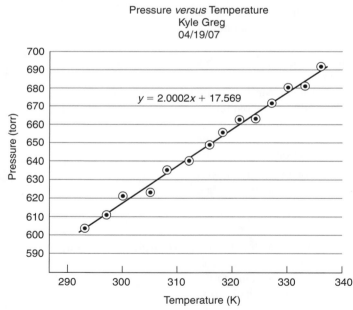

Figure C.5 Using Microsoft Excel the equation for the trendline provides the slope and y-intercept for the pressure vs. temperature data.

Appendix D

Familiar Names of Common Chemicals

Sodium bicarbonate is commonly called baking soda or bicarbonate of soda.

Familiar Name	Chemical Name	Formula
alcohol	ethanol (ethyl alcohol)	C_2H_5OH
aqua regia	mixture of conc nitric and hydrochloric acids	$HNO_3 + 3$ HCl by volume
aspirin	acetylsalicylic acid	$CH_3COOC_6H_4COOH$
baking soda	sodium bicarbonate	$NaHCO_3$
banana oil	amyl acetate	$CH_3COOC_5H_{11}$
bauxite	hydrated aluminum oxide	$Al_2O_3 \cdot xH_2O$
bleaching powder	calcium chloride hypochlorite	$Ca(ClO)_2, Ca(ClO)Cl$
blue vitriol	copper(II) sulfate pentahydrate	$CuSO_4 \cdot 5H_2O$
borax (tincal)	sodium tetraborate decahydrate	$Na_2B_4O_7 \cdot 10H_2O$
brimstone	sulfur	S_8
calamine	zinc oxide	ZnO
calcite	calcium carbonate	$CaCO_3$
Calgon	polymer of sodium metaphosphate	$(NaPO_3)x$
calomel	mercury(I) chloride	Hg_2Cl_2
carborundum	silicon carbide	SiC
caustic soda	sodium hydroxide	$NaOH$
chalk	calcium carbonate	$CaCO_3$
Chile saltpeter	sodium nitrate	$NaNO_3$
copperas	iron(II) sulfate heptahydrate	$FeSO_4 \cdot 7H_2O$
cream of tartar	potassium hydrogen tartrate	$KHC_4H_4O_6$
DDT	dichlorodiphenyltrichloroethane	$(C_6H_4Cl)_2CHCCl_3$
dextrose	glucose	$C_6H_{12}O_6$
Epsom salt	magnesium sulfate heptahydrate	$MgSO_4 \cdot 7H_2O$
fool's gold	iron pyrite	FeS_2
Freon	dichlorodifluoromethane	CCl_2F_2
Glauber's salt	sodium sulfate decahydrate	$Na_2SO_4 \cdot 10H_2O$
glycerin	glycerol	$C_3H_5(OH)_3$
green vitriol	iron(II) sulfate heptahydrate	$FeSO_4 \cdot 7H_2O$
gypsum	calcium sulfate dihydrate	$CaSO_4 \cdot 2H_2O$
hypo	sodium thiosulfate pentahydrate	$Na_2S_2O_3 \cdot 5H_2O$
invert sugar	mixture of glucose and fructose	$C_6H_{12}O_6 + C_6H_{12}O_6$
laughing gas	nitrous oxide	N_2O
levulose	fructose	$C_6H_{12}O_6$
lye	sodium hydroxide	$NaOH$
magnesia	magnesium oxide	MgO
marble	calcium carbonate	$CaCO_3$
marsh gas	methane	CH_4
milk of lime (limewater)	calcium hydroxide	$Ca(OH)_2$
milk of magnesia	magnesium hydroxide	$Mg(OH)_2$
milk sugar	lactose	$C_{12}H_{22}O_{11}$
Mohr's salt	iron(II) ammonium sulfate hexahydrate	$Fe(NH_4)_2(SO_4)_2 \cdot 6H_2O$
moth balls	naphthalene	$C_{10}H_8$
muriatic acid	hydrochloric acid	$HCl(aq)$
oil of vitriol	sulfuric acid	$H_2SO_4(aq)$

Familiar Name	Chemical Name	Formula
oil of wintergreen	methyl salicylate	$C_6H_4(OH)COOCH_3$
oleum	fuming sulfuric acid	$H_2S_2O_7$
Paris green	double salt of copper(II) acetate and copper(II) arsenite	$Cu(CH_3CO_2)_2 \cdot Cu_3(AsO_3)_2$
plaster of Paris	calcium sulfate hemihydrate	$CaSO_4 \cdot \frac{1}{2}H_2O$
potash	potassium carbonate	K_2CO_3
quartz	silicon dioxide	SiO_2
quicklime	calcium oxide	CaO
Rochelle salt	potassium sodium tartrate	$KNaC_4H_4O_6$
rouge	iron(III) oxide	Fe_2O_3
sal ammoniac	ammonium chloride	NH_4Cl
salt (table salt)	sodium chloride	$NaCl$
saltpeter	potassium nitrate	KNO_3
silica	silicon dioxide	SiO_2
sugar (table sugar)	sucrose	$C_{12}H_{22}O_{11}$
Teflon	polymer of tetrafluoroethylene	$(C_2F_4)_x$
washing soda	sodium carbonate decahydrate	$Na_2CO_3 \cdot 10H_2O$
white lead	basic lead carbonate	$PbCO_3 \cdot Pb(OH)_2$
wood alcohol	methanol (methyl alcohol)	CH_3OH

For a listing of more common chemical names, go to www.chemistry.about.com and www.sciencecompany.com (patinas for metal artists).

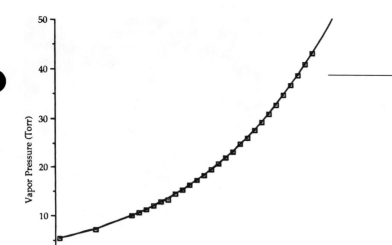

Appendix **E**

Vapor Pressure of Water

Vapor pressure of water as a function of temperature.

Temperature (°C)	Pressure (Torr)
0	4.6
5	6.5
10	9.2
11	9.8
12	10.5
13	11.2
14	12.0
15	12.5
16	13.6
17	14.5
18	15.5
19	16.5
20	17.5
21	18.6
22	19.8
23	21.0
24	22.3
25	23.8
26	25.2
27	26.7
28	28.3
29	30.0
30	31.8
31	33.7
32	35.7
33	37.7
34	39.9
35	42.2
37*	47.1
—	—
100	760

*Body temperature.

Appendix F

Concentrations of Acids and Bases

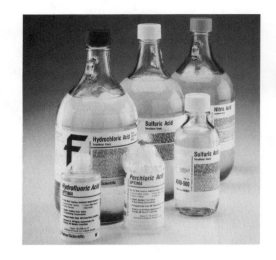

Concentrated laboratory acids and bases.

Reagent	Approximate Molar Concentration	Approximate Mass Percent	Specific Gravity	mL to Dilute to 1 L for a 1.0 M Solution
Acetic acid	17.4	99.5	1.05	57.5
Hydrochloric acid	11.6	36	1.18	86.2
Nitric acid	16.0	71	1.42	62.5
Phosphoric acid	18.1	85	1.70	68.0
Sulfuric acid	18.0	96	1.84	55.6
Ammonia (aq) (ammonium hydroxide)	14.8	28%(NH_3)	0.90	67.6
Potassium hydroxide	13.5	50	1.52	74.1
Sodium hydroxide	19.1	50	1.53	52.4

Caution: *When diluting reagents, add the more concentrated reagent to the more dilute reagent (or solvent). **Never** add water to a concentrated acid!*

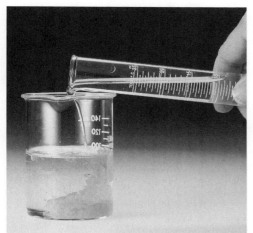

Water Solubility of Inorganic Salts

Many salts, such as cadmium sulfide, have very low solubilities.

Water-Soluble Salts

1. All salts of the chloride ion, Cl^-, bromide ion, Br^-, and iodide ion, I^-, are soluble *except* those of Ag^+, Hg_2^{2+}, Pb^{2+}, Cu^+, and Tl^+. BiI_3 and SnI_4 are insoluble. $PbCl_2$ is three to five times more soluble in hot water than in cold water.

2. All salts of the acetate ion, $CH_3CO_2^-$, nitrate ion, NO_3^-, chlorate ion, ClO_3^-, perchlorate ion, ClO_4^-, and permanganate ion, MnO_4^-, are soluble.

3. All common salts of the Group 1A cations and ammonium ion, NH_4^+, are soluble.

4. All common salts of the sulfate ion, SO_4^{2-}, are soluble *except* those of Ba^{2+}, Sr^{2+}, Pb^{2+}, and Hg^{2+}.

5. All Group 1A and 2A salts of the bicarbonate ion, HCO_3^-, are soluble.

6. *Most* salts of the fluorosilicate ion, SiF_6^{2-}, thiocyanate ion, SCN^-, and thiosulfate ion, $S_2O_3^{2-}$, are soluble. *Exceptions* are the Ba^{2+} and Group 1A fluorosilicates, the Ag^+, Hg_2^{2+}, and Pb^{2+} thiocyanates, and the Ag^+ and Pb^{2+} thiosulfates.

Water-Insoluble Salts

1. All common salts of the fluoride ion, F^-, are insoluble *except* those of Ag^+, NH_4^+, and Group 1A cations.

2. In general, all common salts of the carbonate ion, CO_3^{2-}, phosphate ion, PO_4^{3-}, borate ion, BO_3^{3-}, arsenate ion, AsO_4^{3-}, arsenite ion, AsO_3^{3-}, cyanide ion, CN^-, ferricyanide ion, $[Fe(CN)_6]^{3-}$, ferrocyanide ion, $[Fe(CN)_6]^{4-}$, oxalate ion, $C_2O_4^{2-}$, and the sulfite ion, SO_3^{2-}, are insoluble, *except* those of NH_4^+ and the Group 1A cations.

3. All common salts of the oxide ion, O^{2-}, and the hydroxide ion, OH^-, are insoluble *except* those of the Group 1A cations, Ba^{2+}, Sr^{2+}, and NH_4^+. $Ca(OH)_2$ is slightly soluble. Soluble oxides produce the corresponding hydroxides in water.

4. All common salts of the sulfide ion, S^{2-}, are insoluble *except* those of NH_4^+ and the cations that are isoelectronic with a noble gas (e.g., the Group 1A cations, the Group 2A cations, Al^{3+}, etc.).

5. Most common salts of the chromate ion, CrO_4^{2-}, are insoluble *except* those of NH_4^+, Ca^{2+}, Cu^{2+}, Mg^{2+}, and the Group 1A cations.

6. All common salts of the silicate ion, SiO_3^{2-}, are insoluble *except* those of the Group 1A cations.

Table G.1   Summary of the Solubility of Salts

Anion	Soluble Salts with These Cations	"Insoluble" Salts with These Cations
acetate, $CH_3CO_2^-$	most cations	none
arsenate, AsO_4^{3-}	NH_4^+, Group 1A (except Li^+)	most cations
arsenite, AsO_3^{3-}	NH_4^+, Group 1A (except Li^+)	most cations
borate, BO_3^{3-}	NH_4^+, Group 1A (except Li^+)	most cations
bromide, Br^-	most cations	Ag^+, Hg_2^{2+}, Pb^{2+}, Cu^+, Tl^+
carbonate, CO_3^{2-}	NH_4^+, Group 1A (except Li^+)	most cations
chlorate, ClO_3^-	most cations	none
chloride, Cl^-	most cations	Ag^+, Hg_2^{2+}, Pb^{2+}, Cu^+, Tl^+
chromate, CrO_4^{2-}	NH_4^+, Ca^{2+}, Cu^{2+}, Mg^{2+}, Group 1A	most cations
cyanide, CN^-	NH_4^+, Group 1A (except Li^+)	most cations
ferricyanide, $[Fe(CN)_6]^{3-}$	NH_4^+, Group 1A (except Li^+)	most cations
ferrocyanide, $[Fe(CN)_6]^{4-}$	NH_4^+, Group 1A (except Li^+)	most cations
fluoride, F^-	Ag^+, NH_4^+, Group 1A	most cations
fluorosilicate, SiF_6^{2-}	most cations	Ba^{2+}, Group 1A
hydroxide, OH^-	NH_4^+, Sr^{2+}, Ba^{2+}, Group 1A	most cations
iodide, I^-	most cations	Ag^+, Hg_2^{2+}, Pb^{2+}, Cu^+, Tl^+, Br^{3+}, Sn^{4+}
nitrate, NO_3^-	most cations	none
nitrite, NO_2^-	most cations	none
oxalate, $C_2O_4^{2-}$	NH_4^+, Group 1A (except Li^+)	most cations
oxide, O^{2-}	NH_4^+, Sr^{2+}, Ba^{2+}, Group 1A	most cations
perchlorate, ClO_4^-	most cations	none
permanganate, MnO_4^-	most cations	none
phosphate, PO_4^{3-}	NH_4^+, Group 1A (except Li^+)	most cations
silicate, SiO_3^{2-}	Group 1A	most cations
sulfate, SO_4^{2-}	most cations	Sr^{2+}, Ba^{2+}, Pb^{2+}, Hg^{2+}
sulfide, S^{2-}	NH_4^+, Groups 1A and 2A	most cations
sulfite, SO_3^{2-}	NH_4^+, Group 1A (except Li^+)	most cations
thiocyanate, SCN^-	most cations	Ag^+, Hg_2^{2+}, Pb^{2+}
thiosulfate, $S_2O_3^{2-}$	most cations	Ag^+, Pb^{2+}

Cations	Soluble Salts with These Anions	"Insoluble" Salts with These Anions
ammonium, NH_4^+	most anions	no common anions
Group 1A	most anions	no common anions